GIBBS MEASURES IN BIOLOGY AND PHYSICS

The Potts Model

GIBBS MEASURES IN BIOLOGY AND PHYSICS
The Potts Model

Utkir A. Rozikov

V. I. Romanovskiy Institute of Mathematics,
Tashkent, Uzbekistan

World Scientific

NEW JERSEY • LONDON • SINGAPORE • BEIJING • SHANGHAI • HONG KONG • TAIPEI • CHENNAI • TOKYO

Published by

World Scientific Publishing Co. Pte. Ltd.
5 Toh Tuck Link, Singapore 596224
USA office: 27 Warren Street, Suite 401-402, Hackensack, NJ 07601
UK office: 57 Shelton Street, Covent Garden, London WC2H 9HE

Library of Congress Control Number: 2022021014

British Library Cataloguing-in-Publication Data
A catalogue record for this book is available from the British Library.

GIBBS MEASURES IN BIOLOGY AND PHYSICS
The Potts Model

ISBN 978-981-125-123-8 (hardcover)
ISBN 978-981-125-124-5 (ebook for institutions)
ISBN 978-981-125-125-2 (ebook for individuals)

For any available supplementary material, please visit
https://www.worldscientific.com/worldscibooks/10.1142/12694#t=suppl

Desk Editors: Soundararajan Raghuraman/Kwong Lai Fun

Typeset by Stallion Press
Email: enquiries@stallionpress.com

To my teachers and collaborators

Contents

Preface

This book presents recently obtained mathematical results on Gibbs measures of the q-state Potts model on the integer lattice and on Cayley trees. Moreover, we discuss many applications of the Potts model to real world situations: mainly biology, physics, and some examples of alloy behavior, cell sorting, financial engineering, flocking birds, flowing foams, image segmentation, medicine, sociology etc.

The Potts model is defined by a Hamiltonian (energy) of configurations of spins which take one of q possible values (the $q = 2$ is the well-known Ising model) on vertices of a lattice. The model is used to study the behavior of systems having multiple states (spins, colors, alleles, etc). The structure of the Potts model allows researchers to investigate the internal elements of a system and predict how they will interact with one another to determine the overall behavior of the system. Since the model has a rich mathematical formulation, it has been studied extensively.

The Gibbs measure is one of the important measures in many problems of probability theory and statistical mechanics. It is a measure associated with the Hamiltonian of a (biological, physical) system.

Each Gibbs measure gives a state of the system. More importantly, when the Hamiltonian can be written as a sum of parts (like Potts model), the Gibbs measure has a certain kind of statistical independence (the Markov property), thus leading to its widespread appearance in many problems outside of physics.

The main problem for a given Hamiltonian on a countable lattice is to describe all of its possible Gibbs measures. The existence of some values of parameters at which the uniqueness of Gibbs measure switches to non-uniqueness is interpreted as phase transition.

The set of Gibbs measures for a given Hamiltonian is usually studied by contour methods, algebra and cluster expansion, the FKG inequalities and monotonicity methods, probability and measure theory, methods of Markov random field theory and reconstruction, methods of (thermo)dynamical systems and non-linear analysis. This book discusses the methods and results which were developed recently.

In MathSciNet, there are about 2,700 published works devoted to Gibbs measures, and there are about 2000 works related to the Potts model (the first paper published in 1951). But most of them are written for the audience of physicists with little regard for mathematical rigor. There are a few books written from a mathematical perspective which are devoted to spin systems and Gibbs measures. Most recent book is [75], which gives a concrete mathematical introduction to the theory of Gibbs measures. My book [243] presents all known (obtained until 2013) mathematical results on Gibbs measures on Cayley trees. At that time, the Potts model on trees was not well studied, compared to the Ising model. Even results about translation-invariant (simple) Gibbs measures were not complete and periodic Gibbs measures were not studied.

First complete result about translation-invariant Gibbs measures of the Potts model on Cayley trees appeared in 2014 [138]. Periodic and weakly periodic measures were studied after 2015. These results mainly have applications to physics as thermodynamical systems. In this book, I have collected all these recent results. Moreover, here I have systematically presented very recent results related to Potts models and their Gibbs measures with applications to biology and other sciences.

Thus, the book informs the reader about what has been (mathematically) done in the theory of Gibbs measures of the Potts model and about many-many applications of the Potts model. The main aim of the book is to facilitate the readers (researchers in mathematical biology, statistical physics, applied mathematics, probability and measure theory) in-depth understanding by giving a systematic review of the theory of Gibbs measures of Potts model (developed since 2013) which has wide range of applications in biology, medicine, sociology, physics etc.

The book consists of *three* parts divided into 12 chapters; at the end of each part, there are comments and references related to the part.

Part I of the book has an introductional character. Starting sections of this part present main definitions and related problems of the book. After that, the methods of solutions of the formulated problems are presented. Moreover, this part contains a section as an "excursion" to the applications of the Potts model to several areas of sciences and nature. This section gives many variations of the Potts model, the set of Gibbs measures of which is not completely studied yet. Thus this part gives a clear motivation to read the next chapters of the book. The reader can try to study Gibbs measures of these variations of the Potts model. These measures can be then used to better understanding of the corresponding systems.

Part II presents very recently obtained results related to biological systems (interacting DNA and RNA sets). Using analytic, algebraic and probabilistic methods of statistical physics, we give models of DNA and RNA interacting systems and study their thermodynamic properties depending on Gibbs measures. These results are applied to study distributions of Holliday junctions and branches of DNAs. In case of very high and very low temperatures we give stationary distributions and typical configurations of the Holliday junctions.

Part III is devoted to Gibbs measures of the q-state Potts model with external fields on Cayley trees. In this part, we present recent results (since 2014) about *all* translation-invariant splitting Gibbs measures (TISGMs). In particular, it is shown that at sufficiently low temperatures their maximal number is $2^q - 1$. It is proved that there are $[q/2]$ (where $[a]$ is the integer part of a) critical temperatures at which the number of TISGMs changes and the exact number of TISGMs for each intermediate temperature is calculated. Under some conditions on parameters of q-state Potts model (non-) extremality of these TISGMs in the set of all Gibbs measures are proved. Moreover, for each TISGM μ, we construct the set of boundary conditions such that limiting Gibbs measures with respect to these boundary conditions coincide with μ. To illustrate a contour method on trees, we consider a Potts model with competing interactions. Some periodic and weakly periodic Gibbs measures are studied. It is shown that any periodic Gibbs measure of the Potts model is ether translation-invariant or 2-periodic. In case of the anti-ferromagnetic Potts model for some values of parameters, it is shown that there are several weakly periodic (non-translation-invariant) Gibbs measures. Some explicit formulas for free energies and entropies are given for several boundary conditions including translation-invariant and weakly periodic ones.

In the bibliography of this book, I have collected the possibly maximal list of references which are related to Gibbs measures of the Potts model.

About the Author

Professor Utkir A. Rozikov was born on 20 May 1970 in Bukhara, Uzbekistan. He graduated from the Samarkand State University (1993) with an honors diploma. He is the winner of several Mathematical Olympiads and completed his Ph.D. (1995) and Doctor of Sciences in Physics and Mathematics (2001) degrees from the V.I. Romanovskiy Institute of Mathematics, Uzbekistan Academy of Sciences, Tashkent, where he is currently deputy director.

Rozikov is an internationally renowned expert in probability, mathematical physics and analysis, and specializes in dynamical systems and statistical mechanics on graphs. Rozikov has made significant contributions to the deep analysis of Gibbs states in topical models of statistical mechanics on trees (Ising, Potts, SOS), in particular with random interactions and/or in external field. He has promoted new tools in studying Gibbs measures on graphs, such as group representation theory, information flows, node-weighted random walks, contour methods on trees, and non-linear analysis. Rozikov has also advanced the idea of studying p-adic dynamical systems relevant to multi-scale and/or chaotic systems, applied to statistical mechanics on trees and Cayley graphs of non-amenable groups. His more recent interests are focused on evolution algebras of sex-linked populations, aiming to incorporate thermodynamics in mathematical biology models.

Rozikov is an exceptionally prolific author. Since 1995, he has published over 150 papers in top journals including 12 papers in the flagship *Journal of Statistical Physics*, 26 papers in *Theoretical and Mathematical Physics* and also in *Letters in Mathematical Physics*, *Communications in*

Mathematical Physics, *Journal of Statistical Mechanics: Theory and Experiment*, and *Journal of Mathematical Analysis and Applications*. In 2013 and 2021, he published extensive review papers in *Reviews in Mathematical Physics*, followed by the authoritative research monographs published by World Scientific and Springer, including *Population Dynamics: Algebraic and Probabilistic Approach, An Introduction to Mathematical Billiards, Gibbs Measures on Cayley Trees, Algebra, Complex Analysis, and Pluripotential Theory*. Rozikov was a visiting professor and invited researcher at many universities and research centers in France, Germany, Italy, Spain, UK, Switzerland, Turkey, Pakistan and many others.

Acknowledgments

The book is based on materials collected by the author during several visits to the University of Cambridge, Newton Institute and University of Leeds, UK; ICTP, Trieste, Italy; “La Sapienza” University, Rome, Italy; IHES, Bures-sur-Yvette, France; Universite du Sud Toulon Var, CPT and IMéRA, Marseille, France; University Paris-Est, France; University Santiago de Compostela and University Granada, Spain; University of Bonn and Ruhr University, Bochum, Germany. University of Geneva, Switzerland. I thank all these organizations for their warm hospitality and excellent working conditions.

I thank my coauthors and friends who collaborated with me to develop the theory of Gibbs measures: S. Albeverio, L.V. Bogachev, G.I. Botirov, M. Cassandro, A.C.D. van Enter, S. Friedli, G. Gallavotti, D. Gandolfo, N.N. Ganikhodjaev, B.M. Gurevich, F. Haydarov, F. Henning, Y. Higuchi, O. Hryniv, O. Khakimov, R. Khakimov, S. Kissel, M. Kontsevich, C. Külske, A. LeNy, L. Liao, I. Merola, F.M. Mukhamedov, M. Rahmatullaev, D. Ruelle, J. Ruiz, Ya.G. Sinai, Yu.M. Suhov, S. Shlosman, J.P. Tian, Y. Velenik, V. Zagrebnov and many others. I am grateful to everyone for their help.

I am indebted to my family: wife (Dilfuza) and children (Azamat, Laziz, Sevinch) for their warm attitude to my work.

Utkir A. Rozikov
Tashkent, Uzbekistan
March 2022

PART I

Introduction to thermodynamics and theory of Gibbs measures

Chapter 1

Introduction

In this chapter we give some minimal preliminaries from biology and physics and motivations of studying of the Gibbs measures of the thermodynamic (biological and physical) systems. Moreover, we formulate the main problems in the consideration of such systems. The next chapters of this book will be devoted to solutions of these problems.

1.1 What is a thermodynamic system

Let us give some main notations of this book.

A dynamical system is a system the state of which changes when *time* is increasing. This is very developed field (see for example [120], [250]).

There are some dynamical systems where instead of time parameters, one can consider the changes of a system depending on other parameters, like temperature[1] (making thermodynamic systems) etc.

A thermodynamic system comprises a collection of very large numbers of elements (particles, atoms, bodies, DNAs and electromagnetic field quanta). In general, in addition to being subject to internal effects, a system may also be affected from the outside. Thermodynamic system may be of liquid, solid, gaseous, a biological system, or even some other exotic form.

This field is very developed (see for example: the books [36], [98], [108], [228], [283]).

In this book we consider several thermodynamical systems of natural sciences.

---

[1]Temperature is a measure of the average kinetic energy of the particles in a sample of matter, expressed in terms of units or degrees designated on a standard scale. In this book the temperature is in the Kelvin scale and denoted by T, which is $T \geq 0$ K $= -273.15°$C (degrees Celsius).

The **main problem** of mathematicians for a given (thermo)dynamical system: is to know all changes of states (evolution) of the system when temperature or/and time changes.

For example, an important goal of statistical physics is to determine phase transition temperatures, that is, critical temperatures around which a small change in temperature results in an abrupt, nonanalytic change in various physical properties. This phase transition temperature separates the two phases of the system. In the absence of an external biasing field, for temperatures above this critical temperature, the system exhibits no long-range order, while for temperatures below the critical temperature it does exhibit such order.

For a review about the existing literature related to the field see [75].

1.2 What we do for biology

In biophysics, transduction is the conveyance of energy from one electron (a donor) to another (a receptor), at the same time that the class of energy changes.

Biological thermodynamics (direct quotes from[2]) is the quantitative study of the energy transductions that occur in or between living organisms, structures, and cells and of the nature and function of the chemical processes underlying these transductions. Biological thermodynamics may address the question of whether the benefit associated with any particular phenotypic trait is worth the energy investment it requires.

The field of biological thermodynamics is focused on principles of chemical thermodynamics in biology and biochemistry. Principles covered include the first law of thermodynamics, the second law of thermodynamics, Gibbs free energy, statistical thermodynamics, reaction kinetics, and on hypotheses of the origin of life. Presently, biological thermodynamics concerns itself with the study of internal biochemical dynamics as: protein stability, DNA binding, membrane diffusion, enzyme kinetics, and other such essential energy controlled pathways. In terms of thermodynamics, the amount of energy capable of doing work during a chemical reaction is measured quantitatively by the change in the Gibbs free energy. The physical biologist A.Lotka attempted to unify the change in the Gibbs free energy with evolutionary theory.

[2]See: https://en.wikipedia.org/wiki/Biological_thermodynamics.

Let us give one very interesting example of biological thermodynamics: in biology a *sex-determination system*[3] is a biological system that determines the development of sexual characteristics in an organism.

In many cases, sex determination is *genetic*: males and females have different alleles or even different genes that specify their sexual morphology. In animals, this is often accompanied by *chromosomal* differences.

In other cases, sex is determined by *environmental* variables such as *temperature*. For example, in some species of reptiles, including alligators, some turtles, the tuatara, and a few birds, sex is determined by the temperature at which the egg is incubated during a temperature sensitive period. For some, this is achieved by hotter temperatures being one sex and cooler temperatures being the other. For others, the extreme temperatures are one sex and the middle temperature is the other. This is a very interesting example of thermodynamics in biology.

Environmental sex determination occurred before genetic; it is thought that a temperature — dependent reptile was the common ancestor to sex chromosomes.

As usual a mathematical biologist (as a mathematical physicist) tries to give strong mathematical proof of some known biological (physical) property or law.

It is known[4] that the temperature at which the crocodilian eggs incubate determines the sex of the hatchlings:

- Temperatures above 32°C produce more males.
- Temperatures below 31°C produce more females.

However, to the best of my knowledge there is no any mathematical proof of why for crocodile the critical temperature of sex determination is between 31°C–32°C? What kind of mathematical model can describe this phenomenon?

In [37], [87], [100], [101], [242], [290] it was attempted to introduce thermodynamic methods in biology. Thermodynamic behavior of a set of DNAs and RNAs were studied recently in [246], [248], [252], [253]. Many experimental and theoretical works have brought quantitative insights into DNA base-pairing dynamics that is reviewed in [167].

In the second part of this book we systematically present results of these recently published papers, i.e., we give thermodynamic properties of a set of DNAs and DNA-RNA renaturation. The main model of these

---

[3]https://en.wikipedia.org/wiki/Sex-determination_system.

[4]https://en.wikipedia.org/wiki/Crocodilia.

thermodynamic systems will be the Potts model and we give solutions of the biological problems converting them into description of Gibbs measures (For mathematical definitions of the Potts model and Gibbs measure see Chapter 2).

Biological systems (as macroscopic systems) contain very large number of objects (like DNAs, biomolecules, molecules, particles, organisms etc). For large numbers, the most probable occurrence is overwhelmingly so. As such, the result of any thermodynamic that is, a "macroscopic" measurement is extremely well described by the configuration that refers to the most probable state. Therefore, quite appropriately, thermodynamics focuses primarily on the most probable state of the system. This probability in most natural cases calculated with respect to a Gibbs measure. Therefore the Gibbs measures which we are going to give are very important.

It is known that (see [6]) genetic information is carried in the linear sequence of nucleotides in DNA. Each molecule of DNA is a double helix formed by two complementary strands of nucleotides held together by hydrogen bonds between $G+C$ and $A+T$ base pairs, where $C =$ cytosine, $G =$ guanine, $A =$ adenine, and $T =$ thymine. Duplication of the genetic information occurs by the use of one DNA strand as a template for formation of a complementary strand. The genetic information stored in an organism's DNA contains the instructions for all the proteins the organism will ever synthesize.

Studying of DNA's thermodynamics one wants to know how temperature affects the nucleic acid structure of double-stranded DNA [166]. There are several models of such thermodynamics of DNAs (see for example, [43], [213], [282]).

The book [213] deals with the statistics of DNA, a problem that, recently has attracted the attention of physicists, who attempt to uncover the scaling features of coding (exons) and noncoding (introns) regions of a gene. Although there is still some controversy, the current view is that long-range correlations exist when the complete gene sequence, including introns, is considered and that these correlations are absent when only the exons are read. The book also contains a chapter on the thermodynamics of DNA where several results on variants of the Peyrard-Bishop model for local denaturation or melting of DNA, i.e., the dissociation equilibrium between duplex and single-stranded DNAs, are derived (see [213] and references therein).

In [19] the author introduced an elastic model of large-scale duplex DNA structure. The paper [20] develops this approach in detail, deriving explicit expressions for the elastic equilibrium shapes of stressed DNA and suggesting applications to questions of supercoiling.

In [282] forty thermodynamic parameters were estimated for DNA duplexes with a single bulge loop. To investigate the effect of the type of bulged base and its flanking base pairs, the nearest-neighbor model to DNA sequences with a single bulge loop is applied.

In the second part of this book we give several thermodynamic properties of models of DNAs. *Motivations* of this study are:

- To date, statistical physics models are not available to describe interaction between different DNA molecules.
- This issue has become very interesting over the past years due to the discovery of branched nucleic acid structures, which is known as Holliday junctions [113], which is a branched nucleic acid structure that contains four double-stranded arms joined together. These arms may adopt one of several conformations depending on buffer salt concentrations and the sequence of nucleobases closest to the junction. Holiday junctions play a significant role in the DNA (thermo)dynamics during recombination [150]. An important example is stacking [297], where two of the four participating DNA strands remain helical and two others cross over via Holliday junction.
- Thermodynamics and topology of Holliday junctions have been extensively studied experimentally and computationally, but theoretical understanding remains limited to date.
- To study thermodynamic properties of a model of DNAs we embed them on a Cayley tree. We note that the Cayley tree is an infinite sphere, the center of which is everywhere (i.e., each its vertex can be considered as a center), the circumference nowhere. The consideration of the tree is motivated by the quote of Pascal: "The whole visible world is only an imperceptible atom in the ample bosom of nature. No idea approaches it. We may enlarge our conceptions beyond all imaginable space; we only produce atoms in comparison with the reality of things. It is an infinite sphere, the center of which is everywhere, the circumference nowhere." Moreover, "natural hierarchical" structures in an ultrametric space is such that any point of sphere is its center. This is consistent with Pascal's description the structure of nature. Thus modern science concluding that structure of the things more close to the tree-structure see [130].
- We use theory of Gibbs measures and thermodynamics. Therefore, in terms of thermodynamics, the amount of energy capable of doing work during a chemical reaction can be measured quantitatively by the change in the Gibbs free energy (see [78], [247] for connection between Gibbs measures and free energies on trees).

In a very recent paper [230] the authors used a Gibbs free energy function as a measure of the human brain development. This approach applied to the development of the human brain over the human lifespan: from a prenatal stage to advanced age. The proteomic expression data with the Gibbs free energy used to quantify human brain's proteinprotein interaction networks. A consistent functional dependence of the Gibbs free energies on age for most of the areas and both sexes is found. Moreover, gender data analysis has uncovered the existence of functional differences, showing male Gibbs function values lower than female at prenatal and neonatal ages, which become higher at ages 8 to 40 and finally converging at late adulthood with the corresponding female Gibbs functions.

Using MathSciNet and a Google search one can see that there are few published works devoted to Holliday junctions. But *all* of them are written for biologists audience with little regard for mathematical rigor.

In [302] a model is presented for the mechanochemical-coupling mechanism in a Holliday junction.

In this book we give

- Markov process models of DNA sequence evolution for which the stationary distribution is a Gibbs measure.
- Tree-hierarchy of the set of thermodynamic systems of DNAs.
- Applying translation invariant Gibbs measures of the set of DNAs on the Cayley tree we prove several thermodynamic properties of the DNAs.
- By properties of Markov chains (corresponding to Gibbs measures) we study Holliday junction and branches of DNAs.
- We give conditions (on temperature) of a model of DNA-RNA renaturation.
- In case of very high and very low temperatures we give stationary distributions and typical (most probable) configurations of the systems.

These results were obtained very recently (since 2017).

1.3 What we do for physics

A *physical system* is a portion of the physical universe. In analysis of the physical system the *environment* (i.e., everything outside the system) can be ignored except for its effects on the system.

Statistical physics and thermodynamics describe the behavior of systems on the macroscopic scale. Between of all physical laws, the laws of thermodynamics are perhaps the most universal. In thermodynamics one considers macroscopic systems, each of which comprises very large number of particles. For example, it is known that one gram of hydrogen has approximately $6 \cdot 10^{23}$ atoms. Considering that the age of the Universe is thought to be less than $\approx 10^{18}$ seconds, $6 \cdot 10^{23}$ is a very large number.

Since there is inter-particle interaction, theoretical analysis of most such systems is very complicated [283] and can at best be carried out only approximately. Thermodynamics deals with inter-relationships of observed physical properties of macroscopic systems. This can often help relate easily measurable physical properties to those that are hard to measure.

The main task of thermodynamics is to understand how the total energy of the system can be transformed from one type to the other in order to transform, e.g., heat into mechanical energy (see[5] [36], [98], [75], [132], [283] for detailed explanation of the thermodynamics in physics). An axiomatic foundation of thermodynamics was made in [155], [156] see also references therein.

Let us give some important notations of thermodynamic systems in physics.[6]

Consider a physical system and denote its important quantities by

- P — pressure; T — temperature; U — internal energy;
- V — volume; Q — energy supplied to the system as heat;
- W — thermodynamic work done by the system on its surroundings.

There are several equations in thermodynamics. For example, the *First Law of Thermodynamics* given by the equation $dU = dQ + dW$, which tells us that the increase in the internal energy of a system is equal to the heat supplied to it plus the work done on it.

A *phase* is a chemically homogeneous volume, e.g., solid, liquid or gas, with a boundary separating it from other phases.

The number of intensive state variables that can be varied independently without changing the number of phases in a system is called the number of *degrees of freedom* of the system.

[5]https://en.wikipedia.org/wiki/Thermodynamics.
[6]http://orca.phys.uvic.ca/ tatum/thermod.html.

A *component* is one of a collection of chemically-independent constituents of a system. The *number of components* in a system is the least number of constituents that are necessary to describe the composition of each phase.

The state of a physical system consisting of a single component in a single phase, for example a single gas not a mixture of different gases. In that case it can be described by three intensive state variables, P, V and T. Thus, each state of the system is described by a point in three-dimensional PVT space. However, the intensive state variables are connected by an equation of state $f(P, V, T) = 0$, so that the system is constrained to be on the two-dimensional surface described by this equation. Hence, only two intensive state variables suffice to describe the state of the system, i.e., the system has two degrees of freedom.

The Gibbs phase rule. The *Gibbs phase rule* provides a relation between the number of phases, the number of components and the number of degrees of freedom. It is a general principle governing PVT systems in thermodynamic equilibrium. The rule states that if F is the number of degrees of freedom, C is the number of components and P is the number of phases, then $F = C - P + 2$. The rule assumes the components do not react with each other.

Example. Consider an aqueous solution (in H_2O) of the chlorides and bromides of sodium and potassium co-existing with the crystalline solids NaCl, KCl, NaBr, KBr. In this system there are:

- Five phases: four solid and one liquid.
- Six elements: H, O, Na, K, Cl, Br. But the quantities of each of the elements cannot be varied independently. There are two constraints for number of elements: $n(\mathrm{H}) = 2n(\mathrm{O})$, and $n(\mathrm{Na}) + n(\mathrm{K}) = n(\mathrm{Cl}) + n(\mathrm{Br})$. Therefore, if we know the number of hydrogen atoms, then the number of oxygen atoms is known. If we know the number of any three of Na, K, Cl or Br, then the fourth is known. Thus the number of constituents that can be independently varied is four, i.e., the number of components: four.

Consequently, by Gibbs phase rule, in this example of system there is just one degree of freedom. No more than one intensive state variable can be changed without changing the number of phases.

The macroscopic state of a system is characterized by a distribution on the microstates. The entropy of this distribution is given by the Gibbs entropy formula.[7] Entropy (denoted by S) is a logarithmic measure of the number of states with significant probability of being occupied [148]:

$$S = -k_{\mathrm{B}} \sum_i p_i \log p_i,$$

where k_{B} is the Boltzmann constant. The summation is over all the possible microstates of the system, and p_i is the probability that the system is in the i-th microstate.

Free energy. The (Gibbs) free energy F is defined as

$$F = U + PV - TS.$$

In physics, a *statistical ensemble* is a probability distribution for the state of the system.

A *canonical ensemble* is the statistical ensemble that represents the possible states of a system in thermal equilibrium with a heat bath at a fixed temperature.

It is known that the canonical ensemble or Gibbs measure serves to maximize the entropy of a system, subject to a set of constraints: this is the principle of maximum entropy (see resent book [75] for details).

The *Gibbs measure* is a probability measure which is a generalization of the canonical ensemble to infinite systems. The Gibbs measure gives the probability of the system X being in state x (equivalently, of the random variable X having value x) as

$$\mu(X = x) = \frac{1}{Z(\beta)} \exp(-\beta H(x)),$$

where $H(x)$ is a function from the space of states to the real numbers.[8] The parameter β is (a free parameter) the inverse temperature. The normalizing constant $Z(\beta)$ is the partition function.

Consider an infinite graph G, and let $\Lambda \subset G$ be finite subset. It is convenient to work with reduced free energy $f = -\beta F$, which per unit volume is

$$f(\beta, \Lambda) = \frac{1}{|\Lambda|} \cdot \ln Z(\beta, \Lambda),$$

[7] See https://en.wikipedia.org/wiki/Entropy for variations of this definition of entropy

[8] In physics applications, $H(x)$ is interpreted as the energy of the configuration x or Hamiltonian.

where $Z(\beta, \Lambda)$ is the restriction of the partition function $Z(\beta)$ on the set Λ, by fixing the state of the system outside of Λ.

In physics the reduced free energy encodes the relevant information about the physical system; for example, its first derivative with respect to β yields the internal energy while its second derivative yields, up to a multiplicative prefactor, the specific heat. For a finite graph G, this is clearly a (real) analytic function in β. Some failures of analyticity can only occur in the infinite volume limit:

$$f(\beta, G) = \lim_{\Lambda \nearrow G} \frac{1}{|\Lambda|} \cdot \ln Z(\beta, \Lambda).$$

If the resulting limit of the reduced free energy is an analytic function of β (equivalently, temperature), then the model has no (finite-temperature) phase transition, since U, F, and S will be analytic as well. Finite values of β, or equivalently T, where analyticity fails are finite temperature critical points corresponding to phase transitions.

There can also be zero-temperature critical points. Thus phase transitions are manifest as failures of analyticity in the thermodynamic limit of the reduced free energy, the partition function is very important either to determine these points of non-analyticity or to establish analyticity in some region. If we want to know how the thermodynamic functions behave near a critical temperature, again understanding the partition function is essential.

In case of infinite systems, the total energy is no longer a finite number and cannot be used in the traditional construction of the probability distribution of a canonical ensemble. In statistical physics the limit of intensive properties are studied as the size of a finite system approaches infinity (this limit is called a *thermodynamic limit*).

In general, the Gibbs measures were proposed by Dobrushin, Lanford, and Ruelle to directly study infinite systems, instead of taking the limit of finite systems:

A measure is a Gibbs measure if the conditional probabilities it induces on each finite subsystem satisfy a consistency condition: if all degrees of freedom outside the finite subsystem are frozen, the canonical ensemble for the subsystem subject to these boundary conditions matches the probabilities in the Gibbs measure conditional on the frozen degrees of freedom.

The Hammersley-Clifford theorem[9] states that a probability distribution that has a strictly positive mass or density satisfies one of the Markov

[9] http://www.stat.yale.edu/~pollard/Courses/251.spring04/Handouts/Hammersley-Clifford.pdf.

properties with respect to an undirected graph G if and only if it is a Gibbs random field, that is, its density can be factorized over the complete subgraphs of the graph.

This theorem implies that any probability measure that satisfies a Markov property is a Gibbs measure for an appropriate choice of energy function (see [98], [220]). Therefore, the Gibbs measure applies to widespread problems outside of physics, such as (Hopfield, Markov and Markov logic) networks, and bounded rational potential games in game theory and economics.[10]

The Gibbs measure of an infinite system is not necessarily unique, in contrast to the canonical ensemble of a finite system, which is unique. The existence of more than one Gibbs measure is associated with statistical phenomena such as symmetry breaking and phase coexistence (or transition).

Thus, for a given Hamiltonian on the configuration space (i.e., the space of states) it is very crucial to know all corresponding (to the Hamiltonian) Gibbs measures.

A *lattice model* is a physical model that is defined on a lattice. Lattice models are used in condensed matter physics, where the atoms of a crystal automatically form a lattice. Now lattice models are quite popular and some of them are exactly solvable [14]. Examples of lattice models include the Ising model, the Potts model, the XY model. These models has given insights into the nature of phase transitions, magnetization and scaling behavior, as well as insights into the nature of quantum field theory.

In the third part of this book we give very recently obtained results (since 2014) about Gibbs measures of physical systems energy of which is the Hamiltonian of a q-state Potts model on two lattices: $\mathbb{Z}^d$ and a Cayley tree (this tree is also called Bethe lattice [14]). In particular, the following results are presented in this book:

- The full set of translation-invariant splitting Gibbs measures (TISGMs) is given for the q-state Potts model on a Cayley tree.
- It is shown that at sufficiently low temperatures the maximal number of TISGMs is $2^q - 1$.
- It is proved that there are $[q/2]$ (where $[a]$ is the integer part of a) critical temperatures at which the number of TISGMs changes. The exact number of TISGMs for each intermediate temperature is given.

[10] https://en.wikipedia.org/wiki/Gibbs_measure.

- For the binary tree we give explicit formulae for the critical temperatures and the possible TISGMs.
- We give some regions for the temperature parameter ensuring that a given TISGM is (non-)extreme in the set of all Gibbs measures. In particular we show the existence of a temperature interval for which there are at least $2^{q-1} + q$ extreme TISGMs.
- For each TISGM μ we explicitly give the set of boundary conditions such that limiting Gibbs measures with respect to these boundary conditions coincide with μ.
- For the Potts model with non-zero external field, we give conditions of uniqueness of Gibbs measure. Moreover, the critical lines of the phase transitions are found and their properties are studied.
- For the Potts model with competing one- and two-step interactions on the Cayley tree all periodic (and some non-periodic) ground states are described. By applying a contour method, it is proved that at sufficiently low values of the temperature the model has at least three different limiting Gibbs measures.
- Periodic and weakly periodic Gibbs measures are studied. Under some conditions on the parameters of the Potts model, it is shown that there are weakly periodic (non-periodic) and periodic (non-translation-invariant) Gibbs measures.
- We give some explicit formulas for free energies and entropies in the case of vector-valued boundary conditions. These conditions include translation-invariant and periodic boundary conditions, similar to Dobrushin boundary conditions, as well as boundary conditions corresponding to weakly periodic Gibbs measures.

Chapter 2

Gibbs measures and Potts model

In this chapter we give main definitions and useful facts of the theory of Gibbs measures of lattice spin systems. In the first section we give general definitions. This chapter also contains known results about Gibbs measures of the Potts model on the lattice $\mathbb{Z}^d$ and we give the main method (the theory of Markov random field and the recurrent equations of this theory) of description of Gibbs measures for the Potts model on trees. In the last section we give a review of the applications of the Potts model to medicine, sociology, financial engendering, neural networks, and other sciences. These applications serve to motivate the study of Gibbs measures of the Potts model.

2.1 Definitions

Gibbs measures are familiar subjects in various areas of applied probability. Originally they were defined in the framework of lattice spin systems to describe thermodynamic equilibrium states [75], [98], [220], [267]. In [243] a development of the theory of Gibbs measures on Cayley trees (till 2013) is presented.

2.1.1 *Lattice graphs*

A *graph* is a pair $G = (V, L)$, where V is a set whose elements are called vertices, and $L \subset V \times V$ is a set of two-sets of vertices, whose elements are called edges. The edge connecting $x, y \in V$ is denoted by $l = \langle x, y \rangle$.

A *path* in a graph is a subgraph in which the vertices can be listed in an order $v_1, v_2, \ldots, v_n$ such that the edges are the $\langle v_i, v_{i+1} \rangle$ where $i = 1, 2, \ldots, n-1$.

The distance $d(x, y), x, y \in V$ on the graph is the number of edges of the shortest path from x to y:

$$d(x,y) = \min\{d \,|\, \exists x = x_0, x_1, \dots, x_{d-1}, x_d = y \in V \\ \text{such that } \langle x_0, x_1\rangle, \dots, \langle x_{d-1}, x_d\rangle\}.$$

A (non-empty) set $\Lambda \subset V$ is called *connected* if for any $x, y \in \Lambda$ the path connecting x and y lies in Λ. We denote $\Lambda^c = V \setminus \Lambda$ and the *boundary* of Λ by

$$\partial\Lambda := \{x \in \Lambda^c \colon \exists y \in \Lambda,\ d(x,y) = 1\},$$

and we write $\bar{\Lambda} = \Lambda \cup \partial\Lambda$. The subset of edges in Λ is denoted

$$L_\Lambda := \{\langle x, y\rangle \in L \colon x, y \in \Lambda\}.$$

An unordered pair of vertices $\{x, y\}$ is called connected if a path leads from x to y.

A *connected graph* is an undirected graph in which every unordered pair of vertices in the graph is connected. Otherwise, it is called a disconnected graph.

In this book we consider a *connected* graph $G = (V, L)$, where V is a *countable* set and L is the set of edges.

A *lattice graph* is a graph whose drawing, embedded in some Euclidean space $\mathbb{R}^n$, forms a regular tiling. This implies that the group of bijective transformations that send the graph to itself is a lattice in the group-theoretical sense.

In this book we mainly consider thermodynamic systems on the following two lattice graphs:

- **The lattice $\mathbb{Z}^n$:** the n-dimensional integer lattice (or cubic lattice), which is the lattice in the Euclidean space $\mathbb{R}^n$ whose lattice points are n-tuples of integers. The set $\mathbb{Z}^n$ is an additive group with respect to adding the corresponding components of the vectors.
- **The Cayley tree Γ^k:** The Cayley tree Γ^k of order $k \geq 1$ is an infinite tree, i.e., a graph without cycles, such that exactly $k+1$ edges originate from each vertex. Let $\Gamma^k = (V, L, i)$, where V is the set of vertices Γ^k, L the set of edges and i is the incidence function setting each edge $l \in L$ into correspondence with its endpoints $x, y \in V$. If $i(l) = \{x, y\}$, then the vertices x and y are called the *nearest neighbors*, denoted by $l = \langle x, y\rangle$.

Fix a vertex $x^0 \in V$, interpreted as the *root* of the tree. We say that $y \in V$ is a *direct successor* of $x \in V$ if x is the penultimate vertex on the unique path leading from the root x^0 to the vertex y; that is, $d(x^0, y) = d(x^0, x) + 1$ and $d(x, y) = 1$. The set of all direct successors of $x \in V$ is denoted by $S(x)$.

For a fixed $x^0 \in V$ we set $W_n = \{x \in V | d(x, x^0) = n\}$,

$$V_n = \{x \in V | d(x, x^0) \leq n\}, \quad L_n = \{l = \langle x, y\rangle \in L | x, y \in V_n\}. \tag{2.1}$$

For $x \in W_n$ the set $S(x)$ then has the form

$$S(x) = \{y \in W_{n+1} : \langle x, y\rangle\}. \tag{2.2}$$

For any $x \in V$ denote

$$W_m(x) = \{y \in V : d(x, y) = m\}, \quad m \geq 1.$$

Note that the sequence of balls (V_n) $(n \geq 0)$ is *cofinal* (see [98]), that is, any finite subset $\Lambda \subset V$ is contained in some V_n.

Group representation of the tree.

Let G_k be a free product of $k+1$ cyclic groups of the second order with generators $a_1, a_2, \ldots, a_{k+1}$, respectively, i.e., $a_i^2 = e$, where e is the unit element.

It is known that there exists a one-to-one correspondence between the set of vertices V of the Cayley tree Γ^k and the group G_k.

This correspondence can be given as follows. Fix an arbitrary element $x_0 \in V$ and correspond it to the unit element e of the group G_k. Without loss of generality we assume that the Cayley tree is a planar graph. Using $a_1, \ldots, a_{k+1}$ we numerate nearest-neighbors of element e, moving by positive direction. Now we shall give numeration of the nearest-neighbors of each a_i, $i = 1, \ldots, k+1$ by $a_i a_j$, $j = 1, \ldots, k+1$. Since all a_i have the common neighbor e we give to it $a_i a_i = a_i^2 = e$. Other neighbors are numerated starting from $a_i a_i$ by the positive direction. We numerate the set of all nearest-neighbors of each $a_i a_j$ by words $a_i a_j a_q$, $q = 1, \ldots, k+1$, starting from $a_i a_j a_j = a_i$ by the positive direction. Iterating this argument one gets a one-to-one correspondence (a bijection):

$$\mathfrak{b} : V \to G_k \tag{2.3}$$

(see Chapter 1 of [243] for properties of the group G_k).

Comparing the Cayley tree and $\mathbb{Z}^n$ one can notice that [14]:

(i) The group representation G_k of the Cayley tree is non-commutative, but $\mathbb{Z}^n$ is commutative group.
(ii) The ratio of the number of boundary sites to the number of interior sites of the lattice $\mathbb{Z}^n$ becomes small in the thermodynamic limit of a

large system. For the Cayley tree it does not, since both numbers grow exponentially like k^n. The Cayley tree is a non-amenable graph.

The use of Cayley trees is often motivated by the applications, such as information flows [187] and reconstruction algorithms on networks [59], [169], DNA strands and Holliday junctions [248], [253], evolution of genetic data and phylogenetics [58], bacterial growth and fire forest models [39], or computational complexity on graphs [96]. Crucially, the criticality in such models is governed by phase transitions in the underlying spin systems (in particular, the Potts model).

2.1.2 *Configuration space*

Take a finite set $\Phi = \{1, 2, \ldots, q\}$, $q \geq 2$. This is the set of spin values. For $A \subseteq V$ a spin *configuration* σ_A on A is defined as a function

$$x \in A \to \sigma_A(x) \in \Phi.$$

The set of all configurations coincides with $\Omega_A = \Phi^A$. The cardinality of Ω_A is $|\Omega_A| = |\Phi|^{|A|} = q^{|A|}$.

We denote $\Omega = \Omega_V$ and $\sigma = \sigma_V$. Since V is a countable set, Ω is an uncountable set.

Let the lattice G be a group (like $\mathbb{Z}^n$ or G_k) and G^* be a subgroup of the group G. A configuration $\sigma \in \Omega$ is called G^*-periodic if $\sigma(yx) = \sigma(x)$ for any $x \in G_k$ and $y \in G^*$.

A configuration that is invariant with respect to all shifts is called *translation-invariant.*

In the theory of Gibbs measures the interesting lattices (i.e., graphs on which one considers the measure) are countable and connected. In this book, we mainly consider $\mathbb{Z}^n$ and Γ^k, let V denotes the set of vertices one of these lattices.

In the set $\Omega = \Phi^V$ a notion of convergence for sequences of configurations (i.e., topology) is defined as follows [75]:

Definition 2.1. A sequence $\sigma^{(n)} \in \Omega$ converges to $\sigma \in \Omega$ (written as $\sigma^{(n)} \to \sigma$) if

$$\lim_{n\to\infty} \sigma^{(n)}(x) = \sigma(x), \quad x \in V.$$

Recall sequentially compactness: for every sequence $\{\sigma^{(n)}\}_{n\geq 1} \subset \Omega$, there exists $\sigma^* \in \Omega$ and a subsequence $\{n_m\}_{m\geq 1}$ such that $\sigma^{(n_m)} \to \sigma^*$ when $m \to \infty$.

Since Φ is finite the following fact is known

Proposition 2.1. [75] *With the above notion of convergence, Ω is compact.*

Definition 2.2. A function $f : \Omega \to \mathbb{R}$ is called continuous if $\sigma^{(n)} \to \sigma$ implies $f(\sigma^{(n)}) \to f(\sigma)$. The set of all continuous functions on Ω is denoted by $C(\Omega)$.

2.1.3 *Hamiltonian*

The energy of the configuration $\sigma \in \Omega$ is given by the formal Hamiltonian

$$H(\sigma) = \sum_{\substack{A\subset V:\\ \mathrm{diam}(A)\leq r}} I(\sigma_A), \tag{2.4}$$

where $r \in \mathbb{N}$, $\mathrm{diam}(A) = \max_{x,y\in A} d(x,y)$, $I(\sigma_A) : \Omega_A \to \mathbb{R}$ is a given potential.

For a finite domain $D \subset V$ with the boundary condition φ_{D^c} given on its complement $D^c = V \setminus D$, the conditional Hamiltonian is

$$H(\sigma_D \big| \varphi_{D^c}) = \sum_{\substack{A\subset V: A\cap D\neq\emptyset\\ \mathrm{diam}(A)\leq r}} I(\sigma_A), \tag{2.5}$$

where

$$\sigma_A(x) = \begin{cases} \sigma(x) \text{ if } & x \in A\cap D \\ \varphi(x) \text{ if } & x \in A \cap D^c. \end{cases}$$

A *ground state* of (2.4) is a configuration φ such that $H(\varphi) \leq H(\sigma)$ for all $\sigma \in \Omega$.

2.1.4 *The q-state Potts model*

In the book we mainly consider the Potts model, where the spin at each vertex $x \in V$ can take values in the set $\Phi := \{1, \dots, q\}$. The Potts model with a *nearest-neighbor interaction kernel* $\{J_{xy}\}_{x,y\in V}$ (i.e., such that $J_{xy} = J_{yx}$ and $J_{xy} = 0$ if $d(x,y) \neq 1$) is defined by the formal Hamiltonian

$$H(\sigma) = -\sum_{\langle x,y\rangle\in L} J_{xy}\,\delta_{\sigma(x),\sigma(y)} - \sum_{x\in V} \xi_{\sigma(x)}(x), \qquad \sigma \in \Phi^V, \tag{2.6}$$

where δ_{ij} is the Kronecker delta symbol, i.e.,

$$\delta_{ij} = \begin{cases} 1, & \text{if } i = j \\ 0, & \text{if } i \neq j, \end{cases}$$

and $\xi(x) = (\xi_1(x), \dots, \xi_q(x)) \in \mathbb{R}^q$ is the external (possibly random) field.

Remark 2.1. By the Hamiltonian (2.6) we note that

- The spin-spin interaction is activated only when the neighboring spins are equal, whereas the additive contribution of the external field is provided, at each vertex $x \in V$, by the component of the vector $\xi(x)$ corresponding to the spin value $\sigma(x)$.
- For $q = 2$, the Potts model is equivalent to the Ising model with redefined spins

$$\tilde{\sigma}(x) := 2\sigma(x) - 3 \in \{-1, 1\}, \qquad x \in V,$$

whereby the Hamiltonians in the two models are linked through the relations

$$\delta_{\sigma(x),\sigma(y)} = \frac{\tilde{\sigma}(x)\tilde{\sigma}(y) + 1}{2},$$

$$\xi_{\sigma(x)}(x) = \frac{\xi_2(x) - \xi_1(x)}{2}\tilde{\sigma}(x) + \frac{\xi_1(x) + \xi_2(x)}{2}.$$

2.1.5 *Gibbs measure*

Recall that a probability space is a triple $(\Omega, \mathcal{F}, \mu)$ where:

- Ω is an arbitrary non-empty set,
- $\mathcal{F}$ is a σ-algebra, i.e., a set of subsets of Ω, such that:
  (1) $\Omega \in \mathcal{F}$;
  (2) If $A \in \mathcal{F}$, then also $\Omega \setminus A \in \mathcal{F}$;
  (3) if $A_i \in \mathcal{F}$ for $i = 1, 2, \ldots$, then also

$$\bigcup_{i=1}^{\infty} A_i \in \mathcal{F}.$$

- $\mu : \mathcal{F} \to [0, 1]$ is a probability measure, i.e., is countably additive[1] and $\mu(\Omega) = 1$.

---

[1]If $\{A_i\}_{i=1}^{\infty} \subseteq \mathcal{F}$ is a countable collection of pairwise disjoint sets, then

$$\mu(\bigcup_{i=1}^{\infty} A_i) = \sum_{i=1}^{\infty} \mu(A_i).$$

Any element of $\mathcal{F}$ is called a measurable set and the pair $(\Omega, \mathcal{F})$ is called a measurable space.

Let $A \subset V$ be a subset. On this set fix a configuration φ_A then, the following set is called a cylinder set with base A:

$$\mathcal{S}(\varphi_A) = \{\sigma \in \Omega : \sigma\big|_A = \varphi_A\},$$

that is the set of all configurations from Ω which on the set A coincide with φ_A.

The cylinder σ-algebra is the one that is generated[2] by cylinder sets.

The σ-algebra generated by cylinders with base contained in A is denoted by $\mathcal{F}_A$, this consists of all the events that depend only on the spins inside A.

Denote by $\mathcal{M}_1(\Omega, \mathcal{F})$ the set of all probability measures on the measurable space $(\Omega, \mathcal{F})$. It is known (see [75]) that probability measures on a measurable space $(\Omega, \mathcal{F})$ are uniquely determined by their values on cylinders. Moreover, if $\mu, \nu \in \mathcal{M}_1(\Omega, \mathcal{F})$, then $\mu = \nu$ iff $\mu(\mathcal{S}) = \nu(\mathcal{S})$ for all cylinders $\mathcal{S}$.

In this book, we mainly consider the probability space $(\Omega, \mathcal{B}, \mu)$ where

- Ω is the set of all configurations, i.e. $\Omega = \Phi^V$;
- $\mathcal{B}$ is the cylinder σ-algebra of Ω;
- μ is a *Gibbs measure* which is a probability measure which satisfies (with Hamiltonian H) the Dobrushin-Lanford-Ruelle (DLR) equation (see [60], [144]): for all finite $D \subset V$ and $\sigma_D \in \Omega_D$:

$$\mu\left(\{\omega \in \Omega : \omega\big|_D = \sigma_D\}\right) = \int_\Omega \mu(\mathrm{d}\varphi)\nu_\varphi^D(\sigma_D), \tag{2.7}$$

where ν_φ^D is the conditional probability:

$$\nu_\varphi^D(\sigma_D) = \frac{1}{Z_{D,\varphi}} \exp\left(-\beta H\left(\sigma_D\big|\, \varphi_{D^c}\right)\right). \tag{2.8}$$

Here $\beta = \frac{1}{T}, T > 0-$ temperature and $Z_{D,\varphi}$ stands for the partition function in D, with the boundary condition φ:

$$Z_{D,\varphi} = \sum_{\widetilde{\sigma}_D \in \Omega_D} \exp\left(-\beta H\left(\widetilde{\sigma}_D \,\big|\, \varphi_{D^c}\right)\right).$$

[2]Let F be an arbitrary family of subsets of Ω. Then there exists a unique smallest σ-algebra which contains every set in F (even though F may or may not itself be a σ-algebra). It is, in fact, the intersection of all σ-algebras containing F. This σ-algebra is called the σ-algebra generated by F.

Gibbs measure for the Potts model. For each finite subset $\Lambda \subset V$ ($\Lambda \neq \emptyset$) and any fixed sub-configuration $\eta \in \Phi^{\Lambda^c}$ (called the *configurational boundary condition*), the *Gibbs measure* γ_Λ^η is a probability measure in Φ^Λ defined by the formula

$$\gamma_\Lambda^\eta(\varsigma) = \frac{1}{Z_\Lambda^\eta(\beta)} \exp\left\{ -\beta H_\Lambda(\varsigma) + \beta \sum_{x\in\Lambda} \sum_{y\in\Lambda^c} J_{xy}\, \delta_{\varsigma(x),\eta(y)} \right\}, \qquad \varsigma \in \Phi^\Lambda, \tag{2.9}$$

where H_Λ is the restriction of the Hamiltonian (2.6) to configurations in Λ,

$$H_\Lambda(\varsigma) = -\sum_{\langle x,y\rangle \in L_\Lambda} J_{xy}\, \delta_{\varsigma(x),\varsigma(y)} - \sum_{x\in\Lambda} \xi_{\varsigma(x)}(x), \qquad \varsigma \in \Phi^\Lambda, \tag{2.10}$$

and $Z_\Lambda^\eta(\beta)$ is the partition function (or normalizing constant)

$$Z_\Lambda^\eta(\beta) = \sum_{\varsigma\in\Phi^\Lambda} \exp\left\{ -\beta H_\Lambda(\varsigma) + \beta \sum_{x\in\Lambda} \sum_{y\in\Lambda^c} J_{xy}\, \delta_{\varsigma(x),\eta(y)} \right\}.$$

In the Potts model, the thermodynamic functions (internal energy, specific heat, entropy, and free energy) on the volume Λ can be derived from the partition function:

- The internal energy, which is the sum of the potential and kinetic energy and is defined as
$$U = -\frac{\partial \ln Z_\Lambda^\eta(\beta)}{\partial \beta};$$
- Specific heat, or the energy required to raise a unit amount of material one temperature increment, is
$$C = \frac{\partial U}{\partial T};$$
- Entropy, a measure of the randomness and disorder in a system, is
$$S = \beta U + k_B \ln Z_\Lambda^\eta(\beta);$$
- The total free energy is
$$F = k_B \cdot T \cdot \ln Z_\Lambda^\eta(\beta).$$

Using the nearest-neighbor interaction, formula (2.9) can be rewritten as

$$\gamma_\Lambda^\eta(\varsigma) = \frac{1}{Z_\Lambda^\eta(\beta)} \exp\left\{ -\beta H_\Lambda(\varsigma) + \beta \sum_{x\in\Lambda} \sum_{y\in\partial\Lambda} J_{xy}\, \delta_{\varsigma(x),\eta(y)} \right\}, \qquad \varsigma \in \Phi^\Lambda. \tag{2.11}$$

Now a measure $\mu = \mu_{\beta,\xi}$ on Φ^V is called a *Gibbs measure* if, for any non-empty finite set $\Lambda \subset V$ and any $\eta \in \Phi^{\Lambda^c}$,

$$\mu(\sigma_\Lambda = \varsigma \,|\, \sigma_{\Lambda^c} = \eta) = \gamma_\Lambda^\eta(\varsigma), \qquad \varsigma \in \Phi^\Lambda. \tag{2.12}$$

The main problem is to study the structure of the set $\mathcal{G}(H)$ of all Gibbs measures corresponding to a given Hamiltonian H.

The existence of Gibbs measures for a wide class of Hamiltonians was established in the ground-breaking work of Dobrushin (see, e.g., [98], [75], [267]). However, a complete analysis of the set of Gibbs measures for a given Hamiltonian is often a difficult problem.

Usually, for small values of β (i.e., high temperatures) a Gibbs measure is unique (see [98], [220], [267]) which reflects a physical fact that at high temperatures there is no phase transition. The analysis for low temperatures requires specific assumptions on the form of the Hamiltonian.

Let $\mu_1, \mu_2 \in \mathcal{M}_1(\Omega, \mathcal{F})$ and $\lambda \in [0, 1]$. Then convex combination of these measures is defined by

$$(\lambda\mu_1 + (1 - \lambda)\mu_2)(A) = \lambda\mu_1(A) + (1 - \lambda)\mu_2(A), \quad \forall A \in \mathcal{F}.$$

A set $M \subset \mathcal{M}_1(\Omega, \mathcal{F})$ is called convex if for any $\mu_1, \mu_2 \in M$ and any $\lambda \in [0, 1]$ one has $\lambda\mu_1 + (1 - \lambda)\mu_2 \in M$.

For continuous Hamiltonians (see Definition 2.2) which we consider in this book it is known that for each $\beta > 0$ the Gibbs measures form a non-empty convex compact set $\mathcal{G}(H)$ in the space of all probability measures on $\Omega = \Phi^V$ endowed with the weak topology (see, e.g., [98]).

A measure $\mu \in \mathcal{G}(H)$ is called *extreme* (also called pure state) if it cannot be expressed as $\mu = \lambda\mu_1 + (1 - \lambda)\mu_2$ for some $\mu_1, \mu_2 \in \mathcal{G}(H)$ with $\mu_1 \neq \mu_2$.

The set of all extreme measures in $\mathcal{G}(H)$ denoted by ex$\mathcal{G}(H)$ is a *Choquet simplex*, in the sense that any $\mu \in \mathcal{G}(H)$ can be represented as $\mu = \int_{\mathrm{ex}\mathcal{G}(H)} \nu\, \rho(\mathrm{d}\nu)$, with some probability measure ρ on ex$\mathcal{G}(H)$.

Thus to describe the set $\mathcal{G}(H)$ one has to describe all of its extremal elements.

Extreme Gibbs measures are characterized by the fact that they possess deterministic macroscopic properties, corresponding to the tail σ-algebra defined:

$$\mathcal{T}_\infty = \bigcap_{A \subset V} \mathcal{F}_{V \setminus A},$$

where intersection is taken by finite sets A. Each element of this set is called a tail (or macroscopic) event.

The following theorem characterized the extreme Gibbs measure (see [98] and [75]).

Theorem 2.1. *For each Gibbs measure $\mu \in \mathcal{G}(H)$ the following conditions are equivalent*

1. *μ is extreme in $\mathcal{G}(H)$;*
2. *μ is trivial on the tail σ-algebra $\mathcal{T}_\infty$, i.e., if $A \in \mathcal{T}_\infty$, then $\mu(A)$ is either 1 or 0;*
3. *All $\mathcal{T}_\infty$-measurable functions[3] are μ-almost surely constant;[4]*
4. *μ has short-range correlations: for all $A \in \mathcal{F}$*

$$\lim_{\Lambda \nearrow V} \sup_{B \in \mathcal{F}_{\Lambda^c}} \left|\mu(A \cap B) - \mu(A)\mu(B)\right| = 0.$$

In general, it is very hard to check the criterion mentioned in Theorem 2.1, therefore the problem of extremality in many cases remains open. There are some complete results in the case of Ising model.

The main purpose of this book is to study extreme points of the set of Gibbs measures for the Potts model and apply them to study thermodynamic systems of biology and physics. Extremal Gibbs measures are very important to understanding all possible local behaviors of the (biological and physical) system.

2.2 A review of the Potts model on $\mathbb{Z}^d$

In March of 2022, MathSciNet found more than 2030 entries for "Potts model" in the entire database. By this database the first publication related to the Potts model is [62], which was published in 1951. In just 2020 there were more than 50 publications. Since there are a lot of literature devoted to this model, here I will review only these publications which are considered on $\mathbb{Z}^d$ and mainly devoted to Gibbs measures.

Remark 2.2. To study Gibbs measures for the Hamiltonians on $\mathbb{Z}^d$ one of powerful method is a contour method of the Pirogov-Sinai theory which

---

[3] A Borel set is any subset of $\mathbb{R}$ that can be formed from open sets (or, equivalently, from closed sets) through the operations of countable union, countable intersection, and relative complement. A function $f : \Omega \to \mathbb{R}$ is called $\mathcal{T}_\infty$-measurable (or macroscopic observable) if $f^{-1}(B) \in \mathcal{T}_\infty$, for any Borel subset $B \subset \mathbb{R}$.

[4] A function $f : \Omega \to \mathbb{R}$ is called μ-almost surely constant if there is a constant $c \in \mathbb{R}$ such that $\mu\{\sigma \in \Omega : f(\sigma) = c\} = 1$.

provides a geometric-probabilistic argument to rigorously describe low-temperature Gibbs measures and to formalize the idea that they concentrate on configurations which basically consist in a single configuration perturbed by a small fraction of fluctuations. The latter are defects whose boundary defines a contour and are directly inspired by the classical contours of the argument of Peierls to prove phase transition for the two-dimensional Ising model [92], [267], [307]. There are also Cluster expansions, the FKG inequalities and monotonicity methods [36], [179].

Using the Pirogov-Sinai theory [267] for the q-state Potts model on $\mathbb{Z}^d$ it was proven that there exist critical temperature $T_{\rm cr} \equiv T_{\rm cr}(q,d)$ such that if $T > T_{\rm cr}$ then there is unique Gibbs measure and for any $T < T_{\rm cr}$ there are q distinct extreme Gibbs measures $\mu_\beta^{(i)}$, $i \in \Phi = \{1, \dots, q\}$. Recall that $\beta = 1/T$ and then $\beta_c = 1/T_{\rm cr}$.

Thus the Potts model undergoes a phase transition in infinite volume at the critical temperature $T_{\rm cr}$ in the following sense:

$$\mu_\beta^{(i)}(\sigma(0) = i) = \begin{cases} 1/q & \text{if } \ T > T_{\rm cr} \\ > 1/q & \text{if } \ T < T_{\rm cr} \end{cases} \qquad \text{for any } \ i \in \Phi.$$

The phase transition is called continuous if $\mu_{\beta_c}^{(i)}(\sigma(0) = i) = 1/q$ and discontinuous otherwise.

The main result of the paper [64] states that for $q \in \{1, 2, 3, 4\}$, the Potts model has continuous phase transition. In the case $q = 2$ the result was known for the Ising model. The result was proved by using the random-cluster model (see [2], [17], [52], [65], [106]) with cluster-weight $q \geq 1$, which is not necessarily an integer.

For the critical temperature, $T_{\rm cr}$, denote

$$\tau = \frac{T - T_{\rm cr}}{T_{\rm cr}},$$

then one has to express any of the thermodynamic functions in the form $g(\tau)$, where $g(\tau) \approx c \cdot |\tau|^p$ on the neighborhood of $\tau = 0$ for some p which is called the critical exponent [16]. The critical exponents fall into a set of discrete universality classes, where universality means that the values of the critical exponents are independent of parameters such as the interaction energy J and the choice of lattice (although not its dimension).

Following [16], [217], [273] we recall that phase transitions are broadly classified as first-order (discontinuous) or second-order (continuous, but nonanalytic). Second-order phase transitions are further classified by the critical exponents, since if the phase transitions are continuous, the leading singular behavior as $\tau \to 0$ of the thermodynamic quantities may normally be written in the form $c \cdot |\tau|^p$, where p may be a positive or negative power (there are also cases where the singularity is non-algebraic).

It is also known that the two-dimensional ferromagnetic (i.e., $J > 0$) q-state Potts model, for the thermodynamic limit of regular two-dimensional lattice graphs, provides a set of several different universality classes associated with second-order phase transitions depending on q (in particular for $q = 2, 3, 4$). For $q \geq 5$, the phase transition of this two-dimensional ferromagnetic Potts model is first order.

In [257] it was proved that the anti-ferromagnetic (i.e., $J < 0$) Potts model with q spins has no finite temperature phase transition on lattices where each vertex has degree less than $q/2$.

The value of the critical temperature for the ferromagnetic Potts model on the infinite volume limit of the square lattice has been rigorously determined to be $T_{\rm cr} = \frac{J}{k_B \ln(1+\sqrt{q})}$, and the critical behavior of the two-dimensional Potts ferromagnetic model is known (see [14], [72]).

In [55] the authors presented several results concerning description of Gibbs measures of the Potts model on $\mathbb{Z}^2$ below the critical temperature. It is proved that all Gibbs measures of the q-state Potts model are convex combinations of the q distinct extreme Gibbs measures $\mu^{(i)}$, $i = 1, \dots, q$. It follows from this result that any Gibbs measure is translation-invariant and the extreme Gibbs measures are the infinite-volume measures with pure boundary conditions (that are $i \in \{1, \dots, q\}$, meaning the configurations $\sigma^{(i)}(x) = i$ for every $x \in V$).

The paper [55] also gives several open problems (which are still open) related to the Gibbs measures of q-state Potts model on $\mathbb{Z}^d$:

Critical 2d Potts models. The behavior of two-dimensional q-state Potts models in the critical regime $\beta = \beta_c(q)$ is open. It is conjectured that there is a unique Gibbs state when $q = 3$ and 4, but that, for $q \geq 5$, there is coexistence at β_c of $q + 1$ pure phases: the q low-temperature ordered pure phases and the high temperature disordered phase. This is known to be true when q is large enough [143], [175]. The extension of the latter result to every $q > 4$ remains a mathematical challenge.

Finite-range 2d models. Recall that the Potts model describes nearest-neighboring interactions. The extension of the results of [55], even in the Ising case (i.e., $q = 2$), to general *finite-range* interactions still an open problem.

The question of quasi-periodicity. There is a general conjecture that two dimensional models should always possess a finite number of extreme Gibbs measures, all of which are periodic. In particular, this would imply that all Gibbs measures are periodic, and thus that a two-dimensional quasi-crystal cannot exist (as an equilibrium state).

Models in higher dimensions. The situation in higher dimensions is very different, due to the existence of translation non-invariant states. Even in the very low-temperature 3-dimensional Ising model, the set of extreme Gibbs measures is not known. But in the case of a d-dimensional Ising model for any $d \geq 3$, it is known that all translation-invariant Gibbs measures are convex combinations of the two pure phases at all temperatures [27]. A similar result also holds for large enough values of q [175].

In [97] various q-dimensional central limit theorems are proved for the q-state Potts model on $\mathbb{Z}^d$, $d \geq 2$ for the temperature T sufficiently far from the critical point $T_{\rm cr}$. In particular, it is shown that for $T > T_{\rm cr}$ sufficiently high, and with respect to the unique infinite volume Gibbs measure, the frequencies of the q types of spins in a large box Λ_n, centered and normalized by $\sqrt{|\Lambda_n|}$, converge to a Gaussian limit as $n \to \infty$. For $T < T_{\rm cr}$ sufficiently low, and with respect to the infinite-volume Gibbs measure arising from free boundary conditions, the limit is shown to be a mixture of Gaussian random variables. In case $d = 2$, and $q = 2$ or $q \geq 26$, the result is valid for each $T \neq T_{\rm cr}$. The proof is based on the random cluster representation for the q-state Potts model.

In [202] the authors proved metastability behavior at zero-temperature of the q-state ferromagnetic Potts model. Namely, the properties of the tunneling time between stable configurations of the Glauber dynamics for the Potts model are studied.

In [88] the author considers an extension of the q-state Potts model on $\mathbb{Z}^d$. There a single-spin configuration space is generalized from the finite set $\{1, \ldots, q\}$ to the set of all non-negative integers. Moreover, the a priori single-spin counting measure $\nu(k) = 1/q$ is replaced by a Poisson measure with arbitrary constant λ:

$$\nu(k) = \exp(-\lambda)\,\frac{\lambda^k}{k!}, \quad k \geq 0.$$

The main result of [88] is that the set of all limit Gibbs measures on $\mathbb{Z}^d$ of this model is nonempty. The proof relies on Dobrushin's theorem [267], which states a sufficient condition for the existence of at least one limit Gibbs measure.

2.3 Gibbs measures on trees

Mathematical foundations of random fields on Cayley trees were laid by an extensive analysis of Gibbs measures and phase transitions (see [98] and [243]).

There are several approaches to derive the equation solutions of which describe the Gibbs measures for lattice models on the Cayley tree. One approach is based on properties of Markov random fields on Cayley tree [220] and [271]. Another approach is based on recurrent equations for partition functions [93], [131].

In this section we mainly follow [28] and [139] to explain the method of Markov random field theory and its recurrent equations.

Remark 2.3. 1. Note that a contour method (as in the Pirogov-Sinai theory) is also developed for spin systems defined on trees [33], [192], [237], [240], [245]. Results of these papers systematically presented in Chapter 10 of [243].

2. Another approach is based on the Lee-Yang theory which links the study of phase transitions to the localization of zeros in the complex fugacity plane. This theory initialed by the Lee-Yang theorem which states that if partition functions of certain models in statistical field theory with ferromagnetic interactions are considered as functions of an external field, then all zeros are purely imaginary (or on the unit circle after a change of variable). The first version was proved for the Ising model by T. D. Lee and C. N. Yang in [149]. Their result was later extended to more general models (see for example [25], [51], [115], [154], [172], [173]).

2.3.1 *Splitting Gibbs measure*

To construct Gibbs measures on the Cayley tree Γ^k it is convenient to use a version of Gibbs distributions on the balls (V_n) defined via auxiliary fields encapsulating the interaction with the exterior of the balls. Namely, for a vector field $V \ni x \mapsto \mathbf{h}(x) = (h_1(x), \dots, h_q(x)) \in \mathbb{R}^q$ and each $n \in \mathbb{N}_0 = \{0, 1, \dots\}$, define a probability measure in Φ^{V_n} by the formula

$$\mu_n^h(\sigma_n) = \frac{1}{Z_n} \exp\left\{ -\beta H_n(\sigma_n) + \beta \sum_{x \in W_n} h_{\sigma_n(x)}(x) \right\}, \qquad \sigma_n \in \Phi^{V_n}, \tag{2.13}$$

where $Z_n = Z_n(\beta, \mathbf{h})$ is the normalizing factor and $H_n := H_{V_n}$, that is (see (2.10)),

$$H_n(\sigma_n) = -\sum_{\langle x,y\rangle \in L_n} J_{xy}\, \delta_{\sigma_n(x),\sigma_n(y)} - \sum_{x\in V_n} \xi_{\sigma_n(x)}(x), \qquad \sigma_n \in \Phi^{V_n}. \quad (2.14)$$

The vector field $\{\mathbf{h}(x)\}_{x\in V}$ in (2.13) is called *generalized boundary conditions* (GBC).

One says that the probability distributions (2.13) are *compatible* (and the intrinsic GBC $\{\mathbf{h}(x)\}$ are *permissible*) if for each $n \in \mathbb{N}_0$ the following identity holds,

$$\sum_{\omega\in\Phi^{W_{n+1}}} \mu^h_{n+1}(\sigma_n \vee \omega) \equiv \mu^h_n(\sigma), \qquad \sigma_n \in \Phi^{V_n}, \quad (2.15)$$

where the symbol $\vee$ stands for concatenation of sub-configurations.

By Kolmogorov's extension theorem (see, e.g., [266] and more suitable for our setting is [75]), the compatibility condition (2.15) ensures that there exists a unique measure $\mu^h = \mu^h_{\beta,\xi}$ on Φ^V such that, for all $n \in \mathbb{N}_0$,

$$\mu^h(\sigma_{V_n} = \sigma_n) \equiv \mu^h_n(\sigma_n), \qquad \sigma_n \in \Phi^{V_n}, \quad (2.16)$$

or more explicitly (substituting (2.13)),

$$\mu^h(\sigma_{V_n} = \sigma_n) = \frac{1}{Z_n} \exp\left\{-\beta H_n(\sigma_n) + \beta \sum_{x\in W_n} h_{\sigma_n(x)}(x)\right\}, \qquad \sigma_n \in \Phi^{V_n}. \quad (2.17)$$

Definition 2.3. Measure μ^h satisfying (2.16) is called a *splitting Gibbs measure* (*SGM*).

Remark 2.4. Note that adding a constant $c = c(x)$ to all coordinates $h_i(x)$ of the vector $\mathbf{h}(x)$ does not change the probability measure (2.13) due to the normalization Z_n. The same is true for the external field $\xi(x)$ in the Hamiltonian (2.14). Therefore, without loss of generality consider *reduced GBC* $\check{\mathbf{h}}(x)$, for example defined as

$$\check{h}_i(x) = h_i(x) - h_q(x), \qquad i = 1, \ldots, q-1.$$

The same remark also applies to the external field ξ and its reduced version $\check{\xi}(x)$, defined by

$$\check{\xi}_i(x) := \xi_i(x) - \xi_q(x), \qquad i = 1, \ldots, q-1.$$

Such a reduction can equally be done by subtracting any other coordinate,

$$_\ell\check{h}_i(x) := h_i(x) - h_\ell(x), \qquad {}_\ell\check{\xi}_i(x) := \xi_i(x) - \xi_\ell(x) \qquad (i \neq \ell).$$

Therefore, when working with vectors and vector-valued functions and fields it will often be convenient to pass from a generic vector $\boldsymbol{u} = (u_1, \ldots, u_q)$ to a "reduced vector" $\check{\boldsymbol{u}} = (\check{u}_1, \ldots, \check{u}_{q-1}) \in \mathbb{R}^{q-1}$ by setting $\check{u}_i := u_i - u_q$ $(i = 1, \ldots, q-1)$.

The following statement describes a criterion for the GBC $\{\boldsymbol{h}(x)\}_{x\in V}$ to guarantee compatibility of the measures $\{\mu_n^h\}_{n\in\mathbb{N}_0}$.

Theorem 2.2. *The probability distributions $\{\mu_n^h\}_{n\in\mathbb{N}_0}$ defined in (2.13) are compatible if and only if the following vector identity holds*

$$\beta\check{\boldsymbol{h}}(x) = \sum_{y\in S(x)} \boldsymbol{F}\big(\beta\check{\boldsymbol{h}}(y) + \beta\check{\boldsymbol{\xi}}(y); \mathrm{e}^{\beta J_{xy}}\big), \qquad x \in V, \tag{2.18}$$

where $\check{\boldsymbol{h}}(x) = (\check{h}_i(x), \ldots, \check{h}_{q-1}(x))$, $\check{\boldsymbol{\xi}}(x) = (\check{\xi}_i(x), \ldots, \check{\xi}_{q-1}(x))$,

$$\check{h}_i(x) := h_i(x) - h_q(x), \qquad \check{\xi}_i(x) := \xi_i(x) - \xi_q(x), \qquad i = 1, \ldots, q-1, \tag{2.19}$$

and the map $\boldsymbol{F}(\boldsymbol{u};\theta) = (F_1(\boldsymbol{u};\theta), \ldots, F_{q-1}(\boldsymbol{u};\theta))$ *is defined for* $\boldsymbol{u} = (u_1, \ldots, u_{q-1}) \in \mathbb{R}^{q-1}$ *and* $\theta > 0$ *by the formulas*

$$F_i(\boldsymbol{u};\theta) := \ln \frac{(\theta - 1)\,\mathrm{e}^{u_i} + 1 + \sum_{j=1}^{q-1} \mathrm{e}^{u_j}}{\theta + \sum_{j=1}^{q-1} \mathrm{e}^{u_j}}, \qquad i = 1, \ldots, q-1. \tag{2.20}$$

Proof. *Necessity.* Denote $\boldsymbol{\zeta}(x) := \boldsymbol{h}(x) + \boldsymbol{\xi}(x)$. Assume that the compatibility condition (2.15) holds. Substituting (2.13), it is easy to see that (2.15) can be rewritten as

$$\begin{aligned}\prod_{x\in W_n} \prod_{y\in S(x)} \sum_{\omega(y)\in\Phi} &\exp\big\{\beta\big(J_{xy}\,\delta_{\sigma_n(x),\,\omega(y)} + \zeta_{\omega(y)}(y)\big)\big\}\\ &= \frac{Z_{n+1}}{Z_n} \prod_{x\in W_n} \exp\big\{\beta h_{\sigma_n(x)}(x)\big\},\end{aligned} \tag{2.21}$$

for any $\sigma_n \in \Phi^{V_n}$.

Consider now the equality (2.21) on configurations $\sigma_n^1, \sigma_n^2 \in \Phi^{V_n}$ that coincide everywhere in V_n except at vertex $x \in W_n$, where $\sigma_n^1(x) = i \le q-1$ and $\sigma_n^2(x) = q$. Taking the log-ratio of the two resulting relations, we obtain

$$\sum_{y\in S(x)} \ln \frac{\exp\big\{\beta\big(J_{xy} + \zeta_i(y)\big)\big\} + \sum_{j\ne i} \exp\big\{\beta\zeta_j(y)\big\}}{\exp\big\{\beta\big(J_{xy} + \zeta_q(y)\big)\big\} + \sum_{j=1}^{q-1} \exp\big\{\beta\zeta_j(y)\big\}} = \beta\big(h_i(x) - h_q(x)\big),$$

which is readily reduced to (2.18) in view of the notation (2.19) and (2.20).

Sufficiency. Using (2.19) and (2.20), equation (2.18) can be rewritten in the coordinate form as follows

$$\prod_{y\in S(x)} \sum_{j=1}^{q} \exp\{\beta(J_{xy}\delta_{ij}+\zeta_j(y))\} = a(x)\exp\{\beta h_i(x)\}, \quad i=1,\dots,q-1, \tag{2.22}$$

where we denote

$$a(x) := \exp\{\beta h_q(x)\} \prod_{y\in S(x)} \sum_{j=1}^{q} \exp\{\beta(J_{xy}\delta_{qj}+\zeta_j(y))\}, \qquad x\in V.$$

Hence, using (2.22) and setting $A_n := \prod_{x\in W_n} a(x)$, we get

$$\begin{aligned}\sum_{\omega\in\Phi^{W_{n+1}}} &\mu_{n+1}^h(\sigma_n\vee\omega)\\ &= \frac{\exp\{-\beta H_n(\sigma_n)\}}{Z_{n+1}} \prod_{x\in W_n}\prod_{y\in S(x)}\sum_{j=1}^{q}\exp\{\beta(J_{xy}\delta_{\sigma_n(x),j}+\zeta_j(y))\}\\ &= \frac{A_n}{Z_{n+1}}\exp\left\{-\beta H_n(\sigma_n)+\beta\sum_{x\in W_n} h_{\sigma_n(x)}(x)\right\} = \frac{A_n Z_n}{Z_{n+1}}\mu_n^h(\sigma_n). \end{aligned}\tag{2.23}$$

We have that

$$\sum_{\sigma_n\in\Phi^{V_n}}\sum_{\omega\in\Phi^{W_{n+1}}}\mu_{n+1}^h(\sigma_n\vee\omega) = \sum_{\sigma_{n+1}\in\Phi^{V_{n+1}}}\mu_{n+1}^h(\sigma_{n+1}) = 1,$$

whereas from the right-hand side of (2.23) the same sum is given by

$$\sum_{\sigma_n\in\Phi^{V_n}}\frac{A_n Z_n}{Z_{n+1}}\mu_n^h(\sigma_n) = \frac{A_n Z_n}{Z_{n+1}}.$$

Hence, $A_n Z_n/Z_{n+1} = 1$ and formula (2.23) yields (2.15), as required. This completes the proof. □

By virtue of Theorem 2.2, if the GBC $\{\boldsymbol{h}(x)\}$ and the external field $\{\boldsymbol{\xi}(x)\}$ satisfy the functional equation (2.18) for some $\beta>0$ then there is a (unique) measure $\mu_{\beta,\xi}^h$. This measure is a *Gibbs measure* (see (2.12); since the family (V_n) is cofinal, by [98] it suffices to check that, for each $n\in\mathbb{N}_0$ and any $\eta\in\Phi^{V_n^c}$,

$$\mu^h(\sigma_{V_n}=\sigma_n\,|\,\sigma_{V_n^c}=\eta)\equiv\gamma_n^\eta(\sigma_n), \qquad \sigma_n\in\Phi^{V_n}, \tag{2.24}$$

where γ_n^η is the Gibbs distribution in V_n with configurational boundary condition η (cf. (2.9)). Indeed, denote $\omega := \eta_{W_{n+1}}\in\Phi^{W_{n+1}}$, then, due to the nearest-neighbor interaction in the Hamiltonian (2.6) and according to (2.16), we have

$$\mu^h(\sigma_{V_n} = \sigma_n \,|\, \sigma_{V_n^c} = \eta) = \mu^h(\sigma_{V_n} = \sigma_n \,|\, \sigma_{W_{n+1}} = \omega)$$
$$= \frac{\mu^h(\sigma_{V_{n+1}} = \sigma_n \vee \omega)}{\mu^h(\sigma_{W_{n+1}} = \omega)}$$
$$= \frac{\mu^h_{n+1}(\sigma_n \vee \omega)}{\mu^h_{n+1}(\omega)}.$$

Now by the definitions (2.11) and (2.13), using the proportionality symbol $\propto$ to indicate omission of factors not depending on σ_n, we get

$$\mu^h(\sigma_{V_n} = \sigma_n \,|\, \sigma_{V_n^c} = \eta) \propto \mu^h_{n+1}(\sigma_n \vee \omega)$$
$$\propto \exp\Big\{-\beta H_{n+1}(\sigma_n \vee \omega) + \beta \sum_{x \in W_{n+1}} h_{\omega(x)}(x)\Big\}$$
$$\propto \exp\Big\{-\beta H_n(\sigma_n) + \beta \sum_{x \in W_n} \sum_{y \in S(x)} J_{xy}\, \delta_{\sigma_n(x),\, \omega(y)}\Big\}$$
$$\propto \gamma^{\eta}_{V_n}(\sigma_n), \tag{2.25}$$

and since both the left- and the right-hand sides of (2.25) are probability measures on Φ^{V_n}, the relation (2.24) follows.

The term *splitting* was firstly used[5] in [238] to emphasize that, in addition to the Markov property (see [98]), such measures have the following factorization property: conditioned on a fixed subconfiguration $\sigma_n \in \Phi^{V_n}$, the values $\{\sigma(x)\}_{x \in W_{n+1}}$ are independent under the law μ^h. Indeed, using (2.13) and (2.16), it is easy to see that, for each $n \in \mathbb{N}_0$ and any $\omega \in \Phi^{W_{n+1}}$,

$$\mu^h(\sigma_{W_{n+1}} = \omega \,|\, \sigma_{V_n} = \sigma_n)$$
$$\propto \prod_{x \in W_{n+1}} \exp\big\{\beta J_{x'x}\, \delta_{\sigma_n(x'),\, \omega(x)} + \beta \xi_{\omega(x)}(x) + \beta h_{\omega(x)}(x)\big\}$$
$$\propto \prod_{x \in W_{n+1}} \mu^h\big(\sigma(x) = \omega(x) \,|\, \sigma_{V_n} = \sigma_n\big),$$

where the proportionality symbol $\propto$ indicates omission of factors not depending on ω, and $x' = x'(x) \in W_n$ is the unique vertex such that $x \in S(x')$.

Remark 2.5. Note that $\boldsymbol{F}(\mathbf{0}; \theta) = \mathbf{0}$ for any $\theta > 0$. Thus, for the Potts model with zero external field we have that $\check{\boldsymbol{h}}(x) \equiv \mathbf{0}$ is solution to (2.18) independently of values of other parameters. The SGM corresponding to this zero solution is called disordered (also known as unordered, free, free boundary) Gibbs measure and denoted by μ_0 (see [86], [243], [269]).

---

[5] But the notion of "splitting measure" was used in [174] too.

2.3.2 Boundary laws

Here following [28] and [139] we give a correspondence between the SGM construction given above and an alternative classical approach to defining Gibbs measures on tree-like graphs (including Cayley trees), as presented in the book by Georgii [98].

Definition 2.4. [98] A family $\{\boldsymbol{z}(x,y)\}_{\langle x,y\rangle\in L}$ of positive vectors is called a boundary law if for any $x, y \in V$ such that $\langle x,y\rangle \in L$, and for all $i \in \Phi$ it holds

$$z_i(x,y) = c(x,y) \prod_{v\in\partial\{x\}\setminus\{y\}} \sum_{j\in\Phi} z_j(v,x)\exp\{\beta J_{xv}\delta_{ij} + \beta\xi_i(x) + \beta\xi_j(v)\}, \tag{2.26}$$

where $c(x,y) > 0$ is an arbitrary constant (not depending on $i \in \Phi$).

Proposition 2.2. *The family of permissible GBC $\{\boldsymbol{h}(x)\}_{x\in V}$ defines a* boundary law $\{\boldsymbol{z}(x,y)\}_{\langle x,y\rangle\in L}$.

Proof. For any $y \in V$ and $x \in S(y)$, set

$$z_i(x,y) := \exp\{\beta h_i(x)\}, \qquad i \in \Phi, \tag{2.27}$$

which defines the values of the boundary law on ordered edges $\langle x,y\rangle$ pointing to the root x^0. This definition is consistent, in that the equation (2.26) is satisfied (for such edges) due to the assumed permissibility of the GBC $\{\boldsymbol{h}(x)\}_{x\in V}$ (see Theorem 2.2).

The values $z_i(x,y)$ on the edges $\langle x,y\rangle$ pointing away from the root x^0 (i.e., such that $d(x^0,y) = d(x^0,x)+1$) can be identified inductively (up to proportionality constants) using formula (2.26). The base of induction is set out by choosing $x = x^0$ and $y \in \partial\{x^0\} = S(x^0)$. Then for all $v \in S(x^0)$ one has $z_i(v,x^0) = \exp\{\beta h_i(v)\}$ (see (2.27)), and equation (2.26) yields

$$z_i(x^0,y) = c(x^0,y) \prod_{v\in S(x^0)\setminus\{y\}} \sum_{j\in\Phi} \exp\{\beta h_j(v) + \beta J_{x^0,v}\delta_{ij} + \beta\xi_i(x^0) + \beta\xi_j(v)\},$$

which defines $z_i(x^0,y)$ ($i \in \Phi$) up to an unimportant constant factor. If $z_i(x,y)$ is already defined for all $x \in V_n$ and $y \in S(x)$, then for $x \in W_{n+1}$ and $y \in S(x)$ we have

$$\partial\{x\}\setminus\{y\} = \big(S(x)\setminus\{y\}\big)\cup\{x'\},$$

where $x' \in W_n$ is the unique vertex such that $x \in S(x')$. Noting that the values $z_j(x',x)$ ($j \in \Phi$) are already defined by the induction hypothesis and that $z_j(v,x) = \exp\{\beta h_j(v)\}$ for all $v \in S(x)$, formula (2.26) yields $z_i(x,y)$ (again, up to a proportionality constant), which completes the induction step. □

We recall that a tree-indexed Markov chain is defined as follows. Suppose we are given a tree $\mathcal{T}$ with vertex set V, a probability measure ν, and a transition matrix $\mathbb{P} = (P_{ij})_{i,j\in\Phi}$ on the finite set $\Phi = \{1, \dots, q\}$. We can obtain a tree indexed Markov chain $X : V \to \Phi$ by choosing $X(x^0)$ according to ν and choosing $X(v)$, for each vertex $v \neq x^0$, using the transition probabilities given the value of its parent, independently of everything else. See Definition 12.2 in [98] for a detailed definition.

Remark 2.6. According to the link between GBC $\{\boldsymbol{h}(x)\}_{x\in V}$ and boundary laws $\{\boldsymbol{z}(x,y)\}_{\langle x,y\rangle\in L}$, note that

- The compatibility criterion (2.18) is but a reformulation of the consistency property (2.26) of the boundary law.
- It follows from [98] that any SGM μ^h determines a unique *Markov chain* μ, and vice versa, each Markov chain μ defines a unique SGM μ^h.

Assume $J_{xy} \equiv J$ (i.e., independent on the edges of tree). Then there are translation-invariant boundary laws: $z(x,y) \equiv z$. The SGM corresponding to them are called translation-invariant (TI) SGM.

For marginals on the two-site volumes which consist of two adjacent sites x, y, assuming a translation-invariant boundary law z, in the case of the Potts model, we have

$$\mu(\sigma(x) = i, \sigma(y) = j) = \frac{1}{Z} z_i \exp(J\beta\delta_{ij}) z_j.$$

From this the relation between the boundary law and the transition matrix for the associated tree-indexed Markov chain (SGM) is immediately obtained from the formula of the conditional probability. Indeed, the corresponding transition matrix giving the probability to go from a state i at site x to a state to j at the neighbor y is then

$$P_{ij} = \frac{z_j \exp(J\beta\delta_{ij})}{\sum_{r=1}^{q} z_r \exp(J\beta\delta_{ir})}. \tag{2.28}$$

In the language of GBC: Once we have $\tilde{h}_x$ which is independent of x, i.e., $\tilde{h} = (\tilde{h}_1, \dots, \tilde{h}_q)$, then the tree-indexed Markov chain with states $\{1, 2, \dots, q\}$ is given by the transition matrix $\mathbb{P} = (P_{ij})$ with

$$P_{ij} = \frac{\exp\left(J\beta\delta_{ij} + \tilde{h}_j\right)}{\sum_{r=1}^{q} \exp\left(J\beta\delta_{ir} + \tilde{h}_r\right)}.$$

Remark 2.7. As it was mentioned above the set $\mathcal{G}(H)$ of all Gibbs measures (for a given hamiltonian H) is a non-empty convex compact set $\mathcal{G}(H)$ in the space of all probability measures on Φ^V. Now, using [98] and Remark 2.6, we are ready to make the **crucial observations**, which will be instrumental throughout the book:

- *Any extreme measure* $\mu \in \mathrm{ex}\mathcal{G}(H)$ *is SGM*; therefore, the question of uniqueness of the Gibbs measure is reduced to that in the SGM class. Moreover, for each given temperature, the description of the set $\mathcal{G}(H)$ is equivalent to the full description of the set of all extreme SGMs. Therefore, in this book we only interested to SGM on the Cayley trees.
- *Any SGM corresponds* to a solution of (2.18) given in Theorem 2.2. Thus our main problem is reduced to solving the functional equation (2.18) and to check when a SGM corresponding to a solution is extreme.
- In the next subsection we give some methods for determination of extremality of SGM.

By using of the boundary law $\{\boldsymbol{z}(x,y)\}_{\langle x,y\rangle\in L}$, formula (2.17) can be extended to more general subsets in V [28]. Indeed, fix a vertex x^0 of the tree as the root, then according to [98] adapted to our notation, for any finite *connected* set $\emptyset \neq \Lambda \subset V$ (and $\bar{\Lambda} = \Lambda \cup \partial\Lambda$),

$$\mu^h(\sigma_{\bar{\Lambda}} = \varsigma) = \frac{1}{Z_{\bar{\Lambda}}} \exp\left\{-\beta H_{\bar{\Lambda}}(\varsigma) + \beta \sum_{x\in\partial\Lambda} h^{\dagger}_{\varsigma(x)}(x, x_\Lambda)\right\}, \qquad \varsigma \in \Phi^{\bar{\Lambda}}, \tag{2.29}$$

where $Z_{\bar{\Lambda}} = Z_{\bar{\Lambda}}(\beta, \boldsymbol{h})$ is the normalizing factor, x_Λ denotes the unique neighbor of $x \in \partial\Lambda$ belonging to Λ, and

$$h^{\dagger}_i(x,y) := \beta^{-1} \ln z_i(x,y), \qquad i \in \Phi. \tag{2.30}$$

In particular, if $x \in S(y)$ then (combining (2.30) with (2.27))

$$h^{\dagger}_i(x,y) = h_i(x), \qquad i \in \Phi. \tag{2.31}$$

To link the general expression (2.29) with formula (2.17) for balls V_n, consider part of the boundary $\partial\Lambda$ defined as

$$\partial\Lambda^{\downarrow} := \{x \in \partial\Lambda \colon S(x) \cap \Lambda = \emptyset\}. \tag{2.32}$$

In other words, $\partial\Lambda^{\downarrow}$ consists of the points $x \in \partial\Lambda$ such that the corresponding vertex $x_\Lambda \in \Lambda$ is closer to the root x^0 than x itself. In view of the definition (2.27), for $x \in \partial \Lambda^{\downarrow}$ we get $h_i^{\dagger}(x, x_\Lambda) \equiv h_i(x)$. Clearly, if $x^0 \in \Lambda$ then $\partial\Lambda^{\downarrow} = \partial\Lambda$, but if $x^0 \notin \Lambda$ then the set $\partial\Lambda \setminus \partial\Lambda^{\downarrow}$ is non-empty and, moreover, it contains exactly one vertex, which we denote by $\breve{x}$. Note that $\breve{x} \in \partial\Lambda$ is closer to the root x^0 than $\breve{x}_\Lambda \in \Lambda$, and in this case $h_i^{\dagger}(\breve{x}, \breve{x}_\Lambda)$ is only expressible through the GBC $\{\boldsymbol{h}(x)\}$ via a recursive procedure, as explained above.

Thus, formula (2.29) can be represented more explicitly as follows,

$$\mu^h(\sigma_{\bar{\Lambda}} = \varsigma) = \frac{1}{Z_{\bar{\Lambda}}} \exp\left\{-\beta H_{\bar{\Lambda}}(\varsigma) + \beta \sum_{x \in \partial\Lambda^{\downarrow}} h_{\varsigma(x)}(x) + \beta \sum_{x \in \partial\Lambda \setminus \partial\Lambda^{\downarrow}} h^{\dagger}_{\varsigma(x)}(x, x_\Lambda)\right\}. \tag{2.33}$$

In fact, the last sum in (2.33) includes at most one term, which corresponds to $x = \breve{x}$; more precisely, the latter sum is vacuous whenever $x^0 \in \Lambda$, in which case the first sum in (2.33) is reduced to the sum over all $x \in \partial\Lambda$. In particular, the formula (2.17) is consistent with (2.33) by picking the set $\Lambda = V_{n-1}$ ($n \geq 1$), with boundary $\partial\Lambda = W_n$. For a graphical illustration of the sets involved in formula (2.33), see Fig. 2.1 (for a single-vertex set $\Lambda = \{v\}$ with $v \neq x^0$).

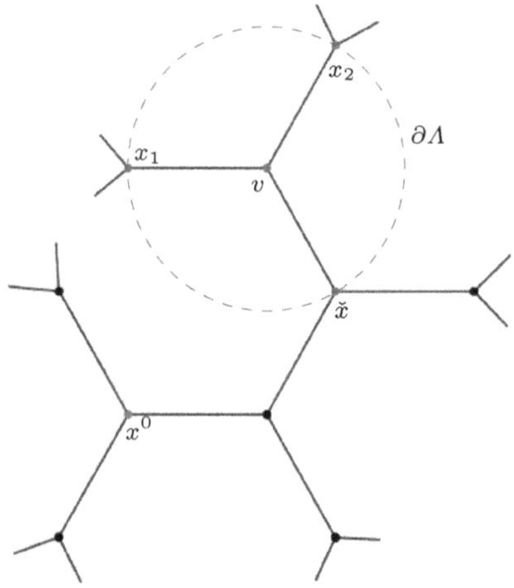

Fig. 2.1 Illustration of the sets in formula (2.33) relative to the root x^0: $\Lambda = \{v\}$, $\partial\Lambda = \{x_1, x_2, \breve{x}\}$ (here $\breve{x}_\Lambda = v$), $\bar{\Lambda} = \{v, x_1, x_2, \breve{x}\}$, and $\partial\Lambda^{\downarrow} = \{x_1, x_2\}$ (see (2.32)).

2.3.3 *Translation-invariant measures*

Here we follow [28] to give main definitions and some results of the transitions on trees. Consider the family of *left* shifts $T_g\colon G_k \to G_k$ $(g \in G_k)$ defined by

$$T_g(a) := ga, \qquad a \in G_k.$$

By the bijection $\mathfrak{b}$ defined in (2.3), this determines conjugate translations on V,

$$\widetilde{T}_z := \mathfrak{b}^{-1} \circ T_{\mathfrak{b}(z)} \circ \mathfrak{b}, \qquad z \in V. \tag{2.34}$$

If $\langle x, y\rangle \in L$ and $y \in S(x)$ (so that $\mathfrak{b}(y) = \mathfrak{b}(x)a_j$, with some generator a_j) then, according to (2.34), $x' := \widetilde{T}_z(x)$ and $y' := \widetilde{T}_z(y) = \mathfrak{b}^{-1}(\mathfrak{b}(z)\mathfrak{b}(y)) = \mathfrak{b}^{-1}(\mathfrak{b}(x')a_j)$ are nearest neighbors, $\langle x', y'\rangle \in L$. Thus, $\widetilde{T}_z$ is an automorphism of V preserving the nearest-neighbor relation. (see Fig. 2.2).

The map (2.34) induces shifts on configurations $\sigma \in \Phi^V$, as

$$(\widetilde{T}_z\sigma)(x) := \sigma(\widetilde{T}_z^{-1}x), \qquad x \in V. \tag{2.35}$$

Definition 2.5. A splitting Gibbs measure $\mu^h = \mu^h_{\theta,\xi}$ is called *translation invariant* (with respect to the group of shifts $(\widetilde{T}_z)$) if for each $z \in V$ the measures $\mu^h \circ \widetilde{T}_z^{-1}$ and μ^h coincide; that is, for any (finite) $\Lambda \subset V$ and any configuration $\varsigma \in \Phi^\Lambda$

$$\mu^h\big(\sigma_{\widetilde{T}_z(\Lambda)} = \widetilde{T}_z(\varsigma)\big) = \mu^h(\sigma_\Lambda = \varsigma). \tag{2.36}$$

Proposition 2.3. *An SGM $\mu^h = \mu^h_{\theta,\xi}$ is translation-invariant under the group of shifts $(\widetilde{T}_z)_{z\in V}$ if and only if the following conditions are satisfied.*

(i) *The reduced field $\{\check{\boldsymbol{\xi}}(x)\}$ is constant over the tree,*

$$\check{\boldsymbol{\xi}}(x) = \check{\boldsymbol{\xi}}(x^0), \qquad x \in V. \tag{2.37}$$

(ii) *The reduced field $\{\check{\boldsymbol{h}}^\dagger(x,y)\}$ (see Section 2.3.2) is symmetric,*

$$\check{\boldsymbol{h}}^\dagger(x,y) = \check{\boldsymbol{h}}^\dagger(y,x), \qquad \langle x,y\rangle \in L. \tag{2.38}$$

(iii) *For any $z \in V$,*

$$\check{\boldsymbol{h}}^\dagger(x,y) = \check{\boldsymbol{h}}^\dagger\big(\widetilde{T}_z(x), \widetilde{T}_z(y)\big), \qquad \langle x,y\rangle \in L. \tag{2.39}$$

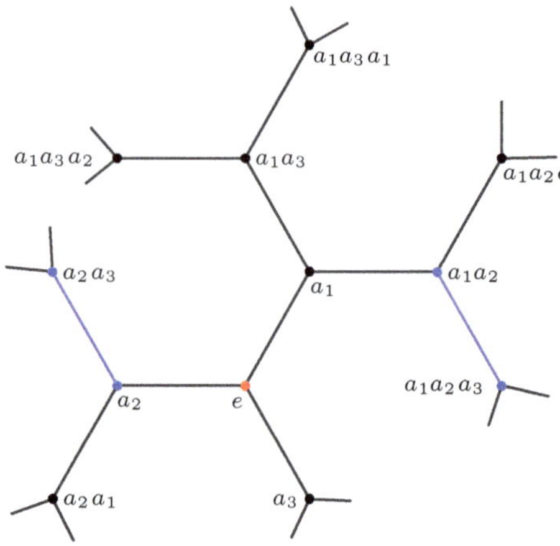

Fig. 2.2 A fragment of the Cayley tree Γ^2 ($k = 2$), with vertices represented (one-to-one) by elements of the free group G_2 with generators $\{a_1, a_2, a_3\}$ (of order 2 each). The identity element $e \in G_2$ designates the root $x^0 \in V$. Starting from e, the elements $a \in G_2$ are inductively expanded along the tree via right-multiplication by one of the generators. Translations on G_2 are defined as left shifts, $T_g \colon a \to ga$ ($g, a \in G_2$). For example, under the shift T_{a_1} the edge $\langle a_2, a_2a_3\rangle$ is mapped to the edge $\langle a_1a_2, a_1a_2a_3\rangle$.

Proof. For $x \neq x^0$, denote by x' the unique vertex such that $x \in S(x')$. Then, according to (2.31) and (2.39),

$$\check{\boldsymbol{h}}^{\dagger}(x, x') = \check{\boldsymbol{h}}(x) = \check{\boldsymbol{h}}(\widetilde{T}_{x'}^{-1}(x)). \tag{2.40}$$

Note that $\widetilde{T}_{x'}^{-1}(x) \in \partial\{x^0\} = W_1$. Thus, there are $k+1$ (vector) values $\check{\boldsymbol{h}}(x_j)$ ($x_j \in W_1$) that determine a translation-invariant SGM μ^h. By translations (2.40), these values (which can be pictured as $k+1$ "colors") are propagated to all vertices $x \in V$ in a periodic "chessboard" tiling, except the root $x = x^0$ where the value $\check{\boldsymbol{h}}(x^0)$ is calculated separately, according to the compatibility formula (2.18).

Necessity. Assume that the measure μ^h is translation invariant, that is, the condition (2.36) holds. Pick the set $\Lambda \subset V$ to be the unit ball

$$V_1 \equiv V_1(x^0) = \{x^0, x_1, \dots, x_{k+1}\}$$

with the center at the root x^0, with $\partial\{x^0\} = \{x_1, \dots, x_{k+1}\}$, where the numbering in (x_j) is consistent with the bijection $\mathfrak{b} \colon V \to G_k$, that is, $x_j = \mathfrak{b}^{-1}(a_j)$. For $z \in V$, the shifted set $V_1(z) := \widetilde{T}_z(V_1) = \{z, z_1, \dots, z_{k+1}\}$ (where $z_j = \widetilde{T}_z(x_j)$, $j = 1, \dots, k+1$) is the unit ball with the center at $z = \widetilde{T}_z(x^0)$, so that $\partial\{z\} = \{z_1, \dots, z_{k+1}\}$. Consider an arbitrary configuration $\varsigma \in \Phi^{V_1}$, with spin values

$$\varsigma(x^0) = i_0, \qquad \varsigma(x_j) = i_j, \qquad j = 1, \dots, k+1.$$

Note that if $y = \widetilde{T}_z(x)$ then, according to (2.35),

$$(\widetilde{T}_z\varsigma)(y) = \varsigma\big(\widetilde{T}_z^{-1}(y)\big) = \varsigma(x). \tag{2.41}$$

Hence, the property (2.36) of translation invariance of μ^h specializes as follows,

$$\begin{aligned} &\mu^h\big(\sigma \in \Phi^V\colon \sigma(z) = i_0,\, \sigma(z_1) = i_1, \dots, \sigma(z_{k+1}) = i_{k+1}\big) \\ &\quad = \mu^h\big(\sigma \in \Phi^V\colon \sigma(x^0) = i_0,\, \sigma(x_1) = i_1, \dots, \sigma(x_{k+1}) = i_{k+1}\big). \end{aligned} \tag{2.42}$$

Using (2.10) and (2.29), and canceling the common term $\beta\sum_{j=1}^{k+1} J\delta_{i_0,i_j}$, the equality (2.42) is reduced to

$$\begin{aligned} &\sum_{j=1}^{k+1}\big[\xi_{i_j}(z_j) + h^\dagger_{i_j}(z_j, z)\big] + \xi_{i_0}(z) \\ &\quad = \sum_{j=1}^{k+1}\big[\xi_{i_j}(x_j) + h^\dagger_{i_j}(x_j, x^0)\big] + \xi_{i_0}(x^0) + \frac{1}{\beta}\ln\frac{Z_{V_1(z)}}{Z_{V_1(x^0)}}. \end{aligned} \tag{2.43}$$

Varying $i_0 \in \Phi$ in (2.43) (whilst keeping all other i_j fixed) shows that the difference $\xi_{i_0}(z) - \xi_{i_0}(x^0)$ does not depend on i_0. More precisely, on subtracting from (2.43) the same equality with $i_0 = q$, we obtain

$$\xi_i(z) - \xi_q(z) = \xi_i(x^0) - \xi_q(x^0), \qquad i = 1, \dots, q-1. \tag{2.44}$$

Since (2.44) holds for any $z \in V$, this proves (2.37) in view of the notation (2.19).

Furthermore, by virtue of (2.37) the equality (2.43) becomes

$$\begin{aligned} \sum_{j=1}^{k+1} h^\dagger_{i_j}(z_j, z) = \sum_{j=1}^{k+1} h^\dagger_{i_j}(x_j, x^0) &+ \sum_{x\in V_1(x^0)} \xi_q(x) \\ &- \sum_{y\in V_1(z)} \xi_q(y) + \frac{1}{\beta}\ln\frac{Z_{V_1(z)}}{Z_{V_1(x^0)}}. \end{aligned} \tag{2.45}$$

Similarly, varying the values $i_1, \dots, i_{k+1}$ in (2.45) (one at a time) yields, for each $j = 1, \dots, k+1$,

$$\check{h}^\dagger_i(z_j, z) = \check{h}^\dagger_i(x_j, x^0), \qquad i = 1, \dots, q-1. \tag{2.46}$$

On the other hand, fix $j \in \{1, \dots, k+1\}$ and consider the shift $\widetilde{T}_{z_j}$, resulting in

$$\widetilde{T}_{z_j}(x^0) = z_j, \qquad \widetilde{T}_{z_j}(x_j) = z.$$

The latter equality follows by recalling the definition of conjugate translations (see (2.34)) and noting that $z_j = \widetilde{T}_z(x_j) = \mathfrak{b}^{-1}(\mathfrak{b}(z)\mathfrak{b}(x_j)) = \mathfrak{b}^{-1}(\mathfrak{b}(z)a_j)$ and, therefore,

$$\widetilde{T}_{z_j}(x_j) = \mathfrak{b}^{-1}(\mathfrak{b}(z_j)\mathfrak{b}(x_j)) = \mathfrak{b}^{-1}(\mathfrak{b}(z)a_j^2) = \mathfrak{b}^{-1}(\mathfrak{b}(z)) = z,$$

because $a_j^2 = e$. Hence, for $\widetilde{T}_{z_j}$ the result (2.46) transforms into

$$\check{h}_i^\dagger(z, z_j) = \check{h}_i^\dagger(x_j, x^0), \qquad i = 1, \dots, q-1. \tag{2.47}$$

Comparing (2.46) and (2.47), we conclude that $\check{h}_i^\dagger(z_j, z) = \check{h}_i^\dagger(z, z_j)$, and the claim (2.38) follows. Let now $y = \widetilde{T}_v(z)$ and $y_j = \widetilde{T}_v(z_j)$, for some $v \in V$. Note that

$$y = \widetilde{T}_y(x^0), \qquad y_j = \widetilde{T}_y(x_j). \tag{2.48}$$

The first equality in (2.48) is clear; to check the second one, note that

$$\begin{aligned}
\mathfrak{b}(y_j) &= \mathfrak{b}(v)\,\mathfrak{b}(z_j) && [y_j = \widetilde{T}_v(z_j)] \\
&= \mathfrak{b}(v)\,(\mathfrak{b}(z)\,\mathfrak{b}(x_j)) && [z_j = \widetilde{T}_z(x_j)] \\
&= (\mathfrak{b}(v)\,\mathfrak{b}(z))\,\mathfrak{b}(x_j) && \\
&= \mathfrak{b}(y)\,\mathfrak{b}(x_j) && [y = \widetilde{T}_v(z)].
\end{aligned}$$

That is, $\mathfrak{b}(y_j) = \mathfrak{b}(y)\,\mathfrak{b}(x_j)$ and (2.48) follows. Thus, formula (2.46) applied to the edge $\langle y_j, y\rangle$ gives

$$\check{h}_i^\dagger(y_j, y) = \check{h}_i^\dagger(x_j, x^0), \qquad i = 1, \dots, q-1.$$

Combined with (2.46), this implies

$$\check{h}_i^\dagger(z_j, z) = \check{h}_i^\dagger(y_j, y) = \check{h}_i^\dagger(\widetilde{T}_v(z_j), \widetilde{T}_v(z)),$$

and (2.39) follows. This completes the "only if" part of the proof.

Sufficiency. Suppose that the conditions (2.37), (2.38) and (2.39) are satisfied. It suffices to verify formula (2.36) for the balls V_n $(n \ge 1)$. For $z \in V$, denote $\Lambda := \widetilde{T}_z(V_{n-1})$, then $\bar{\Lambda} = \widetilde{T}_z(V_n)$ and $\partial\Lambda = \widetilde{T}_z(W_n)$. Furthermore, for $\varsigma \in \Phi^{V_n}$ set $\varsigma_z := \widetilde{T}_z(\varsigma) \in \Phi^{\bar{\Lambda}}$ (see (2.35)). Observe that if $y = \widetilde{T}_z(x)$ then, according to (2.41),

$$\varsigma_z(y) = \varsigma(\widetilde{T}_z^{-1}(y)) = \varsigma(x). \tag{2.49}$$

Hence, recalling (2.10), we have

$$\begin{aligned}
H_{\bar{\Lambda}}(\varsigma_z) &= -\sum_{\langle y, y'\rangle \in E_{\bar{\Lambda}}} J\delta_{\varsigma_z(y),\varsigma_z(y')} - \sum_{y \in \bar{\Lambda}} \xi_{\varsigma_z(y)}(y) \\
&= -\sum_{\langle x, x'\rangle \in E_n} J\delta_{\varsigma(x),\varsigma(x')} - \sum_{x \in V_n} \xi_{\varsigma(x)}(\widetilde{T}_z(x)) \\
&= H_n(\varsigma) + \sum_{x \in V_n} (\xi_q(x) - \xi_q(\widetilde{T}_z(x))),
\end{aligned}$$

where at the last step we used the property (2.37). Thus, from formula (2.29) we obtain, omitting factors not depending on ς,

$$\mu^h(\sigma_{\bar{\Lambda}} = \varsigma_z) \propto \mu^h(\sigma_{V_n} = \varsigma) \cdot \exp\left\{\beta \sum_{y\in\partial\Lambda} h^\dagger_{\varsigma_z(y)}(y, y_\Lambda) - \beta \sum_{x\in W_n} h_{\varsigma(x)}(x)\right\}, \tag{2.50}$$

where y_Λ is the unique neighbor of $y \in \partial\Lambda$ in Λ. Note that if $y = \widetilde{T}_z(x)$, with $x \in \partial V_{n-1} = W_n$, then $y_\Lambda = \widetilde{T}_z(x')$, where $x' \in W_{n-1}$ is the unique vertex such that $x \in S(x')$. Thus, using (2.49), (2.39) and (2.40), we can write

$$\begin{aligned}
\exp\left\{\beta \sum_{y\in\partial\Lambda} h^\dagger_{\varsigma_z(y)}(y, y_\Lambda)\right\} &= \exp\left\{\beta \sum_{x\in W_n} h^\dagger_{\varsigma(x)}(\widetilde{T}_z(x), \widetilde{T}_z(x'))\right\} \\
&\propto \exp\left\{\beta \sum_{x\in W_n} h^\dagger_{\varsigma(x)}(x, x')\right\} \\
&\propto \exp\left\{\beta \sum_{x\in W_n} h_{\varsigma(x)}(x)\right\}.
\end{aligned}$$

Returning to (2.50), this gives $\mu^h(\sigma_{\bar{\Lambda}} = \varsigma_z) \propto \mu^h(\sigma_{V_n} = \varsigma)$, and since

$$\sum_{\varsigma\in\Phi^{V_n}} \mu^h(\sigma_{\bar{\Lambda}} = \varsigma_z) = 1 = \sum_{\varsigma\in\Phi^{V_n}} \mu^h(\sigma_{V_n} = \varsigma),$$

it follows that $\mu^h(\sigma_{\bar{\Lambda}} = \varsigma_z) = \mu^h(\sigma_{V_n} = \varsigma)$, and the proof of the "if" part is complete. □

The criterion of translation invariance given by Proposition 2.3 and its proof firstly appears in [28]. In Georgii [98], a version of this result is established (in the language of boundary laws) for *completely homogeneous* SGM, that is, assuming the invariance under the group of *all automorphisms* of the tree Γ^k. Note that the latter group is generated by the group of (left) shifts $(\widetilde{T}_z)$ and pairwise inversions between vertices $x_j, x_\ell \in \partial\{x^0\}$ [164].

Consequently, we have the following

Corollary 2.1. *An SGM $\mu^h = \mu^h_{\theta,\xi}$ is completely homogeneous if and only if $\check{\boldsymbol{\xi}}(x) \equiv \check{\boldsymbol{\xi}}^0$ for all $x \in V$ and $\check{\boldsymbol{h}}(x) \equiv \check{\boldsymbol{h}}^0$ for all $x \neq x^0$.*

Remark 2.8. In this book (as in [80], [138], [243]), we use the term "translation invariant" (and the abbreviation TISGM) having in mind just completely homogeneous SGM. The translation-invariance characterized in Proposition 2.3 is very interesting but technically more challenging, so it is not addressed in this book. We will see below that the subclass of completely homogeneous SGM in the Potts model is very rich.

2.3.4 *Extreme Gibbs measures*

As it was mentioned above checking the criterion of extremality (Theorem 2.1) for a Gibbs measure is difficult. But for some concrete models, there is some alternative criterion which is more simpler one.

To check the extremality of a Gibbs measure we apply arguments used for the reconstruction on trees [73], [122], [171], [186], [187].

Consider Markov chains with states $\{1, 2, \dots, q\}$ and matrix of transition probabilities $\mathbb{P} = (P_{ij})$.

Condition of non-extremality. Let μ be a Gibbs measure on the Cayley tree of order $k \geq 2$.

It is known that a sufficient condition for non-extremality (which is equivalent to solvability of the associated reconstruction) of a Gibbs measure μ corresponding to the matrix $\mathbb{P}$ is that $k\lambda_2^2 > 1$, where λ_2 is the second largest (in absolute value) eigenvalue of $\mathbb{P}$ [122].

Condition of extremality. Martin in [171] gives the following condition for extremality (non-reconstructibility) in case of $q = 2$:

$$k\left(\sqrt{P_{11}P_{22}} - \sqrt{P_{12}P_{21}}\right)^2 \leq 1.$$

More general conditions are given by the reconstruction insolvability on trees. Let us give the main notations of the reconstruction problems.

An information flow is a process on a tree $T = (V; E)$ in which information is transmitted from the root of the tree to all the nodes of the tree [185], [186], [187].

More precisely, the information flow on each edge is given by a channel on a finite alphabet $\Phi = \{1, \dots, q\}$.

Let $\mathbf{M}_{i,j}$ be the transition probability from i to j; M be the random function (or channel) which satisfies for all i and j that $\mathbf{P}[M(i) = j] = \mathbf{M}_{i,j}$, and $\lambda_2(M)$ be the eigenvalue of $\mathbf{M}$ which has the second largest absolute value.

Assume that M defines an ergodic Markov chain (irreducible and aperiodic). At the root ρ one of the symbols of Φ is chosen according to some initial distribution. Denote this (random) symbol by σ_ρ. This symbol is then propagated in the tree in the following way.

Given that the parent of v, denoted v', has value $\sigma_{v'}$, the probability that σ_v is j is given by $\mathbf{M}_{\sigma_{v'},j}$. That is, for each vertex v having v' as a parent, we let $\sigma_v = M_{v',v}(\sigma_{v'})$, where $\{M_{v',v}\}$ are independent copies of M.

Equivalently, let $\Gamma(v)$ be the set of all vertices which are connected to ρ through paths which do not contain v. Then the process satisfies:

$$\mathbf{P}[\sigma_v = j \mid (\sigma_w)_{w\in\Gamma(v)}] = \mathbf{P}[\sigma_v = j \mid \sigma_{v'}] = \mathbf{M}_{\sigma_{v'},j}.$$

Let $d(\cdot,\cdot)$ be the graph-metric distance on T, and $L_n = \{v \in V : d(\rho, v) = n\}$ be the nth level of the tree. Denote by $\sigma_n = (\sigma_v)_{v\in L_n}$ the symbols (configuration) at the n-th level of the tree.

Note that $(\sigma_n)_{n=1}^{\infty}$ is a Markov chain.

Definition 2.6. The reconstruction problem for T and M is solvable if there exist $i, j \in \Phi$ for which

$$\lim_{n\to\infty} |\mathbf{P}_n^i - \mathbf{P}_n^j| > 0, \tag{2.51}$$

where $\mathbf{P}_n^l$ denotes the conditional distribution of σ_n given that $\sigma_\rho = l$ and $|\cdot|$ denotes the total variation norm.

Thus, the question of reconstructibility is whether the conditional distributions of configurations on the leaves given different configurations at the root will be distinguishable in total variation distance when the distance of leaves from the origin goes to infinity.

This problem arose independently in biology, information theory and statistical physics. Moreover, the reconstruction is known to be equivalent to non-extremality of Gibbs measure of the models of statistical physics (like Ising, Potts models) [187].

In this book we use a result of [176] (see also [177]) to establish a bound for reconstruction insolvability corresponding to the matrix (channel) of a solution to (2.15).

Let us first give some necessary definitions from [176].

For $k \geq 2$, let $\mathbb{T}^k$ denote a *half tree*, i.e., the infinite rooted k-ary tree (in which every vertex has k children). Consider an *initial finite complete subtree* $\mathcal{T}$, that is a tree of the following form: in the rooted tree $\mathbb{T}^k$, take all vertices at distance $\leq d$ from the root, plus the edges joining them, where d is a fixed constant. We identify subgraphs of $\mathcal{T}$ with their vertex sets and write $E(A)$ for the edges within a subset A and ∂A for the boundary of A, i.e., the neighbors of A in $(\mathcal{T} \cup \partial\mathcal{T}) \setminus A$.

In [176] the key ingredients are two quantities, κ and γ, which bound the probabilities of percolation of disagreement down and up the tree, respectively. Both are properties of the collection of Gibbs measures $\{\mu_{\mathcal{T}}^{\tau}\}$, where the boundary condition τ is fixed and $\mathcal{T}$ ranges over all initial finite complete subtrees of $\mathbb{T}^k$.

For a given subtree $\mathcal{T}$ of $\mathbb{T}^k$ and a vertex $x \in \mathcal{T}$, we write $\mathcal{T}_x$ for the (maximal) subtree of $\mathcal{T}$ rooted at x that is a tree given by $\mathcal{T} \cap \mathbb{T}^k_x$, with $\mathbb{T}^k_x$ the half tree with root x.

We draw the trees with the root at the top and the leaves at the bottom. When x is not the root of $\mathcal{T}$, let $\mu^s_{\mathcal{T}_x}$ denote the (finite-volume) Gibbs measure in which the parent of x has its spin fixed to s and the configuration on the bottom boundary of $\mathcal{T}_x$ (i.e., on $\partial\mathcal{T}_x \setminus \{\text{parent of } x\}$) is specified by τ.

For two measures μ_1 and μ_2 on Ω, $\|\mu_1 - \mu_2\|_x$ denotes the variation distance between the projections of μ_1 and μ_2 onto the spin at x, i.e.,

$$\|\mu_1 - \mu_2\|_x = \frac{1}{2}\sum_{i=1}^{q} |\mu_1(\sigma(x) = i) - \mu_2(\sigma(x) = i)|.$$

Denote by $\Omega^\tau_{\mathcal{T}}$ the set of configurations σ given on $\mathcal{T} \cup \partial\mathcal{T}$ that agree with τ on $\partial\mathcal{T}$, i.e., τ specifies a boundary condition on $\mathcal{T}$. For any $\eta \in \Omega^\tau_{\mathcal{T}}$ and any subset $A \subseteq \mathcal{T}$, the Gibbs distribution on A conditional on the configuration outside A being η is denoted by μ^η_A.

Let $\eta^{x,s}$ be the configuration η with the spin at x set to s.

Let $\mathbb{P} = (P_{ij})$ be a channel on $\mathbb{T}^k$, and μ its associated Gibbs measure. Following [176] define

$$\kappa = \frac{1}{2}\max_{i,j}\sum_{l} |P_{il} - P_{jl}|; \tag{2.52}$$

$$\gamma \equiv \gamma(\mu) = \sup_{A \subset \mathbb{T}^k} \max \|\mu_A^{\eta^{y,s}} - \mu_A^{\eta^{y,s'}}\|_x,$$

where the supremum is taken over all subsets $A \subset \mathbb{T}^k$, the maximum is taken over all boundary conditions η, all sites $y \in \partial A$, all neighbors $x \in A$ of y, and all spins $s, s' \in \{1, \dots, q\}$.

Recall that a channel $\mathbb{P} = (P_{ij})^q_{i,j=1}$ is called

- *Ergodic* if it is an irreducible and aperiodic Markov chain (here irreducible and aperiodic refers to the probability kernel $\mathbb{P}$, i.e., the normal Markov chain[6] with this transition kernel is irreducible and aperiodic). Therefore the chain has a unique stationary distribution $\pi = (\pi_1, \dots, \pi_q)$ with $\pi_i > 0$ for all i.
- *Permissive* if for arbitrary finite A and boundary condition outside A being η the conditioned Gibbs measure on A, corresponding to the channel is positive for at least one configuration on A.

[6]Where "time" does not have tree structure.

As the main ingredient we apply [176] , which is

Theorem 2.3. *For an arbitrary ergodic and permissive channel* $\mathbb{P} = (P_{ij})_{i,j=1}^q$ *on a tree, the reconstruction of the corresponding tree-indexed Markov chain is impossible if* $k\kappa\gamma < 1$.

We note that the above mentioned results of [176] are given for a rooted Cayley tree (half tree $\mathbb{T}^k$). The following lemma of [139] says that the results can be extended for unrooted Cayley tree (full tree Γ^k):

Lemma 2.1. *Consider a tree-indexed Markov chain with a given fixed transition matrix* (P_{ij}) *on two different graphs: First, a Cayley tree where every vertex has* $k+1$ *nearest neighbors. Second, a rooted tree where only the root has* k *nearest neighbors and all other vertices have* $k+1$ *neighbors. Then the chain on the rooted tree is reconstructable if and only if the chain on the Cayley tree is reconstructable.*

Proof. Let us denote by ρ_n^i the joint probability distribution of the spins $\{\sigma^i(v) : v \in V,\, d(x^0, v) = n\}$ which are at the boundary at distance n to the root x^0 on the rooted tree, this is obtained by sending a symbol i from the root x^0 using the transition matrix of the Markov chain. It is well known that the chain is non-reconstructable if and only if, for any pair of different symbols $i, j \in \{1, \dots, q\}$, the relative entropy[7] vanishes in the large-n limit, that is we have

$$\lim_{n\to\infty} S(\rho_n^i | \rho_n^j) = 0.$$

(For more background on this, see Propositions 14 and 15 on page 295 of [185].)

We will show that non-reconstructability on the rooted tree (the root has k adjacent vertices, the half tree) implies non-reconstructability on the Cayley tree (the root has $k+1$ adjacent vertices, the full tree). The other direction is obvious. To do so, we consider the Cayley tree and single out a vertex x^0 which we will call the root. Let us label its nearest neighbors by $v_1, \dots, v_k, v_{k+1}$. The boundary at distance n for the Cayley tree decomposes into a first part which is connected to the root x^0 via the vertices $v_1, \dots, v_k$

---

[7] For discrete probability distributions μ and ν defined on the same probability space, $\mathcal{X}$, the relative entropy from ν to μ is defined to be

$$S(\mu \,|\, \nu) = \sum_{x\in\mathcal{X}} \mu(x) \log\left(\frac{\mu(x)}{\nu(x)}\right).$$

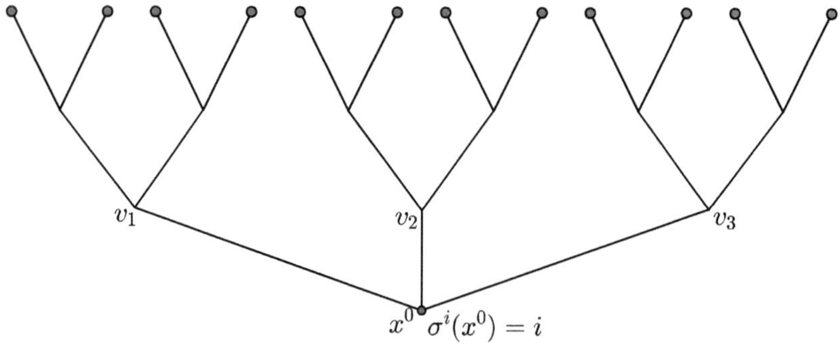

Fig. 2.3 The full tree for $k = 2$ and the root is x^0. The boundary at distance $n = 3$ is divided to two parts (blue and red points).

(see the blue dots in Fig. 2.3) and a second part of those vertices which is connected to x^0 via v_{k+1} (see the red dots in Fig. 2.3).

We denote the spin distributions corresponding to this decomposition of boundary vertices by $\rho_n^{i,<}$ and $\rho_n^{i,>}$, and their joint distribution by $\rho_n^{i,\text{full}}$. By the Markov property of the chain the spins on these two parts of the boundary are mutually independent and we can write $\rho_n^{i,\text{full}} = \rho_n^{i,<} \times \rho_n^{i,>}$. As a consequence we have for the relative entropy between measures on the full boundary which are obtained by sending $i \neq j$ that

$$S(\rho_n^{i,\text{full}}|\rho_n^{j,\text{full}}) = S(\rho_n^{i,<}|\rho_n^{j,<}) + S(\rho_n^{i,>}|\rho_n^{j,>}). \tag{2.53}$$

We recognize that $\rho_n^{i,<}$ is the distribution of boundary sites on the rooted tree where the root has k children. Noting that $\rho_n^{i,>}$ is the distribution of boundary sites on a tree where the root has one child, and all other sites have k children, we use the recursion from root to boundary to write $\rho_n^{i,>} = \sum_{a=1}^q P_{i,a}\rho_{n-1}^{a,<}$ as a sum over the spin-value a of the single vertex v_{k+1} at distance 1 to the root. Here $\rho_{n-1}^{a,<}$ is the distribution of the boundary sites at distance $n-1$ from v_{k+1} obtained by sending the letter a. By convexity of the relative entropy in both arguments we get

$$S(\rho_n^{i,>}|\rho_n^{j,>}) \leq \sum_{a,b} P_{i,a}P_{j,b}S(\rho_{n-1}^{a,<}|\rho_{n-1}^{b,<}). \tag{2.54}$$

By (2.53) and (2.54) we see now that non-reconstruction on the smaller tree, and hence $\lim_{n\to\infty} S(\rho_n^{i,<}|\rho_n^{j,<}) = 0$ for all $i \neq j$, implies that

$$\lim_{n\to\infty} S(\rho_n^{i,\text{full}}|\rho_n^{j,\text{full}}) = 0$$

for all $i \neq j$, and hence non-reconstruction on the larger tree. □

It is easy to see that the channel $\mathbb{P}$ corresponding to a TISGM of the Potts model is ergodic and permissive. Thus by Theorem 2.3 and Lemma 2.1 the criterion of *extremality* of a TISGM is $k\kappa\gamma < 1$.

The constant γ does not have a clean general formula, but can be estimated in specific models (as Ising, Hard-Core etc.). For example, if $\mathbb{P}$ is the channel of the Potts model then (see Theorem 8.1 in page 160 of [176]):

$$\gamma \leq \frac{\theta - 1}{\theta + 1}. \tag{2.55}$$

In the Part III we shall use these conditions of (non-)extremality to check extremality of many SGMs.

2.4 Applications of the Potts model

In this section the Hamiltonian of the Potts model is extended to encode forces in addition to simple nearest neighbor interactions, but the Gibbs measure is still generally defined.

The most studied model of statistical mechanics is the Ising model (i.e., 2-state Potts model), in MathSciNet there are about 6,000 papers devoted to the problems related to Ising model. In particular, this model plays a very special role in statistical mechanics and gives the simplest nontrivial example of a system undergoing phase transitions. In [198] (see also [13]) some physical motivations why the Ising model on a Cayley tree is interesting are given.

The *Potts model* was introduced by R. B. Potts [219] as a lattice system with $q \geq 2$ spin states and nearest-neighbor interaction, to generalize of the Ising model. The Potts model has been quickly picked up by a host of research in diverse areas, such as statistical mechanics [14], [300], as one of few "exactly soluble" models demonstrating a phase transition [59], [68], [134], [143], [175], probability [106], algebra [172], graph theory [16], conformally invariant scaling limits [231], computer science [96], statistics [107], [200], biology [104], medicine [276], [278], sociology [259], [261], financial engineering [229], [279], computational algorithms [38], [76], technological processes [257], [285], and many others.

Here we give brief (not complete[8]) review of some interesting application of the Potts model.

---

[8]See: https://www.sciencedirect.com/topics/physics-and-astronomy/pott-model https://worldwidescience.org/topicpages/q/q-state+potts+model.html for a list and abstracts of works related to Potts model.

The detailed results concerning to some biological and physical systems will be presented in the next chapters.

2.4.1 *Alloy behavior*

An alloy is[9] a combination of metals or metals combined with one or more other elements. Examples are combining the metallic elements gold and copper produces red gold, gold and silver becomes white gold, and silver combined with copper produces sterling silver.

By modeling of microstructural evolution one can study many metals processing companies because the alloys are usually designed computationally by tailoring their microstructural features (see [1] and references therein). These features entail grain size, particle/precipitate content, recrystallization fraction, and crystallographic texture, among others [218]. Moreover, in many industrial thermo-mechanical processes, various annealing phenomena (for example: recrystallization and grain growth) are incompletely understood. One of the computer simulation methods used to study such phenomena is the Monte Carlo Potts model (MCPM). This model has been used to simulate annealing phenomena such as grain growth in single- and two-phase polycrystalline materials, directional grain growth, particle pinning, static and dynamic recrystallization, microstructure, abnormal and nanocrystalline grain growth. The MCPM method has also demonstrated its applicability to modeling recrystallization in aluminum alloys.

To simulate recrystallization in a particle containing alloy, the paper [222] have coupled the finite element crystal plasticity method with MCPM.

In [1] a MCPM was used to model the primary recrystallization and grain growth in cold rolled single-phase Al alloy. The general vision of such modeling is to be able to optimize the microstructural features (grain size, recrystallization fraction, and crystallographic texture) during recrystallization and grain growth computationally in three dimensions in single-phase Al alloys. This simulation provides beneficial tools for understanding annealing and related phenomena in thermal treatments of rolled structures.

2.4.2 *Anomalies*

Anomalies are typically defined in terms of deviation from some expected behavior. There are numerous definition of anomalies relating to various

[9] https://en.wikipedia.org/wiki/Alloy.

fields and areas of science and technology (see [29], [210] and the references therein for a review of general, non-network based anomaly detection defined anomalies as "patterns in data that do not conform to a well defined notion of normal behavior"). Anomalies in online social networks can signify irregular, and often illegal behavior. Detection of such anomalies has been used to identify malicious individuals, including spammers, sexual predators, and online fraudsters ([47], [258]).

In recent paper [210], based on the Potts model in the context of IT networks, the author discussed the occurrence of anomalies affecting the process of phase transitions.

To define the phase transition in network structures, following [210] take a network structure $G = \{V, E\}$, where V is the set of vertices (nodes) and E the set of edges. Each of the nodes can take one of the q values from the set $\{1, 2, \ldots, q\}$ of available states. However, the system remains in one of the global S states (stable or unstable) corresponding to a given phase. Then, the phase transition indicating the transition of the network structure G from the state S to the state S'. As a result of the transition of the network structure from the S state to the S' state, macroscopic changes occur, which are closely related to changes in the G structure properties. It should be noted that, very rarely, a change in a single node or edge condition directly contributes to the phase transition of the entire structure.

The considered issue was examined from the perspective of phase transitions in network structures, particularly in IT networks, Internet of Things and Internet of Everything. A classification of anomalies in relation to the states of particular nodes in the network structure is given. Considered anomalies included homogeneous, heterogeneous, individual and cyclic disorders. The results of tests and simulations showed the impact of anomalies on the phase transitions in the network structures (see, for example, Fig. 2.4).

This study is motivated by possible applications in modeling the processes occurring in network structures, particularly in IT networks.

2.4.3 *Cell sorting*

In biology the cell (meaning "small room") is the basic structural, functional, and biological unit of all known organisms. A cell is the smallest unit of life.

Cell migration is a central process in the development and maintenance of multicellular organisms. Tissue formation during embryonic

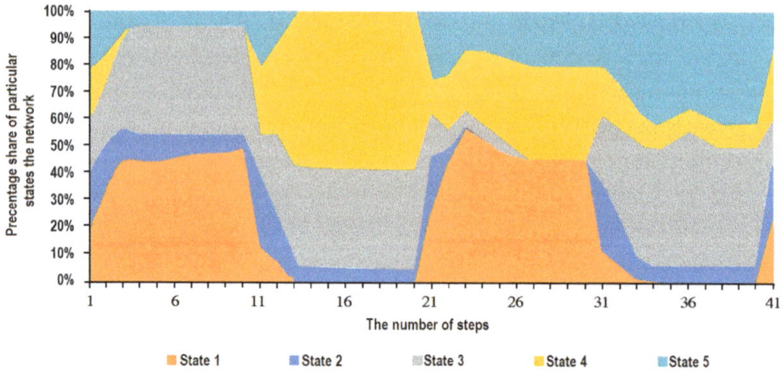

Fig. 2.4 An example of the impact of anomalies on the phase transition in the network structures [210].

development, wound healing and immune responses all require the orchestrated movement of cells in particular directions to specific locations. Cells often migrate in response to specific external signals, including chemical signals and mechanical signals [165].

Cellular Potts models (CPMs) (see [291] and references therein) are asynchronous probabilistic cellular automata developed specifically to model interacting cell populations. They are used to the field of cell and tissue biology. In particular, when the details of intercellular interaction are essentially determined by the shape and the size of the individual cells as well as the length of the contact area between neighboring cells.

Following [291] let us give the mathematical foundation of CPMs: Consider a lattice (usually $\mathbb{Z}^2$) then a CPM assigns a state $\sigma(x)$ from the set $\Phi = \{0, 1, \dots, q\}$ to each site x of a finite set S of the lattice. The set Φ contains the elementary states, called cell indices in the context of CPMs, where $q \in \mathbb{N}$ is the number of cells.

The state of the system as a whole is described by configurations $\sigma \in \Phi^S$. Given a configuration σ, a cell is the set of all points in S with the same cell index, i.e.,

$$\text{cell}_i := \{x \in S : \sigma(x) = i\}, \quad 0 \neq i \in \Phi.$$

The value $\sigma(x) = 0$ is assigned to a given node x if this node is not occupied by a cell. These nodes are interpreted to be filled by extracellular medium.

Thus each cell is of a certain cell type, which determines the migration and interaction properties of the cell.

Let Λ be the set of all possible cell types.

Denote by $\tau : \Phi \to \Lambda$ the map that assigns to each cell its cell type.

A cell with index $i \in \Phi$ has volume

$$V_i(\sigma) = \sum_{x \in S} \delta_{i,\sigma(x)},$$

and surface length

$$M_i(\sigma) = \sum_{\substack{\text{interfaces } (x,y) \in S^2: \\ |x-y|=1, \sigma(x) \neq \sigma(y)}} \delta_{i,\sigma(x)},$$

A CPM is a discrete-time Markov chain with state space Φ^S, where the transition probabilities are specified with Gibbs measure corresponding to a Hamiltonian.

The Hamiltonian to control single aspects of the cells interdependence structure contains several terms. The standard CPM uses the following two terms, $H = H_s + H_v$:

First, a surface interaction term

$$H_s(\sigma) = \frac{1}{2} \sum_{\text{interfaces } (x,y) \in S^2} J(\tau(\sigma(x)), \tau(\sigma(y))), \quad \sigma \in \Phi^S,$$

Here, $J : \Lambda^2 \to \mathbb{R}$ is a symmetric matrix of surface energy coefficients.

Second, a volume constraint

$$H_v(\sigma) = \sum_{i=1}^{q} \lambda_{\tau(i)} (V_i(\sigma) - v_{\tau(i)})^2, \quad \sigma \in \Phi^S,$$

where v_τ the target volume, and λ_τ, the strength of the volume constraint, are cell-type-specific parameters, $\tau \in \Lambda$. Moreover, depending on the phenomenon under investigation, further summands can be included. For instance, a constraint can be imposed on the surface length.

In [7] (see also the references therein) a CPM is considered, to simulate single cell migration over flat substrates with variable stiffness.

In the migration the cells sense their surroundings and respond to different types of signals. Cells preferentially crawl from soft to stiff substrates by reorganizing their cytoskeleton from an isotropic to an anisotropic distribution of actin filaments.

The following configurations are studied [7]:

1. A substrate including a soft and a stiff region.
2. A soft substrate including two parallel stiff stripes.
3. A substrate made of successive stripes with increasing stiffness to create a gradient.
4. A stiff substrate with four embedded soft squares.

For each case it is evaluated the morphology of the cell, the distance covered, the spreading area and the migration speed.

Cell sorting is the process of taking cells from an organism and separating them according to their type. The cells are labeled and tagged to identify areas of interest and their effect. They are separated based on differences in cell size, morphology (shape), and surface protein expression [235].

Following [104] we simulate cell sorting in two-dimensional aggregates using the Potts model defined on the lattice $\mathbb{Z}^2$, which constrains cell size and allows for different surface energies between different cell types.

The Potts model describes a collection of q cells by defining q degenerate spins, $\sigma(x) \in \Phi = \{1, \ldots, q\}$, where $x \in \mathbb{Z}^2$. Thus a cell is a configuration on the lattice.

The cells need not be connected. Mismatched bonds between different cells have energy 1 and bonds between like spins have energy 0. The Hamiltonian is

$$H(\sigma) = \sum_{\langle x,y \rangle} (1 - \delta_{\sigma(x)\sigma(y)}).$$

Consider Gibbs measures associated with this hamiltonian. Phase transitions (or non-uniqueness of Gibbs measure) has meaning that there is a critical temperature T_{cr}, above which each cell dissociates. Below T_{cr} the spins coalesce into compact cells, though for finite temperatures isolated mismatched spins are possible. At $T = 0$ the cells grow by relaxation, as the pattern minimizes its total surface area. The pattern's length scale increases as cells disappear. The dynamics and statistics agree well with experiments on surface-energy-driven grain growth, e.g., in soap froth [102].

In the cell sorting while the surface-energy-driving mechanism is the same as for grain growth, biological cells have generally a fixed range of sizes. Thus the pattern cannot lose energy by coarsening, since cells cannot disappear, similar constrained evolution occurs in bubbles in magnetic films [298].

The differences in contact energies between cells of different types (differential adhesion) cause cell motion which reduces the pattern's energy. To take this into account the authors of [104] added an elastic-area constraint to the Hamiltonian of the Potts model: introduce a symbol τ for the cell type. In the model there are three cell types, "light" $= l$, "dark" $= d$ and "medium" $= M$ that is $\tau \in \{l, d, M\}$. The surface energy between two cells then depends on the types of the cells. Each cell still has a unique spin $\sigma \in \Phi$, and consists of all lattice sites with that spin, but there may be many cells of each type, i.e., with the same τ.

The extended Potts model is defined by the following

$$H(\sigma) = \sum_{\langle x,y \rangle} J(\tau(\sigma(x)), \tau(\sigma(y)))[1 - \delta_{\sigma(x)\sigma(y)}] + \lambda \sum_{\text{spin types } \sigma} \left(a(\sigma) - A_{\tau(\sigma)}\right)^2 \theta(A_{\tau(\sigma)}),$$

where $\tau(\sigma)$ is the type associated with the cell σ and $J(\tau, \tau')$ is the surface energy between types τ and τ'. λ is a Lagrange multiplier specifying the strength of the area constraint, $a(\sigma)$ the area of a cell σ, and A_τ, the target area for cells of type τ.

Because of the surface energy, each cell usually contains slightly fewer than A_τ, lattice sites. Moreover the biological aggregates are usually surrounded by a fluid medium (i.e. $\tau = M$), e.g., culture solution, substrate, or extracellular matrix, which is defined as a single cell with associated type, interaction energies, and unconstrained volume. Assume the target area A_M of the medium to be negative and to suppress the area constraint include

$$\theta(x) = \begin{cases} 0 & \text{if } x < 0 \\ 1 & \text{if } x > 0. \end{cases}$$

It is interesting to determine whether a model of this type exhibits biologically reasonable cell sorting. In coarsening, patterns are usually characterized by their side and area distributions and their moments, as well as the exponent describing the average rate of area growth. In cell sorting, the areas are approximately fixed so the area information is not useful.

In [104] the authors used Potts-model dynamics, with one Monte Carlo time step defined to be 16 times the number of spins in the array, but they suppress the nucleation of heterogeneous spins by requiring that a lattice site flip only to a spin belonging to one of its neighbors. It is found a long-distance cell movement leading to sorting with a logarithmic increase in the length scale of homogeneous clusters. Sorted clusters then round. Moreover two successive phases are found: a rapid boundary-driven creation of a low-cohesivity cell monolayer around the aggregate, and a slower boundary-independent internal rearrangement.

2.4.4 *Financial markets*

In a data finding groups of alike elements is of great interest in all quantitative sciences. For multivariate data, where the objects are characterized by a vector of attributes, a number of efficient and well understood clustering algorithms exist (see [229] and the references therein). They allow to find clusters of similar objects based on a metric between the attribute vectors.

The work [229] combines the early idea of [77] for graph bi-partitioning with a modified Ising Hamiltonian and the Potts model clustering of multivariate data of [23]. By this way the authors of [229] have mapped the communities of a network onto the magnetic domains in the ground state or in local minima of a suitable Hamiltonian. Based on a q-state Potts model a fast community detection algorithm is presented. Communities in networks (groups of densely interconnected nodes that are only loosely connected to the rest of the network) are found to coincide with the domains of equal spin value in the minima of a modified Potts spin glass Hamiltonian. Comparing global and local minima of the Hamiltonian allowed for the detection of overlapping (fuzzy) communities and quantifying the association of nodes to multiple communities as well as the robustness of a community.

In [279], (see also [280]) a 3-state model based on the Potts model is proposed to simulate financial markets. The three states are assigned to buy, sell and inactive states. The model shows the main stylized facts observed in the financial market: fat-tailed distributions of returns and long time correlations in the absolute returns.

The exponential distributions of returns are observed as the inactivity increases. At low inactivity rate, this model effectively reduces to the two-state Bornholdt model [31].

Financial markets can be regarded [303] (see also the references therein) as a complex interactive system that contains a lot of interactive particles, which have some similarity with certain particle physics systems. Therefore, some of the statistical physics methods have been applied to the financial markets, to measure the nonlinear behaviors of the stock price and stock indexes, for instance, the multi-fractal of stock markets, the entropy of return time series, and the correlation dimension of financial time series is calculated.

The work [294] uses the Potts model to simulate and characterize the time evolution of a market time series. A two-dimensional 3-state Potts model is used to develop a stock price time series model.

This financial model imitates:

(i) Traders taking a selling position;
(ii) Traders taking a buying position; and
(iii) Traders taking no trading position.

which are classified as type 1, type 2, and type 3, respectively.

It is assumed (in [294]) that stock price behavior is strongly affected by the number of traders $\omega^{(1)}(t)$ (traders of type 1), $\omega^{(2)}(t)$ (traders of type 2), and $\omega^{(3)}(t)$ (traders of type 3).

Consider a single stock and assume that there are L^2 traders of this stock who are located in a square lattice: $L \times L \subset \mathbb{Z}^2$. Moreover, assume that each trader can trade a unit number of stock at each time $t \in \{1, 2, \ldots, T\}$. At each time t, the fluctuation of stock price process is strongly influenced by the number of traders who take buying positions and the number of traders who take selling positions. When the number of traders in selling positions is smaller than the number of traders in buying positions, the stock price is considered low by market participants, and the stock price gradually increases. The similar is true in the opposite case. In this proposed financial model, the clusters of parallel spins in the square-lattice Potts model are designated groups of market traders acting together.

In [295] high-frequency financial data, the return intervals between price volatilities and the probability density function of return intervals follows a scaling function are studied. The paper [294] investigated dynamic behavior of the return intervals in the Potts financial model and in real stock markets.

The work [303] considers a new microscopic Potts interaction financial price model on $\mathbb{Z}^3$, to investigate the nonlinear complexity behaviors of stock markets and to explain the interaction behavior among the agents. Then the composite multi-scale entropy method is applied to the intrinsic mode functions and the corresponding shuffled data to study the complexity behaviors. The empirical results indicated that the there-dimensional financial model is feasible.

In [305] permutation entropy and sample entropy are developed to the fractional cases, weighted fractional permutation entropy and fractional sample entropy. The effectiveness of these entropies as complexity measures is analyzed by application to the logistic map, which is a typical one-dimensional map creating chaos in some range of a parameter, such as the time series of some stock market indices and the price evolution of an artificial stock market using the Potts model. Moreover, the numerical research on nonlinear complexity behaviors is compared between the returns series of Potts financial model and the actual stock markets.

In very recent paper [32] the author considered a q-spin Potts model version of a simple Ising market model to represent the dynamics of a stock market index in a spin model. A self-organized gain-loss asymmetry in the time series of an index variable composed of stocks in this market is observed.

2.4.5 *Flocking birds*

Flocking[10] is the behavior exhibited when a group of birds, called a flock, are foraging or in flight. There are parallels with the shoaling behavior of fish, the swarming behavior of insects, and herd behavior of land animals.

Examples: starlings are known for aggregating into huge flocks of hundreds to thousands of individuals, murmur at ions, which when they take flight altogether, render large displays of intriguing swirling patterns in the skies above observers.

Mathematical models used to study the flocking behaviors of birds can also generally be applied to the "flocking" behavior of other species.

In mathematical modeler, "flocking" is the collective motion by a group of self-propelled entities and is a collective animal behavior exhibited by many living beings. The flocking transition is an out-of-equilibrium phenomenon and abundant in nature (see [48] and references therein): from human crowds, mammalian herds, bird flocks, fish schools to unicellular organisms such as amoebae, bacteria, collective cell migration in dense tissues, and sub-cellular structures including cytoskeletal filaments and molecular motors.

The physics of flocking is also prevalent in nonliving substances such as rods on a horizontal surface agitated vertically, self propelled liquid droplets, liquid crystal hydrodynamics, and rolling colloids.

In basic models of flocking behavior the following three simple rules are taken into account:

- Separation — avoid crowding neighbors (short range repulsion).
- Alignment — steer toward average heading of neighbors.
- Cohesion — steer toward average position of neighbors (long range attraction).

With these rules, the flock moves in an extremely realistic way, creating complex motion and interaction that would be extremely hard to create otherwise.

[10]https://en.wikipedia.org/wiki/Flocking_(behavior).

In [270] the authors studied the active Ising model, a stochastic lattice gas where collective motion emerges from the spontaneous breaking of a discrete symmetry. Active particles of this model on $\mathbb{Z}^2$ undergo a diffusion biased in one of two possible directions (left and right) and align ferromagnetically their direction of motion, hence yielding a minimal flocking model. It is shown that the transition to collective motion amounts in this model to a bona fide liquid-gas phase transition in the canonical ensemble.

In [48] the 4-state active Potts model (APM) is considered, which has four internal states corresponding to four motion directions and is defined on a two-dimensional lattice with coordination number 4. Its two main ingredients leading to flocking are the local alignment interactions and self propulsion via biased hopping to neighboring sites without repulsive interactions. Thus, the paper [48] generalizes the flocking transition of [270] by the APM when particles can move in more than two directions. The AMP is defined as follows.

Consider N particles defined on a two-dimensional lattice with coordination number q (square lattice for $q = 4$ and triangular lattice for $q = 6$). Each particle is in one of q discrete internal states corresponding to a movement in one of the q lattice directions. Each site can be occupied by multiple particles and each particle can either flip to a different spin state or hop to the nearest neighbor site. The alignment (i.e., flip) probability for particles on site i are defined by the local Hamiltonian

$$H_i = \frac{J}{2\rho_i} \sum_{k=1}^{\rho_i} \sum_{l \neq k} (q\delta_{\sigma_k \sigma_l} - 1),$$

where the double sum runs over all particle pairs (k, l) in site i, $\sigma_k \in \{1, \ldots, q\}$ denotes the state of the k-th particle on site i, and J is the strength of the coupling between different particles on the same site. The number of particles on site i is

$$\rho_i = \sum_{j=1}^{q} \sum_{k} \delta_{\sigma_k, j},$$

the number of particles in state j.

In [48] the authors studied the 4-state APM on the square lattice in which active particles have four internal states corresponding to the four directions of motion. A local alignment rule inspired by the ferromagnetic 4-state Potts model and self-propulsion via biased diffusion according to the internal particle states leads to flocking at high densities and low noise. The phase diagram of the APM is computed and the flocking dynamics in the region is explored, in which the high-density (liquid) phase coexists with the low-density (gas) phase and forms a fluctuating band of coherently moving particles. Moreover, as a function of the particle self-propulsion velocity, a novel reorientation transition of the phase-separated profiles from transversal to longitudinal band motion is revealed, which is absent in the active Ising model.

2.4.6 *Flowing foams*

Foams (see [54]) are complex fluids composed of gas bubbles tightly packed in a surfactant solution. In spite they generally consist only of Newtonian fluids, foam flow obeys nonlinear laws. This can result from non-affine deformations of the disordered bubble packing as well as from a coupling between the surface flow in the surfactant mono-layers and the bulk liquid flow in the films, channels, and nodes. A similar coupling governs the permeation of liquid through the bubble packing that is observed when foams drain due to gravity.

In [256] the CPM successfully simulates drainage and shear in foams. The CPM is used to investigate instabilities due to the flow of a single large bubble in a dry, mono-disperse two-dimensional flowing foam. It is shown that as in experiments in a Hele-Shaw cell,[11] above a threshold velocity the large bubble moves faster than the mean flow. These simulations reproduce analytical and experimental predictions for the velocity threshold and the relative velocity of the large bubble, demonstrating the utility of the CPM in foam rheology studies.

2.4.7 *Image segmentation*

In digital imaging, a pixel (or picture element) is a physical point in a raster[12] image. The pixel is the smallest controllable element of a picture represented on the screen.

More pixels typically provide more accurate representations of the original image. In color imaging systems, a color is typically represented by three or four component intensities such as red, green, and blue, or cyan, magenta, yellow, and black (which can be assigned to states of the Potts model).

---

[11] The term Hele-Shaw cell is commonly used for cases in which a fluid is injected into the shallow geometry from above or below the geometry, and when the fluid is bounded by another liquid or gas.

[12] A raster is a dot matrix data structure that represents a generally rectangular grid of colored points in a display medium.

An image segmentation is[13] the process of partitioning a digital image into multiple segments (sets of pixels, also known as image objects).

The segmentation is needed to simplify and/or change the representation of an image into something that is more meaningful and easier to analyze. This is the process of assigning a label to every pixel in an image such that pixels with the same label share certain characteristics.

In image analysis the problems involving incomplete data arise when some part of the data is missing or unobservable [45]. In this problems one wants to recover an original image which is hidden.

In [45] a Markov model-based image segmentation is studied, which involves hidden Markov random fields.

Let S be the set of pixels with a neighborhood system defined on it and $N(i)$ be the set of pixels considered as neighbors of pixel i.

The q-state (q-color) Potts model for the hidden data is defined as follows. Each pixel $i \in S$ takes one of q states, which can represent q different class assignments, represented as vectors $z_i \in \mathbb{R}^q$, such that one coordinate is 1, all others are equal to 0.

Then it is easy to see that the Kronecker delta has the representation $\delta_{z_i,z_j} = z_i^t z_j$.

The parameter β accounts for strength of the inter-pixel interaction.

The Gibbs measure of the configuration $\mathbf{z} = \{z_i, i \in S\}$ is defined by

$$\mu_\beta(\mathbf{z}) = \frac{1}{Z(\beta)} \exp\left(\frac{\beta}{2} \sum_{i\in S} \sum_{j\in N(i)} z_i^t z_j\right).$$

The parameter β is needed for control of spatial dependencies between pixels and leads to satisfying results in many cases (see [18], [45]).

This model has very nice symmetry (with respect to permutations of colors), it follows from this symmetry that the different colors in the image have the same probability (with respect to Gibbs measure of the Potts model). Therefore, to have distinct probabilities of the colors one can use the Potts model with external field which involves q-dimensional vector $v = (v_1, \dots, v_q)^t$ and the measure is

$$\mu_{\beta,v}(\mathbf{z}) = \frac{1}{Z(\beta, v)} \exp\left(\frac{\beta}{2} \sum_{i\in S} \sum_{j\in N(i)} z_i^t z_j + \sum_{i\in S} z_i^t v\right),$$

where v interpreted as weights for the different values of z_i.

---

[13] https://en.wikipedia.org/wiki/Image_segmentation.

In [45] the external fields are used to better adequacy especially when the color proportions are very unbalanced in the images to be recovered.

As usual the pixels set S in image analysis is $\mathbb{Z}^2$ with the second order neighborhood system: for each site, the neighbors are the eight sites surrounding it.

The image segmentation involves observed and unobserved data to be recovered.

In [45] the unobserved data is modeled as a discrete Markov random field, where a commonly used distribution for the random field is the q-color Potts model with a non-zero external field. Moreover, an illustration of the advantages of introducing the external field is given with numerical experiments.

2.4.8 *Medicine*

Cancer is a group of diseases characterized by the uncontrolled growth of tumor[14] cells that can occur anywhere in the body.

The paper [119] contains a survey of mathematical models that explicitly take into account the spatial architecture of three-dimensional tumors and address tumor development, progression and response to treatments. In particular, it discusses Potts models of epithelial acini, multicellular spheroids, normal and tumor spheroids and organoids, and multi-component tissues.

Here, following [26], [153], [276], [278], we give a review of the application of Potts model to cancer diseases.

In [26] the authors discussed CPM's application in developmental biology, focusing on the development of blood vessels, a process called vascular morphogenesis. A range of models focusing on uncovering the basic mechanisms of vascular morphogenesis are introduced: network formation and sprouting and then show how these models are extended with models of intracellular regulation and with interactions with the extracellular micro-environment. The integration of models of vascular morphogenesis in several examples of organ development in health and disease, including development, cancer, and age-related macular degeneration are reviewed. The computational efficiency of the CPM and the available strategies for the validation of CPM-based simulation models are presented.

Digital pathology imaging of tumor tissues, which captures histological details in high resolution, is fast becoming a routine clinical procedure.

[14]Uncontrolled cells may form a mass called a tumor.

In [153] (see also references therein) the authors considered the problem of modeling a pathology image with irregular locations of three different types of cells: lymphocyte, stromal, and tumor cells. A novel Bayesian hierarchical model is given, which incorporates a hidden Potts model to project the irregularly distributed cells to a square lattice and a Markov random field prior model (Gibbs measures of the Potts model) to identify regions in a heterogeneous pathology image. The model allowed to quantify the interactions between different types of cells, some of which are clinically meaningful.

The development of a tumor is initiated as the genomes of individual cells in an organism become destabilized, which usually kills cells, but in rare cases it modifies the properties of the cell in a way that allows it to proliferate and form a tumor. Despite a growing wealth of available molecular data, the growth of tumors, invasion of tumors into healthy tissue, and response of tumors to therapies are still poorly understood [278].

In [278] the authors review attempts to develop theoretical frameworks for collective cell behavior during tumor development. Mathematical descriptions of tumor growth and development range from continuum-level descriptions of gene-regulatory networks or tumor cell populations, to detailed, spatial models of individual and collective cell behavior. A cellular phenotype determines the success of a cancer cell in competition with its neighbors, irrespective of the genetic mutations or physiological alterations that gave rise to the altered phenotype. The CPM widely used to study the question what phenotypes can make a cell "successful" in an environment of healthy and cancerous cells. In particular, [278] reviews use of the CPM for modeling tumor growth, tumor invasion, and tumor progression.

In [9] a multi-scale, individual-based simulation environment is given that integrates CompuCell3D for lattice based modeling on the cellular level and Bionet[15] solver for intracellular modeling. CompuCell3D or CC3D provides an implementation of the lattice-based CPM and a Monte Carlo method based on the metropolis algorithm for system evolution. The integration of CC3D for cellular systems with Bionet solver for subcellular systems enables to develop a multi-scale mathematical model and to study the evolution of cell behavior due to the dynamics inside of the cells, capturing aspects of cell behavior and interaction that is not possible using continuum approaches. This multi-scale modeling technique is applied to a model of cancer growth.

[15]https://en.wikipedia.org/wiki/BIOSCI.

In [276], to simulate the spatiotemporal evolution of the tumor, a mathematical model is constructed. Using both Potts model and nutrient competition a good visual simulation of tumor growth was provided, reproducing experimental results which also shows that the tumor growth is sensitive to the nutrient environment. It was noted that the results may have some medical signification:

(i) Tumors grow exponentially in the beginning. The tumor migrated toward the nutrient.
(ii) There are some differences in the dependence on nutrient between malignant cells and healthy cells. It may allow to control the nutrient environment in human host factitiously to avoid or cure cancer.
(iii) It may exist several critical nutrients which play a very different even opposite role in the growth process of malignant cells.

2.4.9 *Neural network*

A neural network is a network or circuit of neurons, or in a modern sense, an artificial neural network, composed of artificial neurons or nodes. Thus a neural network is either a biological neural network, made up of real biological neurons, or an artificial neural network, for solving artificial intelligence problems.

The neural network is defined by its node characteristics, the learning rules, and the network topology. The learning rules control the improvement of the network performance through appropriate adaptive changes of the weights of the links. Many papers are devoted to studies on neural networks in the fields of pattern recognition, system identification, control, and other interdisciplinary areas.

In the network theory the *Hebbian learning rule* is a learning rule that specifies how much the weight of the connection between two units should be increased or decreased in proportion to the product of their activation. The rule builds on Hebbs's learning rule which states that the connections between two neurons might be strengthened if the neurons fire simultaneously.

For networks of formal two-state neurons the sequential learning problem is considered [157]: a set of patterns has been memorized by adjusting a matrix of the synaptic efficacious. Then an extra pattern is presented to the network and should be stored in addition to previous patterns in such a way that the latter are not used. This problem has been solved in the framework of the pseudo-inverse learning rules. The resulting synaptic matrix has the same value as for other variants of the pseudo-inverse learning rule if memorizing of the previous patterns was performed by some pseudo-inverse rule.

The q-state Potts-glass model is a symmetric feedback neural network with neurons of q states [160], [304]. Each neuron can be modeled with a Potts spin σ_i, where subscript i indicates neuron $i \in \{1, \ldots, q\}$, that may represent the color or the shade of grey of a pixel in a pattern.

In [11] the critical properties of the Potts model with $q = 3$ and 8 states in one-dimension on directed small-world networks are investigated. The Potts model on these networks presents a second-order phase transition with a new set of critical exponents for $q = 3$. For $q = 8$ the system exhibits only a first-order phase transition.

The authors of [160] have studied the q-states Potts generalization of the Hopfield neural network evolving in parallel. Precise results are obtained on the asymptotic number of stored independent and identically distributed patterns so that either the patterns are fixed points of the dynamics or corrupted patterns get attracted to the original ones after one or several steps of the dynamics.

In [304] the authors have studied the q-state Potts-glass neural network with the pseudo-inverse rule. It is found that there exists a critical point of $q_c = 14$, below which the storage capacity and the retrieval quality can be greatly improved by introducing the pseudo-inverse rule. It was shown that the dynamics of the neural networks constructed with the two learning rules respectively are quite different; but however, regardless of the learning rules, in the q-state Potts-glass neural networks with $q \geq 3$ there is a common novel dynamical phase in which the spurious memories are completely suppressed. This property has never been noticed in the symmetric feedback neural networks. Free from the spurious memories implies that the multi-state Potts-glass neural networks would not be trapped in the meta-stable states, which is a favorable property for their applications.

In very recent paper [281] using the techniques of neural networks, the three-dimensional, 5-state ferromagnetic Potts model on the cubic lattice and two-dimensional 3-state anti-ferromagnetic Potts model on the square lattice are studied. The whole or part of the theoretical ground state configurations of the studied models are considered as the training sets.

The results of the three-dimensional Potts model imply that the neural networks approach is as efficient as the traditional method since the signal of a first order phase transition, namely tunneling between two channels, determined by the neural networks method is as strong as that calculated with the Monte Carlo technique. Furthermore, the outcomes associated with the considered two-dimensional Potts model indicate even little partial information of the ground states which can lead to conclusive results regarding the studied phase transition. These results demonstrate that the performance of neural networks, using the theoretical ground state configurations as the training sets, is impressive.

2.4.10 *Phase transitions*

In fact, the Potts model firstly appeared in statistical physics [14; 300]. In Part III of this book we give strong mathematical presentation of this model for physical point of view.

It is known[16] that as a model of a physical system, the Potts model is simple, but useful as a model system for the study of phase transitions. For example, the ferromagnetic Potts model on $\mathbb{Z}^2$ has a first order transition if $q > 4$. In case $q = 4$ a continuous transition is observed, as in the Ising model where $q = 2$. Other use of Potts model is found through the model's relation to percolation problems and the Tutte and chromatic polynomials found in combinatorics [257], [300].

For integer values of $q \geq 3$, the model displays the phenomenon of 'interfacial adsorption' with intriguing critical wetting properties when fixing opposite boundaries in two different states.

The q-state Potts model with $q \geq 3$ and an external field $\alpha \in \mathbb{R}$ on Cayley tree was considered in [214], [215]) using physical argumentation. In particular, they identified the critical temperature point T_c and also suggested an explicit critical boundary in the phase diagram for $\alpha \geq 0$.

It should be stressed that the phase transition occurring at these critical boundaries is not of type "uniqueness/non-uniqueness", with which we are concerned in the present book, but in fact the so-called "order/disorder" phase transition. The latter was studied rigorously in [96] in connection with the computational complexity of approximating the partition function of the Potts model. The useful classification of critical points deployed in [96] is based on the notion of *dominant phase*; in particular, a critical point is determined, from this point of view as a threshold beyond which only ordered phases are dominant.

In the Part III of this book, we do not investigate the thermodynamical nature of phase transitions, but following [28], [138], [244] and other works, we study the number of splitting Gibbs measures.

---

[16] https://en.wikipedia.org/wiki/Potts_model.

2.4.11 *Political trends*

In [203] the Potts model is applied to Twitter data related to political trends. Twitter is a micro blogging environment where users post small messages, or Twitts, depicting their likes and dislike toward a certain topic, e.g. candidates to the next political elections.

The following general scheme is used (see [203] and the references therein): For each electoral event there are q main candidates, each candidate is represented by a label $\sigma = 1, \ldots, q$, and the activity of a large number of users (order 10^5 or more) is followed for D days (or D minutes) prior or after the election day (or election minute), or partially prior and after. The days (minutes) of observation are represented by the integer $t = 1, \ldots, D$, while the corresponding collected number of messages (order 10^6 or more) is represented by $N(t)$. At each t, the activity of the ith message, $i = 1, \ldots, N(t)$, is then analyzed by looking for key words providing the political preference of the user toward one specific candidate among the q ones: $\sigma_i(t)$.

The q percentages (a normalized mean number of mentions)

$$x_\sigma(t) = \frac{1}{N(t)} \sum_{i=1}^{N(t)} \delta_{\sigma_i(t),\sigma},$$

are tracked till the final time D. Taking into account that $\sum_\sigma x_\sigma(t) = 1$, these data constitute a set of $(q-1)D$ independent information. The statistical model used for these data corresponds to the generalized Potts model defined as follows. Consider a system of N messages $\sigma_1, \ldots, \sigma_N$, where σ_i can take one over q possible values $1, \ldots, q$. Define

$$H(\sigma) = -\sum_i \sum_\sigma h_{\sigma,i} \delta_{\sigma_i \sigma} - \sum_{i<j} \sum_{\sigma,\, \sigma'} J_{\sigma,\sigma';i,j} \delta_{\sigma_i \sigma} \delta_{\sigma' \sigma_j},$$

where $h_{\sigma,i}$ and $J_{\sigma,\sigma';i,j}$ are two sets of parameters, called external fields and couplings, which may have some constrains and also depend on time. On the base of several electoral events and assuming a stationary regime, the authors of [203] found the following:

(i) the maximization of the entropy singles out a microscopic model (single-Twitt-level) that coincides with a q-state Potts model having suitable couplings and external fields to be determined via an inverse problem from the two sets of data;

(ii) correlations decay as $1/N_e$, where N_e is a small fraction of the mean number of Twitts;

(iii) the simplest statistical models that reproduce these correlations are the multinomial distribution, characterized by q external fields, and the mean field Potts model, characterized by one coupling;
(iv) this coupling turns out to be always close to its critical value.

2.4.12 *Protein family*

A protein is a polypeptide chain consists of a sequence of amino acids. This sequence contains amino acid units and (one additional state) gaps or empty spaces. The gap state must be available for an amino acid to move when they fold into three dimensional structures to form domains.

A protein family[17] is a group of evolutionary-related proteins. Usually a protein family has a corresponding gene family, in which each gene encodes a corresponding protein with a 1:1 relationship. The evolutionary history of a protein family is typically represented by a phylogenetic tree.[18]

Proteins in a family descend from a common ancestor and typically have similar three-dimensional structures, functions, and significant sequence similarity. Two segments of a DNA may have shared ancestry because of three phenomena: either a speciation event, or a duplication event, or else a horizontal gene transfer event. Sequence homology is the biological homology between DNA, RNA, or protein sequences, defined in terms of shared ancestry in the evolutionary history of life.

Potts model is used to describe the sequence variability of sets of evolutionarily related protein sequences (i.e., homologous protein families). The notion of homologous protein families implies that present sequences derive from a common ancestor (see [233] and references therein).

This model gives a probability $\mu(A) = \frac{1}{Z}\exp(H(A))$, (with respect to Gibbs measure) to each aligned amino acid sequence $A = (A_1, \ldots, A_L)$ of length L, with $A_i \in \mathcal{A} = \{A, C, \ldots, Y, -\}$ being either one of the 20 amino acids, or an alignment gap, '$-$', representing amino acid insertions or deletions. The total alphabet size is $q = |\mathcal{A}| = 21$. The Hamiltonian is defined by

$$H(A) = -\sum_{1\le i<j\le L} J_{ij}(A_i, A_j) - \sum_{i=1}^{L} h_i(A_i),$$

with strong statistical couplings J_{ij} between different positions. This have been found to be indicative of contacts of the corresponding amino acids

[17] https://en.wikipedia.org/wiki/Protein_family.
[18] https://en.wikipedia.org/wiki/Phylogenetic_tree.

in the three-dimensional protein fold, thereby facilitating protein structure prediction from sequence information.

This statistical energy landscape (i.e., $H(A)$) around a sequence has been found to be informative about the effects of mutations on the protein's functionality (or fitness) [152].

In [233] the authors gave a principled way to correct for phylogenetic effects in the inference of Potts models from sequence data. Although the standard technique to account for these effects in co-evolutionary analysis relies on an empirical re-weighting of sequences. The method of [233] aims at doing so using the phylogenetic tree as well as an evolutionary model. The global nature of Potts models implies that the evolutionary model used should depend on the full sequence. An inference scheme is given that takes the phylogeny of a protein family into account in order to correct biases in estimating the frequencies of amino acids. Using artificial data, it is shown that a Potts model inferred using these corrected frequencies performs better in predicting contacts and fitness effect of mutations.

2.4.13 *Protein folding*

Protein folding is the physical process meaning that a protein chain acquires its native three-dimensional structure, a conformation that is usually biologically functional, in an expeditious and reproducible manner.

This subsection is based on [268].

A Hamiltonian representing interacting amino acids on a lattice is given by

$$H(\sigma) = -\frac{1}{2}\sum_{i\neq j} J_{ij}\delta_{\sigma_i,\sigma_j} - \sum_i h_i \delta_{\sigma_i,\sigma_{q_0}},$$

where the external field h_i favors variables to align in σ_{q_0}; that is forming domains of amino acid σ_{q_0}. Here J_{ij} is the pairwise interaction strength or the exchange parameter between variables σ_i and σ_j at two different lattice sites i and j. Note that the model allows each lattice site to have one out of q different states.

The Kronecker delta function and the negative sign in front of the first term of the Hamiltonian favor to have same amino acids at sites i and j. Due to this pairwise attractive interaction between same residues, domains of same amino acids is expected to be formed by folding the amino acid chain into a three dimensional structure. By the q-state Potts model, this structural formation is characterized by "ordering" of spin variables σ in a specific state.

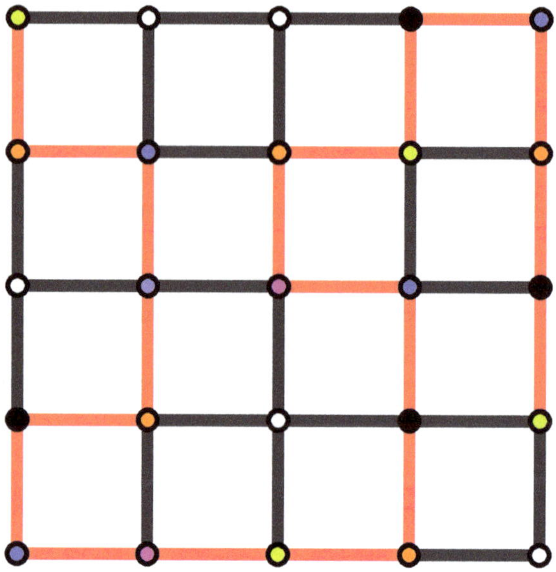

Fig. 2.5 The two-dimensional protein is represented as a chain of amino acids (colored dots) connected by peptide bonds (red lines). The different colored dots occupied at lattice sites represent the different amino acids, the white dots represent the empty sites on the two dimensional lattice.

In case when the entropy dominates at higher temperatures, the chain stretches inside the lattice (see Fig. 2.5) by moving amino acids to empty sites. When the energy dominates at lower temperatures, favored by the Potts model, same color amino acids cluster together by moving them closer. This clustering and stretching of same colors due to the competition between the energy and the entropy represents the protein folding and un-folding.

The general investigation of protein folding problem requires to solve the inverse Potts model, i.e., finding the coupling constants J_{ij} and h_i from the data base.

The main protein folding problem is understanding the question of how a protein's amino acid sequence dictates its structure. The various interaction parameters inside the protein and the local environment are directly responsible for the folding. In [268] the authors studied the thermodynamic properties of the protein folding (the Potts model) using a statistical mechanics approach. Namely, converting the interacting amino acids into an effectively non-interacting model using a mean-field theory, the Helmholtz free energy is found. Then by investigating the Helmholtz free energy, the properties of protein folding transition is qualitatively studied. It is shown that the protein folding phase transition is a strongly first order.

Potts models of protein sequence co-variation are statistical models constructed from the pair correlations observed in a multiple sequence alignment of a protein family. In ([152] see also references therein) the Potts models are reviewed to predict protein structure and sequence-dependent conformational free energy landscapes, to survey protein fitness landscapes and to explore the effects of epistasis on fitness.

2.4.14 *Signal reconstruction*

Piecewise constant signals and images are an important kind of data. Typical examples include bar code signals, logos, cartoons, Quick response codes, and text images, which are widely used in both general commercial and automotive industry use [296].

The Potts model has applications in signal reconstruction.[19]

Assume that we are given noisy observation of a piecewise constant signal g in $\mathbb{R}^n$. To recover g from the noisy observation vector f in $\mathbb{R}^n$, one seeks a minimizer of the corresponding inverse problem, the L^p-Potts functional $P_\gamma(u)$ which is defined by

$$P_\gamma(u) = \gamma\|\nabla u\|_0 + \|u - f\|_p^p = \gamma \sum_i \left(1 - \delta_{u_i u_{i+1}}\right) + \sum_{i=1}^{n} |u_i - f_i|^p,$$

where the jump penalty $\|\nabla u\|_0$ forces piecewise constant solutions and the data term $\|u - f\|_p^p$ couples the minimizing candidate u to the data f. The parameter $\gamma > 0$ controls the trade-off between regularity and data fidelity. There are fast algorithms for the exact minimization of the L^1 and the L^2-Potts functional (see [76]). The algorithms are accompanied by a series of illustrative examples from molecular biology.

In [299] it is shown that each sequence of minimizers of the discrete L^1-Potts functionals has an accumulation point, which is a minimizer of the continuous model. Moreover, an algorithm to compute an exact minimizer of the discrete L^1-Potts functional is given.

In image processing, the Potts functional is related to the segmentation problem. However, in two dimensions the problem is NP-hard (see [38] for details).

[19] https://en.wikipedia.org/wiki/Potts_model#cite_note-1.

2.4.15 *Smectic phase*

The liquid crystals phases are often called (see [204]) mesophases, which (according to the degrees of symmetry and the physical properties of liquid crystals with respect to their molecular arrangement) include four kinds of structure: nematic, smectic, cholesteric and columnar liquid crystals. The smectic phase is almost a crystalline solid for which molecules are ordered in equidistant layers.

It is known (see [204]) that the smectic phase, as the other liquid crystals phases, has different microscopic origins. Though their orderings belong to the same classification, namely the smectic phase, each smectic liquid crystal may have properties different from the others, for example the temperature-dependence of physical quantities and the reaction to an applied electric field.

In [12] it is shown that the smectic and nematic phases can be obtained directly by cooling the system with appropriate Hamiltonian using a mobile Potts model.

2.4.16 *Sociology*

In [16], [259], [261] an application of Potts model to studies of human behavior is given: Nobel laureate Thomas Schelling published a seminal paper [259] which in addition to organized and economic explanations considers the possibility of micro-motive explanations for racial segregation. The premise is that individual decisions to avoid minority status (or to require being in a minority of some minimum size) could lead to the macro-effect of segregation. Schelling places vacancies, stars, and zeros randomly on a checkerboard and then iteratively considers the happiness of the stars and zeros with their local neighborhoods, moving an unhappy star or zero to the nearest vacant spot that meets their happiness criteria. Schelling works through many different experiments to come up with some very compelling results on segregation. From this model it seems that people do consciously or subconsciously segregate themselves from people who are different than they are.

The work [178] gives a similar premise to Schellings (that micro-motive explanations can lead to immigrant ghettos) with a more Potts-like model where the Hamiltonian measures the happiness of individuals with their neighbors, the temperature is viewed as a social temperature where warmer temperatures reflect facilitation of integration and assimilation, and at each step in the simulation two neighbors are able to exchange places with a probability based on the likelihood of the new state with respect to the current state. The paper [261] extends the work of [178] to address up to seven different ethnic groups.

Following [15] one can construct a Potts model for simulating human behavior in the following way. Use a lattice to depict your neighborhood, city, business, or any other venue in which people interact with one another. To have more groups, we can have elderly people, college roommates, families with teenagers, and families with small children. Assume that members of each of these groups of people are living together in a brand new development. Label the elderly with a 1, the college roommates with a 2, the families with teenagers with a 3, and the families with small children with a 4. The members of these groups have preferences about who they live near. For example, the elderly do not want to live next to the college roommates because of the large parties that they tend to throw. The couples with small children might want to live next to one another so that their kids can play together without going far from home. The Hamiltonian for this experiment would measure overall happiness as opposed to energy. Outside forces might be the price of other houses in other neighborhoods, proximity to work, or how much people like their current house. The Metropolis Algorithm could then be run to develop higher probabilities for lattice states with higher overall happiness. Eventually, we would likely see preferences playing out in the form of segregation. This is just a rough sketch of a Potts model scenario. This may give an appreciation for the versatility of the Potts model when it comes to real world situations.

2.4.17 *Spin glasses*

A glassy system is complex system that exhibits a very slow dynamics that prevents it from reaching the equilibrium state. Examples are real glasses, spin glasses, supercooled liquids, polymers, granular material, colloids, ionic conductors, orientational glasses, and vortex glasses [74].

A spin glass is a model[20] of a certain type of magnet. Magnetic spins are the orientation of the north and south magnetic poles in three-dimensional space. In ferromagnetic solids, component atoms magnetic spins all align in the same direction. Spin glasses are contrasted with ferromagnetism as

[20] https://en.wikipedia.org/wiki/Spin_glass.

"disordered" magnets in which their atoms spins are not aligned in a regular pattern.

The complex internal structures that arise within spin glasses are termed "metastable" because they are "stuck" in stable configurations other than the lowest-energy configuration. The mathematical complexity of these structures is difficult but fruitful to study experimentally or in simulations; with applications to physics, chemistry, materials science and artificial neural networks in computer science.

The Potts model of spin glass is defined by the Hamiltonian [74]:

$$H(\sigma, s) = -qJ \sum_{\langle i,j \rangle} \left[\delta_{\sigma_i \sigma_j}(\epsilon_{ij} s_i s_j + 1) - 2\right],$$

where associated with each lattice site is an Ising spin $s_i \in \{-1, 1\}$ and a q-state Potts spin $\sigma_i \in \{1, \dots, q\}$. The sum is extended over all nearest-neighbor sites, $\epsilon_{i,j} = \pm 1$ is a random quenched variable, and J is the strength of interaction.

The model is a superposition of a ferromagnetic q-state Potts model and a $\pm J$ Ising spin glass model [180].

In [74] it is shown that this model exhibits for all q a spin glass transition at $T_{SG}(q)$ and a percolation transition at higher temperature $T_P(q)$. It is shown that for all values of $q > 1$ at $T_P(q)$ there is a thermodynamic transition in the universality class of a ferromagnetic q-state Potts model. Moreover, the efficiency of the cluster dynamics is compared with that of standard spin-flip dynamics.

There are variations of the Potts spin glass model, which has been studied extensively in the physics literature (see, e.g., [42], [180], [209] and references therein).

In [209] the following Potts spin glass model is considered:

$$H_N(\sigma) = \frac{1}{\sqrt{N}} \sum_{1 \le i,j \le N} g_{ij} \delta_{\sigma_i \sigma_j},$$

where the interaction parameters g_{ij} are independent standard Gaussian random variables and $\sigma_i \in \{1, \dots, q\}$. This model generalizes the Sherrington-Kirkpatrick model (see [265]) to the case when spins take more than two values but their interactions are counted only if the spins are equal. The analogue of the Parisi variational formula for the free energy is obtained.

2.4.18 *Storage capacity*

Storage capacity measures how much data a computer system may contain.

A Potts unit (spin value) can be regarded (see [201]) in the neuroscience context as representing a local subnetwork or cortical patch of real neurons, endowed with its set of dynamical attractors, which span different directions in activity space, and are therefore converted to the states of the Potts unit (which is defined precisely as having states pointing each along a different dimension of a simplex). One can define the model as an auto-associative network of Potts units interacting through tensor connections. The memories are stored in the weight matrix of the network and they are fixed: each memory μ is a vector or list of the states taken in the overall activity configuration by each unit.

Take each Potts unit to have S possible active states, labeled e.g. by the index k, as well as one quiescent state, $k = 0$, when the unit does not participate in the activity configuration of the memory. Using this model in [201] the storage capacity of the Potts network is studied. The storage capacity calculation, performed using replica tools, is limited to fully connected networks, for which a Hamiltonian can be defined. To extend the results to the case of intermediate partial connectivity, the self-consistent signal-to-noise analysis is derived for the Potts network; and the implications for semantic memory in humans is discussed.

In [118] the theory of neural networks is extended to include discrete neurons with q, $q \geq 2$ discrete states. The dynamics of such systems are studied by using Potts model. The maximum number of storage patterns is found to be proportional to $Nq(q-1)$, where q is the number of Potts states and N is the size of the network.

It is known (see [136]) that the capacity to store information in any device, and in particular the capacity to store concepts in the human brain, is limited. In [136] it is shown in a minimal model of semantic memory, and in progressive steps, how one can expect the storage capacity to behave depending on the parameters of the system. It was deduced the minimum requirements of any model of this kind in order to have a high capacity. The calculation specifies that in the Potts model the number of concepts that can be stored is neither linear nor an arbitrary power of the number S of values a feature can take, but quadratic in S.

2.4.19 *Symmetric channels*

Symmetric channels on q symbols have the state space $\{1,\ldots,q\}$ and $q \times q$-matrices:

$$\mathbf{M} = \begin{pmatrix} 1-(q-1)\delta & \delta & \delta \ldots & \delta \\ \delta & 1-(q-1)\delta & \delta \ldots & \delta \\ \ldots & \vdots & \ldots \vdots & \ldots \\ \delta & \delta & \delta \ldots 1-(q-1)\delta \end{pmatrix}, \tag{2.56}$$

with $\lambda_2(M) = 1 - q\delta$.

Depending on the sign of $\lambda_2(M)$ we distinguish between ferromagnetic Potts models where $\lambda_2(M) > 0$, and anti-ferromagnetic models where $\lambda_2(M) < 0$. In case $1-(q-1)\delta = 0$ we obtain the model of proper colorings of the tree:

$$\mathbf{M} = \begin{pmatrix} 0 & (q-1)^{-1} & (q-1)^{-1} \ldots & (q-1)^{-1} \\ (q-1)^{-1} & 0 & (q-1)^{-1} \ldots & (q-1)^{-1} \\ \ldots & \vdots & \ldots \vdots & \ldots \\ (q-1)^{-1} & (q-1)^{-1} & (q-1)^{-1} \ldots & 0 \end{pmatrix}. \tag{2.57}$$

Several bounds for the reconstruction problem (for q-state Potts models on a Cayley tree of order $k \geq 2$) are known (see [185]-[187], [269]):

- If $k\lambda_2^2(M) > 1$ then the reconstruction problem is solvable.
- If $k|\lambda_2(M)| \leq 1$, then the reconstruction problem is unsolvable.
- If $k\lambda_2(M) > 1$ and q is sufficiently large, then the reconstruction problem is solvable.
- If $k\frac{(1-q\delta)^2}{1-(q-2)\delta} \leq 1$ then the reconstruction problem is unsolvable.

As it was mentioned above the most general result on reconstruction is the Kesten-Stigum bound [122] which says that reconstruction holds when $k\lambda_2^2 > 1$.

In [269] the author proved the first exact reconstruction threshold in a non-binary model establishing the Kesten-Stigum bound for the 3-state Potts model on regular trees of large degree. The Kesten-Stigum bound is not tight for the q-state Potts model when $q \geq 5$. Moreover, asymptotics for these reconstruction thresholds are determined.

In genetics (see, e.g., [44], [274]) an information flow can be used to represent propagation of a genetic property from ancestor to its descendants. In communication theory, this process represents a communication network on the tree where information is transmitted from the root of the tree. Moreover, the process was studied in statistical physics, see, e.g., [24], [111]. More precisely, in statistical physics a non-uniqueness of the Gibbs measure means that for all n there exists σ_n such that the distribution of σ_ρ given σ_n has total variation distance at least $\delta > 0$ from uniform. This is a weaker condition than reconstruction solvability and it was studied in statistical physics for Ising and Potts models (see [98], [108], [220]).

The crucial role of the reconstruction problem in Phylogeny was demonstrated in [188] and [189].

2.4.20 *Technological processes*

Technology means activities, because it is in constant development. In general technological processes can be design processes, making processes and processes in the phase of using and assessing technology [292].

Ostwald ripening[21] is a phenomenon observed in solid solutions or liquid sols that describes the change of an inhomogeneous structure over time, i.e., small crystals or sol particles dissolve, and redeposit onto larger crystals or sol particles. The Ostwald ripening is generally found in water-in-oil emulsions, while flocculation is found in oil-in-water emulsions.

Some analytic methods used to model processes such as grain growth and Ostwald ripening by many simplifying assumptions.

In [272] the q-state Potts model was adapted to study grain growth (development). By this model it was treated the next level of complexity by incorporating grain boundary topology. Moreover, the growth kinetics of a two dimensional, connected assembly of mutually interacting grains are described.

Now following [285] we give one more application of the Potts model. Markets often demand small quantities of many different components with highly specialized performance requirements. Therefore, the development cost of each component cannot be amortized over large numbers and becomes prohibitive. Good predictive tools are necessary for reduction of the development cost. It have been proved that the Potts model is a powerful and useful tool for simulating a wide variety of microstructural evolution

[21] https://en.wikipedia.org/wiki/Ostwald_ripening.

problems both during processing and during component use. In [285] the use and application of the Potts model to Ostwald ripening is presented.

Here, to see how the Potts model appears, we give phase equilibria characteristics of [285]:

Consider $\mathbb{Z}^2$ as the lattice of the q-state Potts model. A canonical ensemble of sites, which may be visualized as building blocks of the microstructure, populates a simulation lattice. These sites possess certain energies based on their inherent characteristics and on their interactions with their neighboring domains and evolve to minimize the total free energy of the system.

The evolution mechanism is that each domain can be modified based on the energetics of the modification being considered. Then a digitized microstructures can be represented on $\mathbb{Z}^2$ with a periodic boundary conditions. The components necessary for Ostwald ripening can be incorporated into the Potts model by populating the simulation lattice with a canonical ensemble having two components designated as A and B. The A-sites are the primary building blocks of the grains and the B-sites of the matrix, as shown in Fig. 2.6.

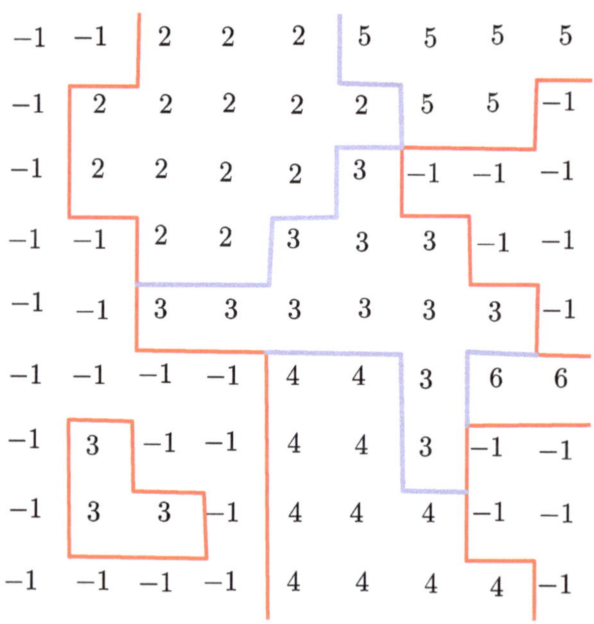

Fig. 2.6 Illustration of the technique used to digitize microstructures. The grains in a liquid with grain boundaries (blue) and solid-liquid interfaces (red) are shown. The B-sites are all the sites labeled "-1" and A-sites are labeled with positive integers.

The two components, A and B, must separate into two phases, grains dispersed in a matrix. The two phases were generated by defining bond energies between the A- and B-sites, so that the components would segregate into two phases. The $A - B$ bonds are assigned higher energies than $A - A$ or $B - B$ bonds. Thus the system minimizes its energy by segregating the A- and B-sites into two phases.

2.4.21 *Wetting transition*

Here we follow [159] (see also references therein). The study of liquid behavior on a surface is a topic that is receiving increasing interest due to its diverse applications.

Examples of application are: microfluidics for biotechnology, textile lubrication, and self-cleaning. The inspiration arose in nature, more precisely from the plant (the lotus), whose leaves have a natural hydrophobicity.

Based on the surface structure of the leaf, similar artificial rough surfaces were produced. A drop of water on this type of surface can exhibit two states of wettability:

- Wenzel state, in which the drop of water penetrates the surface cavities;
- Cassie-Baxter (CB) state, in which the drop remains on the top of the pillars, as in the lotus leaf.

The Wenzel approach considers that the liquid fills up completely the grooves on the rough surface increasing thus the total area of the liquid-solid interface.

The CB approach assumes that air pockets are trapped inside the grooves, and thus, that the liquid only contacts the solid at the top of the asperities. Thus, the drop sits on a composite surface comprised of solid and trapped air.

The energy barrier between the CB-Wenzel state transition is given as a product of the following factors: the pillar height, pillar thickness, and the area that the droplet occupies between the pillars. This energy barrier separating the states may be large enough to prevent a spontaneous transition from occurring.

The work [159] is based on a two-dimensional Potts model of a droplet placed on a regularly patterned surface of pillars of different heights and interspacing.

The model is defined as follows. Consider a square lattice, where to each network site, i, a value $\sigma_i \in \Phi = \{1, \dots, q\}$ is given. The set Φ is labels associated with the different media: air, liquid, solid etc.

The Hamiltonian (the total energy) of the system is

$$H(\sigma) = \frac{1}{2} \sum_{\langle i,j \rangle} \gamma_{\sigma_i \sigma_j} \left(1 - \delta_{\sigma_i \sigma_j}\right) + \lambda \left(A_{\sigma=1} - A_T\right)^2 + mg \sum_i \delta_{1\sigma_i} h_i,$$

where $\sigma = \{\sigma_i : i \in \mathbb{Z}^2\}$ the configuration (the labels on the sites of the lattice) and $\gamma_{\sigma_i \sigma_j}$ is the interfacial energy between neighboring sites belonging to different media [104]. The second term corresponds to an elastic compression energy: $A_{\sigma=1}$ is the drop area, A_T is a target area, and λ is a Lagrange multiplier related to the inverse of the liquid compressibility [114]. This term is required to avoid the droplet disappearance by energy minimization. The last term is the contribution from the gravitational energy, where g is the acceleration of gravity[22] and h_i is the vertical coordinate of a label belonging to the droplet defined as having unit mass ($m_{\sigma=1} = 1$).

In [263] (see also the references therein) the transition and the influence of the energy barrier on the transition are studied through thermodynamics and molecular dynamics models.

In [159] Potts model simulation compared with the studies developed in [263], [264] employing thermodynamics and two-dimensional molecular dynamics.

2.5 Comments and references for Part I

This part of the book has an introduction character.

In each page I have cited to the references of the material and gave briefly the obtained results of the literature. Starting sections of this part presents main definitions and related problems of the book.

Chapter 2 mainly devoted to the tools of solutions of the formulated problems. The previous section is an "excursion" to the applications of the Potts model to several areas of sciences and nature. Moreover, the last section gives many variations of the Potts model, the set of Gibbs measures of which is not completely studied yet. Thus this part gives a clear motivation to read the next chapters of the book. Moreover, the reader can

[22]The gravity can be neglected in case of very small droplets.

try to study Gibbs measures of the variations of the Potts model listed in the previous section. These measures can be then used to better understanding the corresponding systems.

For future reading see recent works [22], [32], [49], [50], [64], [71], [99], [103], [105], [141], [146], [207], [211], [228], [232], [281], [286], [301], [310] and references therein.

PART II

Thermodynamics in biology: The Potts model

Chapter 3

Hierarchy of DNAs and Holliday junctions

In this chapter we consider codon-based model of DNA evolution, where a Gibbs measure is the stationary measure. In a model of interacting DNAs we define a DNA as a sequence of ± 1's and embed it on a path of a Cayley tree of order $k \geq 2$. Using group representation of the Cayley tree, we give a hierarchy of a countable set of DNAs each of which 'lives' on the same Cayley tree. Then we give a Hamiltonian of this set of DNAs as a 3-state Potts model (states are considered as DNA base pairs) on a set of admissible configurations. To study thermodynamic properties of the model of DNAs we describe corresponding translation-invariant Gibbs measures (TIGM) of the model on the Cayley tree of order two ($k = 2$). Each such measure describes a phase of the set of DNAs. We show that there is a critical temperature T_c such that

(i) if temperature $T > T_c$ then there exists unique TIGM;
(ii) if $T = T_c$ then there are two TIGMs;
(iii) if $T < T_c$ then there are three TIGMs.

These results applied to study distributions of Holliday junctions and branches of DNAs. In case of very high and very low temperatures we give stationary distributions and typical configurations of the Holliday junctions.

The case $k \geq 3$ remains open.

3.1 Models and Markov process of DNA sequence evolution

3.1.1 *Definitions*

Recall the following definitions. Deoxyribonucleic acid (i.e., DNA)[1] is a molecule that carries most of the genetic instructions used in the growth, development, functioning and reproduction of all known living organisms and many viruses. Most DNA molecules consist of two biopolymer strands coiled around each other to form a double helix.

The two DNA strands are known as polynucleotides since they are composed of simpler units called nucleotides.

The two strands of DNA in a double helix can be pulled apart like a zipper, either by a mechanical force or high temperature.

In this chapter we shall give a model of this thermodynamics. Each nucleotide is composed of a nitrogen-containing nucleobase — either cytosine (C), guanine (G), adenine (A), or thymine (T) — as well as a sugar called deoxyribose and a phosphate group. The nucleotides are joined to one another in a chain by covalent bonds between the sugar of one nucleotide and the phosphate of the next, resulting in an alternating sugar-phosphate backbone (see Fig. 3.1).

According to base pairing rules (A+T, and C+G), hydrogen bonds bind the nitrogenous bases of the two separate polynucleotide strands to make double-stranded DNA.

Branched DNA can occur if a third strand of DNA is introduced and contains adjoining regions able to hybridize with the frayed regions of the pre-existing double-strand. Although the simplest example of branched DNA involves only three strands of DNA, complexes involving additional strands and multiple branches are also possible (see Fig. 3.2).

Using methods of thermodynamics (of statistical physics) we describe a general structure of such branching process.

A Holliday junction is a branched nucleic acid structure that contains four double-stranded arms joined together. These arms may adopt one of several conformations depending on buffer salt concentrations and the sequence of nucleobases closest to the junction (see Fig. 3.3).

In [112] the Holliday junction considered as an exquisitely sensitive force sensor whose force response is amplified with an increase in its arm lengths, demonstrating a lever-arm effect at the nanometer-length scale. Mechanical

[1] See https://en.wikipedia.org/wiki/DNA. Some colored pictures of this section are taken from internet.

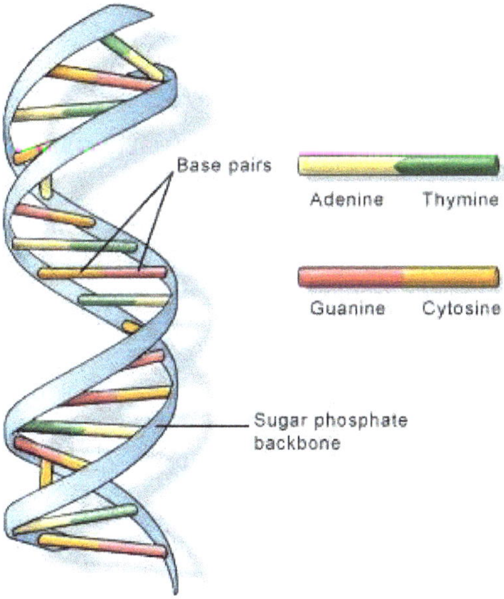

Fig. 3.1 Structure of a DNA.

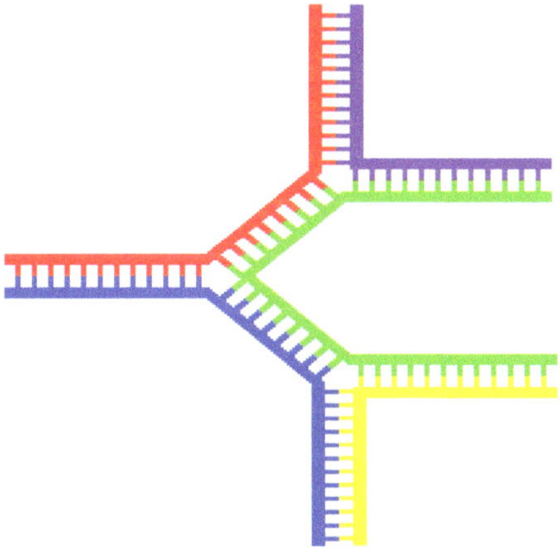

Fig. 3.2 Multiple branched DNA.

interrogation of the Holliday junction in three different directions helped elucidate the structures of the transient species populated during its conformational changes.

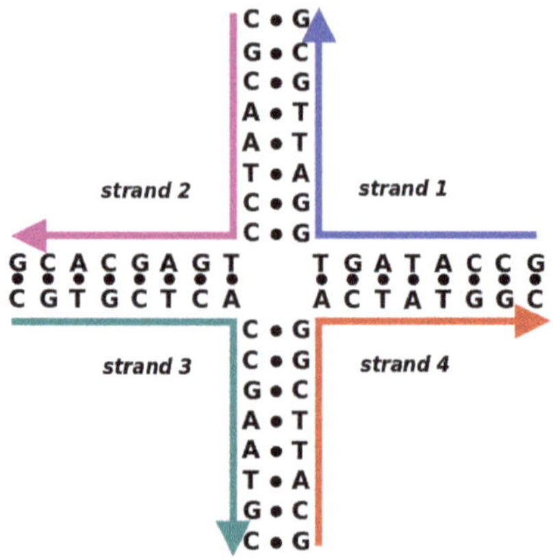

Fig. 3.3 Schematic of a Holliday junction showing the base sequence and secondary structure but not the tertiary structure.

In MathSciNet there is only one paper [302] (except papers of the author of this book) related to Holliday junctions.

In very recent papers [127], [128] the author considered a DNA molecule as a configuration of the Blume-Capel model on paths in a Cayley tree and using Gibbs measures, probability distributions of Holliday junctions in the Blume-Capel model of DNA molecules are studied.

It is known that (see [302] and references therein) the Escherichia coli[2] RuvA and RuvB proteins are involved in the late stages of recombination and the recombinational repair of damaged DNA.

The RuvA protein forms symmetric tetramers and binds to the center of a Holliday junction, unfolding the junction into a square planar configuration and keeping the four single-stranded segments at the center apart.

In [302] a model is presented for the mechanochemical-coupling mechanism in a Holliday junction. Using the model, the relationship between the power-stroke size, the step size of DNA translocation and the ratio of the rotational rate of DNA over that of RuvB relative to RuvA is predicted.

In this chapter we consider mathematical models of DNAs and study their properties (in particular, the Holliday junctions between them).

3.1.2 *A codon-based model*

In this subsection following [117] we give a model where Gibbs measure is the stationary measure.

[2]https://en.wikipedia.org/wiki/Escherichia_coli.

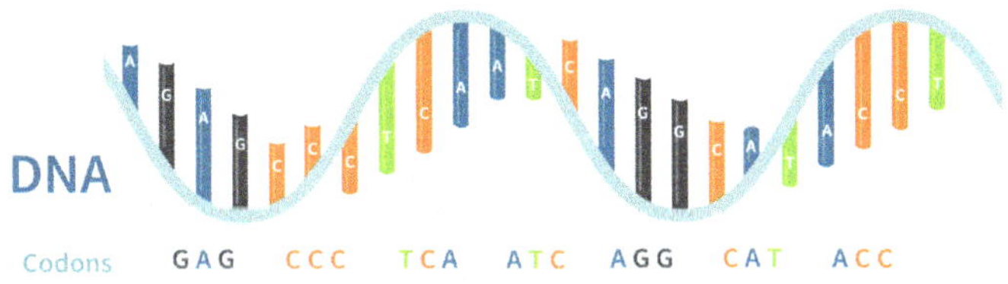

Fig. 3.4 Examples of codons in a DNA.

Some regions of the sequences (the genes) encode proteins, that is, strings of amino acids. In these regions triplets of nucleotides are called codons.[3] The codons are translated into amino acids via the genetic code (see Fig. 3.4).

Since there are 64 codons (for example, ATG, GCT, GCC, TGT), with three being stop codons (TAA, TGA, TAG) that signal end of translation, and only 20 amino acids, the genetic code is degenerate.

In [117] a model is given for the substitution process in lentiviral[4] genes, which is an extension of such a model considered in [212]. It is shown that the stationary measure is a Gibbs measure for the codon sequence of this model. The Gibbs measure can be written as a Markov chain along the sequence of codons. Let us present this model and results. Consider a model where a change in a codon sequence consists in a substitution (change) of one nucleotide only. Write a codon sequence as $z_i, \dots, z_n$, with $z_i = (z_i^1, z_i^2, z_i^3)$, where the upper index j in z_i^j indicates the position within the codon and $z_i^j \in \{A, C, G, T\}$.

Denote by $z_i(j,b)$ the new codon which is identical to z_i except at codon position i, where z_i^j has been replaced by b. Assume that the intensity γ for such a change depends upon z_i as well as the neighbors z_{i-1}^3 and z_{i+1}^1, which has the form

$$\begin{aligned}\gamma\left(z_i(j,b); z_{i-1}^3, z_i, z_{i+1}^1\right) &= M\left(z_i, z_i(j,b)\right)\pi_b^j \\ \times\lambda_{31}^{\mathbf{1}_{\mathrm{CG}}\left(z_{i-1}^3, z_i(j,b)^1\right)-\mathbf{1}_{\mathrm{CG}}\left(z_{i-1}^3, z_i^1\right)}&\lambda_{12}^{\mathbf{1}_{\mathrm{CG}}\left(z_i(j,b)^1, z_i(j,b)^2\right)-\mathbf{1}_{\mathrm{CG}}\left(z_i^1, z_i^2\right)} \\ \times\lambda_{23}^{\mathbf{1}_{\mathrm{CG}}\left(z_i(j,b)^2, z_i(j,b)^3\right)-\mathbf{1}_{\mathrm{CG}}\left(z_i^2, z_i^3\right)}&\lambda_{31}^{\mathbf{1}_{\mathrm{CG}}\left(z_i(j,b)^3, z_{i+1}^1\right)-\mathbf{1}_{\mathrm{CG}}\left(z_i^3, z_{i+1}^1\right)},\end{aligned} \tag{3.1}$$

where $\lambda_{ij} > 0$ are parameters,

$$\mathbf{1}_{\mathrm{CG}}(s_1, s_2) = \begin{cases} 1, & \text{if } (s_1, s_2) = (\mathrm{CG}) \\ 0, & \text{if } (s_1, s_2) \neq (\mathrm{CG}), \end{cases}$$

[3] https://en.wikipedia.org/wiki/DNA_and_RNA_codon_tables.
[4] https://en.wikipedia.org/wiki/Lentivirus.

$$M\left(z_i, z_i(j,b)\right) = K^{\mathbf{1}_{\mathrm{TS}}\left(z_i^j, b\right)} f^{\mathbf{1}_{\text{non-syn}}\left(z_i, z_i(j,b)\right)},$$

with some $K > 0$, $f > 0$ and $\mathbf{1}_{\mathrm{TS}}$ an indicator function for a transition, i.e., the substitution of a purine[5] (A or G) for a purine or a pyrimidine[6] (C or T) for a pyrimidine, and with $\mathbf{1}_{\text{non-syn}}$ an indicator function for a change in the amino acid. The parameters $\pi_b^j \geq 0$ satisfy

$$\sum_b \pi_b^j = 1, \quad \text{for each} \quad j,$$

and we only consider those j, b for which $z_i(j,b)$ is not a stop codon.

In a study of HIV-1 gene sequences the authors of [212] considered the above model with $\lambda_{31} = 1$ and $\lambda_{12} = \lambda_{23}$. When $\lambda_{31} = 1$ the codons evolve independently and the stationary frequency for a codon (s_1, s_2, s_3) is

$$\begin{cases} \kappa\lambda^2 \pi_{s_1}^1 \pi_{s_2}^2 \pi_{s_3}^3 & \text{if } (s_1, s_2) = (\mathrm{CG}) \text{ or } (s_2, s_3) = (\mathrm{CG}) \\ \kappa \pi_{s_1}^1 \pi_{s_2}^2 \pi_{s_3}^3 & \text{otherwise,} \end{cases}$$

where $\lambda = \lambda_{12} = \lambda_{23}$ is the common value and κ is a normalizing constant.

From the stationary frequencies one can see that this model can take into account different nucleotide compositions in the three codon positions and low frequencies of CG pairs at codon positions $(1,2)$ and $(2,3)$. However, in lentiviral gene sequences low frequencies of CGs are observed across codon boundaries, that is, at codon positions $(3,1)$, as well as within codons.

In [212] the authors presented a scenario, where pentets

$$\left(z_{i-1}^3, z_i^1, z_i^2, z_i^3, z_{i+1}^1\right),$$

evolve independently according to the intensities given in (3.1). In such a model the stationary frequencies for pentets are of the form

$$\begin{cases} \kappa\lambda^4 \pi_{s_0}^3 \pi_{s_1}^1 \pi_{s_2}^2 \pi_{s_3}^3 \pi_{s_4}^1 & \text{two CGs in } \tilde{s} \\ \kappa\lambda^2 \pi_{s_0}^3 \pi_{s_1}^1 \pi_{s_2}^2 \pi_{s_3}^3 \pi_{s_4}^1 & \text{one CG in } \tilde{s} \\ \kappa \pi_{s_0}^3 \pi_{s_1}^1 \pi_{s_2}^2 \pi_{s_3}^3 \pi_{s_4}^1 & \text{no CG in } \tilde{s}, \end{cases} \tag{3.2}$$

where $\tilde{s} = (s_0, s_1, s_2, s_3, s_4)$ is the pentet. They established the importance of depression of CGs between codons. In recognition of the fact that overlapping pentets do not evolve independently the authors refrain from performing any evolutionary analysis using the pentet model.

---

[5]A purine is an aromatic heterocycle composed of carbon and nitrogen. Purines include adenine and guanine, which participate in DNA and RNA formation.

[6]Pyrimidine is one of two classes of heterocyclic nitrogenous bases found in the nucleic acids DNA and RNA: in DNA the pyrimidines are cytosine and thymine, in RNA uracil replaces thymine.

For the model defined in (3.1) the stationary measure for the codon sequence is

$$\pi(z) = \frac{1}{Z}\left(\prod_{i=1}^{n}\left(\pi^1_{z_i^1}\pi^2_{z_i^2}\pi^3_{z_i^3}\right)\right)\lambda_{31}^{2\sum_{i=1}^{n+1}\mathbf{1}_{\mathrm{CG}}\left(z_{i-1}^3,z_i^1\right)} \times \lambda_{12}^{2\sum_{i=1}^{n}\mathbf{1}_{\mathrm{CG}}\left(z_i^1,z_i^2\right)}\lambda_{23}^{2\sum_{i=1}^{n}\mathbf{1}_{\mathrm{CG}}\left(z_i^2,z_i^3\right)}, \tag{3.3}$$

where Z is a normalizing constant. This can be proved by showing that

$$\pi(z)\gamma\left(z_i(j,b); z_{i-1}^3, z_i, z_{i+1}^1\right) = \pi\left(z_1,\ldots,z_{i-1},z_i(j,b),z_{i+1},\ldots,z_n\right)\gamma\left(z_i; z_{i-1}^3, z_i(j,b), z_{i+1}^1\right).$$

Note that the stationary measure (3.3) is a Gibbs measure which is specified through an interaction function $\exp(\phi(\cdot,\cdot))$ between z_{i-1}^3 and z_i^1 :

$$\exp(\phi(a,b)) = \left(\lambda_{31}^2\right)^{\mathbf{1}_{\mathrm{CG}}(a,b)}, \tag{3.4}$$

that measures the presence of the pair CG across a codon boundary, and a potential $\exp\left(\phi_0(\cdot)\right)$ for the codon $s=(s_1,s_2,s_3)$ itself:

$$\exp\left(\phi_0(s)\right) = \pi^1_{s_1}\pi^2_{s_2}\pi^3_{s_3}\left(\lambda_{12}^2\right)^{\mathbf{1}_{\mathrm{CG}}(s_1,s_2)}\left(\lambda_{23}^2\right)^{\mathbf{1}_{\mathrm{CG}}(s_2,s_3)}. \tag{3.5}$$

The Gibbs measure is important because the stationary measure becomes a Markov chain along the codon sequence. Since the interaction in (3.4) is between z_{i-1}^3 and z_i^1 only, the conditional distribution of z_i given z_{i-1} depends on z_{i-1}^3 only. Note that the Gibbs structure gives immediately from (3.4) and (3.5), in the case with $\lambda_{12}=\lambda_{23}=\lambda_{31}$, that conditionally on z_{i-1}^3 and z_{i+1}^1 the stationary frequencies for the codons are given by (3.2).

Now the Markov chain along the codon sequence embodied in (3.3) can be defined by

$$\mathrm{P}\left(z_i=(s_1,s_2,s_3)\mid z_{i-1}^3=a\right) = \frac{\exp\left\{\phi\left(a,s_1\right)+\phi_0(s)\right\}}{qh(a)}h\left(s_3\right), \tag{3.6}$$

for some number q and some function $h(\cdot)$.

By (3.4) one can see that it suffices to take $h(\mathrm{A})=h(\mathrm{G})=h(\mathrm{T})=1$ and $h(\mathrm{C})=\tau$. Then τ and q are determined from the two equations

$$q = \sum_{s\in\tilde{S}}\exp\left(\phi_0(s)\right)h\left(s_3\right),$$

$$q\tau = \sum_{s\in\tilde{S}}\left(\lambda_{31}^2\right)^{\mathbf{1}_{\mathrm{G}}(s_1)}\exp\left(\phi_0(s)\right)h\left(s_3\right),$$

where $\tilde{S}$ is the set of non-stop codons.

The usefulness of (3.6) is that this formula is explicit and that simple tests can be designed to test the adequacy of the transition probabilities in (3.6).

Define now the matrix $V(\cdot,\cdot)$ by

$$V(a, s_3) = \sum_{s_1, s_2 : s \in \tilde{S}} \exp\left(\phi(a, s_1) + \phi_0(s)\right), \quad s = (s_1, s_2, s_3).$$

Then h is a right eigenvector of V with eigenvalue q. Using

$$V(\mathrm{A}, \cdot) = V(\mathrm{G}, \cdot) = V(\mathrm{T}, \cdot),$$

one can see that a left eigenvector $l(\cdot)$, normalized such that $l(\mathrm{C}) = 1$, has to be of the form

$$ql(b) = V(\mathrm{C}, b) + V(\mathrm{A}, b)\xi,$$

where $\xi = l(\mathrm{A}) + l(\mathrm{G}) + l(\mathrm{T})$ is determined by

$$q = V(\mathrm{C}, \mathrm{C}) + V(\mathrm{A}, \mathrm{C})\xi.$$

Now the invariant distribution for z_i^3 in the Markov chain (3.6) can be written as

$$\mathrm{P}\left(z_i^3 = a\right) = \frac{l(a)h(a)}{\tau + \xi} = \frac{1}{\tau + \xi}(V(\mathrm{C}, a) + V(\mathrm{A}, a)\xi)\tau^{\mathbf{1}_{\mathrm{C}}(a)}.$$

To generalize the above considered model, following [117], we consider a continuous-time Markov process on the space of codon sequences of length n. Denote the sequence by $z = (z_1, \ldots, z_n)$ with $z_i \in \tilde{S}$, the set of nonstop codons. Assume that we have two boundary codons z_0 and z_{n+1} which are fixed. These could be initiation or stop codons. When the process makes a jump there will only be a change in one codon. A codon can be changed from any value to any other value in $\tilde{S}$, corresponding to all intensities being positive.

The intensity for a change from z_i to y_i is allowed to depend on the two neighboring codons, i.e., the intensity is

$$\gamma\left(y_i; z_{i-1}, z_i, z_{i+1}\right).$$

In case $1 < i < n$ all the codon variables will belong to $\tilde{S}$, whereas when $i = 1$ we allow z_0 to be a stop codon and similarly with z_{n+1}.

We are now interested to a stationary distribution to be of a form that allows explicit calculations. This allows the possibility of designing tests based on one sequence only and simple estimates can be obtained for some of the parameters of the model.

Consider a Gibbs measure for the stationary distribution, i.e. a measure defined through an interaction function $\phi : \tilde{S} \times \tilde{S} \to \mathbb{R}$ and given by

$$\mathrm{P}(z) = \frac{1}{Z} \exp\left\{ \phi_1(z_1) + \sum_{i-2}^{n} \phi(z_{i-1}, z_i) + \phi_n(z_n) \right\}, \tag{3.7}$$

where Z is a normalizing constant. The functions ϕ_1 and ϕ_n are defined on $\tilde{S}$.

We now show that the Gibbs measure allows us to rewrite this measure as a Markov chain along the codon sequence, thereby making the analysis of the measure simple.

Note that the normalizing constant Z is not easy to find directly. To find Z and properties of the Gibbs measure it will be useful to rewrite the measure as a Markov chain along the sequence of codons. Define

$$Q(a,b) = \exp\{\phi(a,b)\}, \quad a, b \in \tilde{S},$$

where $\tilde{S}$ is the set of 61 non-stop codons.

Let q be the largest eigenvalue of Q and let r be a right eigenvector corresponding to q,

$$\sum_b Q(a,b)r(b) = qr(a), \quad \forall a \in \tilde{S}.$$

Define also

$$T(a,b) = \frac{Q(a,b)r(b)}{qr(a)}, \quad a, b \in \tilde{S}.$$

Note that, this is a transition matrix and one has

$$\prod_{i=2}^{n} T(z_{i-1}, z_i) = \frac{r(z_n)}{q^{n-1} r(z_1)} \exp\left\{ \sum_{i=2}^{n} \phi(z_{i-1}, z_i) \right\},$$

and $\mathrm{P}(z)$ from (3.7) becomes

$$\mathrm{P}(z) = \frac{1}{Z} \frac{q^{n-1} r(z_1)}{r(z_n)} \exp\{\phi_1(z_1) + \phi_n(z_n)\} \prod_{i=2}^{n} T(z_{i-1}, z_i).$$

Thus for the normalizing constant we get

$$Z = q^{n-1} \sum_{a,b \in \tilde{S}} \frac{r(a)}{r(b)} \exp\{\phi_1(a) + \phi_n(b)\} T^{n-1}(a,b), \tag{3.8}$$

which can be evaluated numerically, allowing $\mathrm{P}(z)$ to be calculated.

The calculation of T^{n-1} requires of the order of $61^2 n$ calculations.

Frequencies in the limit $n \to \infty$ can be evaluated with the help of the stationary distribution for the transition matrix T. Let l be a left eigenvector for Q with eigenvalue q,

$$\sum_a l(a)Q(a,b) = ql(b), \quad \forall b \in \tilde{S},$$

and normalize l so that

$$\sum_a l(a)r(a) = 1.$$

Then $\{l(a)r(a), a \in S\}$ is the stationary distribution for T. Thus the frequency of a particular codon c will converge to $l(c)r(c)$ as $n \to \infty$, and the frequency of a particular neighboring pair of codons (c_1, c_2) will converge to

$$l(c_1)\, r(c_1)\, T(c_1, c_2) = q^{-1} l(c_1) \exp(\phi(c_1, c_2))\, r(c_2)\,.$$

Using approximation $T^{n-1}(a,b)$ by the stationary probability $l(b)r(b)$, from (3.8) one gets the following approximation of the normalizing constant:

$$Z \approx q^{n-1} \sum_{a,b \in \tilde{S}} r(a) \exp\{\phi_1(a) + \phi_n(b)\}\, l(b)r(b).$$

3.1.3 *A 3-state model of the interacting DNAs*

To define the model, we follow [246], consider a set of DNAs which "live" on a tree-like graph.

For simplicity of notations we denote "A+T" by "-1" and "C+G" by "+1".

We consider a DNA as a very long sequence consisting of -1 and $+1$'s.

Let l be an edge of this graph we have a function $\sigma(l)$ with three possible values $-1, 0, 1$ (an analogue of spin values in physical systems), in case $\sigma(l) = 0$ we say the edge l does not belong to a DNA. If this l separates two DNA then the value $\sigma(l) = 1$ or $\sigma(l) = -1$ means that these two DNA have a Holliday junction.

Let Γ^k be a Cayley tree of order $k \geq 1$ and G_k be its group representation (see Subsection 2.1.1).

We consider a normal subgroup $\mathcal{H}_0 \subset G_k$ of infinite index constructed as follows. Recall that a_i, $i = 1, \dots, k+1$ are generators of the group G_k, with $a_i^2 = e$, for all i.

Let the mapping $\pi_0 : \{a_1, ..., a_{k+1}\} \longrightarrow \{e, a_1, a_2\}$ be defined by

$$\pi_0(a_i) = \begin{cases} a_i, \text{ if } \ i = 1, 2 \\ e, \ \text{ if } \ i \neq 1, 2. \end{cases}$$

Denote by G_1 the free product of cyclic groups $\{e, a_1\}, \{e, a_2\}$. Note that G_1 is a representation of $\mathbb{Z}$-the tree of order 1, i.e., the vertices are labeled by the integer numbers.

Consider

$$f_0(x) = f_0(a_{i_1} a_{i_2} ... a_{i_m}) = \pi_0(a_{i_1})\pi_0(a_{i_2}) \dots \pi_0(a_{i_m}).$$

Then it is easy to see that f_0 is a homomorphism and hence

$$\mathcal{H}_0 = \{x \in G_k : \ f_0(x) = e\},$$

is a normal subgroup of infinite index.

Now we consider the factor group

$$G_k/\mathcal{H}_0 = \{\mathcal{H}_0, \mathcal{H}_0(a_1), \mathcal{H}_0(a_2), \mathcal{H}_0(a_1 a_2), \dots\},$$

where $\mathcal{H}_0(y) = \{x \in G_k : f_0(x) = y\}$.

Introduce the notations

$$\mathcal{H}_n = \mathcal{H}_0(\underbrace{a_1 a_2 \dots}_{n}),$$

$$\mathcal{H}_{-n} = \mathcal{H}_0(\underbrace{a_2 a_1 \dots}_{n}).$$

In this notation, the factor group can be represented as

$$G_k/\mathcal{H}_0 = \{\dots, \mathcal{H}_{-2}, \mathcal{H}_{-1}, \mathcal{H}_0, \mathcal{H}_1, \mathcal{H}_2, \dots\}.$$

We introduce the following equivalence relation on the set G_k: $x \sim y$ if $xy^{-1} \in \mathcal{H}_0$. Then G_k can be partitioned to countably many classes $\mathcal{H}_i$ of equivalent elements. The partition of the Cayley tree Γ^2 w.r.t. $\mathcal{H}_0$ is shown in Fig. 3.5 (the elements of the class $\mathcal{H}_i$, $i \in \mathbb{Z}$, are merely denoted by i).

$\mathbb{Z}$-path. Denote

$$q_i(x) = |W_1(x) \cap \mathcal{H}_i|, \quad x \in G_k,$$

where $|\cdot|$ is the counting measure of a set.

Note that (see [241]) if $x \in \mathcal{H}_m$, then

$$q_{m-1}(x) = 1, \quad q_m(x) = k - 1, \quad q_{m+1}(x) = 1. \tag{3.9}$$

Indeed, if $x \in \mathcal{H}_m$ then $f_0(x) = \underbrace{a_i a_j \dots a_i}_{m}$ (or $a_i a_j ... a_j$ depending on m).

Since the set of all nearest neighbors of x is $\{xa_s, \ s = 1, ..., k+1\}$ we have

$$f_0(xa_s) = f_0(x)f_0(a_s) = a_i a_j \dots a_i f_0(a_s) = \begin{cases} \underbrace{a_i a_j \dots a_j}_{m-1} & \text{if } \ a_s = a_i \\ \underbrace{a_i a_j \dots}_{m} & \text{if } \ a_s \neq a_i, a_j \\ \underbrace{a_i a_j \dots a_i a_j}_{m+1} & \text{if } \ a_s = a_j. \end{cases}$$

This relation gives (3.9).

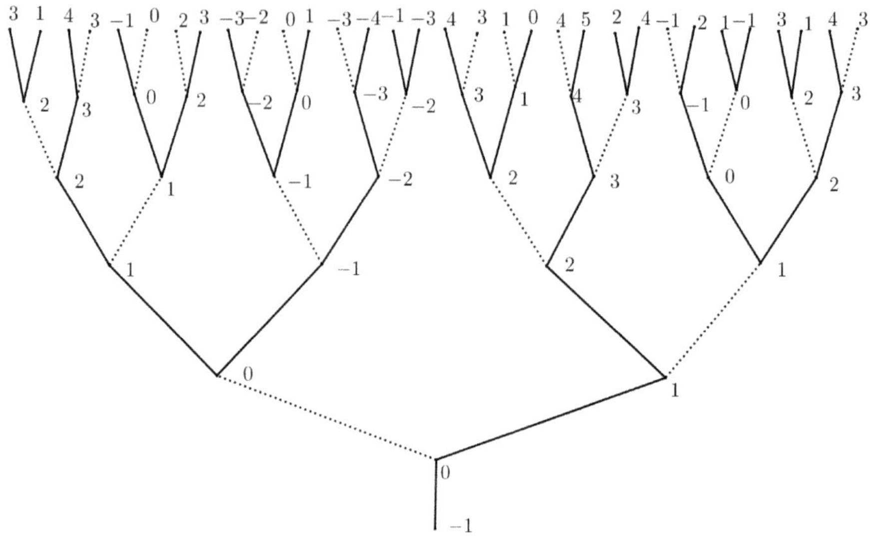

Fig. 3.5 The partition of the Cayley tree Γ^2 w.r.t. $\mathcal{H}_0$, the elements of the class $\mathcal{H}_i$, $i \in \mathbb{Z}$, are denoted by i. $\mathbb{Z}$-pathes are solid lines.

From this fact it follows that for any $x \in V$, if $x \in \mathcal{H}_m$ then there is a unique two-side-path (containing x) such that the sequence of numbers of equivalence classes for vertices of this path in one side are $m, m+1, m+2, \dots$ in the second side the sequence is $m, m-1, m-2, \dots$. Thus the two-side-path has the sequence of numbers of equivalent classes as $\mathbb{Z} = \{\dots, -2, -1, 0, 1, 2, \dots\}$. Such a path is called $\mathbb{Z}$-path. (In Fig. 3.5 one can see the unique $\mathbb{Z}$-paths of each vertex of the tree.) $\mathbb{Z}$-paths are solid pathes.

Since each vertex x has its own $\mathbb{Z}$-path one can see that the Cayley tree considered with respect to normal subgroup $\mathcal{H}_0$ contains infinitely many (countable) set of $\mathbb{Z}$-pathes.

The model. Let L be the set of edges of a Cayley tree. Consider function σ which assigns to each edge $l \in L$, values $\sigma(l) \in \{-1, 0, 1\}$. Value $\sigma(l) = -1$ (resp. $+1$) means that edge l is 'occupied' by $-1 =$"A+T" (resp. $1 =$"C+G"), and $\sigma(l) = 0$ that l is 'vacant'.

A configuration $\sigma = \{\sigma(l),\ l \in L\}$ on edges of the Cayley tree is given by a function from L to $\{-1, 0, 1\}$. The set of all configurations in L is denoted by Ω. Configurations in L_n are defined analogously and the set of all configurations in L_n is denoted by Ω_n.

A configuration $\sigma = \{\sigma(l),\ l \in L\}$ is called *admissible* if $\sigma(l) \neq 0$ for any $l \in \mathbb{Z}$-path.

Remark 3.1. The admissibility condition guarantees that on each $\mathbb{Z}$-path the configuration describes a DNA. This admissibility reminds the admissibility in case of hard-core models (see Chapter 7 of [243] and [275]), but they are different, because in the hard-core models the admissibility is given by conditions for configurations on nearest neighboring vertices.

The restriction of an admissible configuration on a $\mathbb{Z}$-path is called *a DNA*, because it is a sequence of -1 and 1 values (see above the part about DNA).

For an admissible configuration σ we have $\sigma(l) = 0$ iff $l = \langle x, y\rangle$ with $x \sim y$, i.e., $f_0(x) = f_0(y)$. Comparing this with Holliday junction one can see that DNAs corresponding to two $\mathbb{Z}$-paths can have a junction only through the edge $l = \langle x, y\rangle$ with $x \sim y$.

The most common discrete models treat DNA as a collection of rigid subunits representing the base pairs (see Fig. 3.1) [277]. This description has long been used by chemists to characterize DNA crystal structures. Following formulas (4.9) and (4.10) of [277] we consider the following simple model of the energy of the configuration σ of a set of DNAs:

$$H(\sigma) = J \sum_{\langle l,t\rangle \in L\times L} \sigma(l)\sigma(t), \tag{3.10}$$

where $J > 0$ is a coupling constant, $\sigma(l) \in \{-1, 0, 1\}$ and $\langle l, t\rangle$ stands for nearest neighbor edges, i.e. edges which have a common endpoint.

Tree-hierarchy of the set of DNAs.

We shall give a Cayley tree-hierarchy of the set of DNAs, as a DNA crystal.

Given an admissible configuration σ on a Cayley tree, since there are countably many $\mathbb{Z}$-pathes we have a countable many distinct DNAs. We say that two DNA *are neighbors* if there is an edge (of the Cayley tree) such that one its endpoint belongs to the first DNA and another endpoint of the edge belongs to the second DNA. By the construction it is clear (see Fig. 3.5) that such an edge is unique for each neighboring pair of DNAs. This edge has equivalent endpoints, i.e., both endpoints belong to the same class $\mathcal{H}_m$ for some $m \in \mathbb{Z}$.

Moreover these countably infinite set of DNAs have a hierarchy that

- Any two DNA do not intersect.
- Each DNA has its own countably many set of neighboring DNAs.

- For any two neighboring DNAs, say D_1 and D_2, there exists a unique edge $l = l(D_1, D_2) = \langle x, y\rangle$ with $x \sim y$ (of the Cayley tree) which connects DNAs.
- Two DNAs, D_1, D_2 (corresponding to two $\mathbb{Z}$-paths) can have a Holliday junction if and only if they are neighbors, i.e. through the unique edge $l(D_1, D_2)$.
- For any finite $n \geq 1$ the ball V_n has intersection only with finitely many DNAs.

3.2 Finite dimensional distributions and equations

Let Ω_n^a (resp. Ω^a) be the set of all *admissible* configurations on L_n (resp. L).

Denote

$$E_n = \{\langle x, y\rangle \in L : x \in W_{n-1},\ y \in W_n\},$$

$$\Omega_n^{ba} = \text{ the set of admissible configurations on } E_n.$$

For $l \in E_{n-1}$ denote

$$S(l) = \{t \in E_n : \langle l, t\rangle\}.$$

It is easy to see that (see Fig. 3.5)

$$S(l) \cap \mathbb{Z} - \text{path} = \begin{cases} \{l_0, l_1\} \subset L, & \text{if } l \notin \mathbb{Z} - \text{path} \\ \{l_1\} \subset L, & \text{if } l \in \mathbb{Z} - \text{path} \end{cases}.$$

We denote

$$S_0(l) = S(l) \setminus \{l_0, l_1\},\ \ l \notin \mathbb{Z} - \text{path},$$

$$S_1(l) = S(l) \setminus \{l_1\},\ \ l \in \mathbb{Z} - \text{path}.$$

Define a finite-dimensional distribution of a probability measure μ on Ω_n^a as

$$\mu_n(\sigma_n) = Z_n^{-1} \exp\left\{\beta H_n(\sigma_n) + \sum_{l \in E_n} h_{\sigma(l),l}\right\}, \tag{3.11}$$

where $\beta = 1/T$, $T > 0$ is temperature, Z_n^{-1} is the normalizing factor, $\{h_{i,l} \in \mathbb{R}, i = -1, 0, 1, l \in L\}$ is a collection of real numbers and

$$H_n(\sigma_n) = J \sum_{\substack{l,t \in L_n: \\ \langle l,t\rangle}} \sigma(l)\sigma(t).$$

Remark 3.2. We note that the quantities $\exp(h_{i,l})$ define a boundary law (see Subsection 2.3.2 and Chapter 12 of [98]). In this section these quantities mean a boundary law of our biological system of DNAs.

We say that the probability distributions (3.11) are compatible if for all $n \geq 1$ and $\sigma_{n-1} \in \Omega^a_{n-1}$:

$$\sum_{\omega_n \in \Omega_n^{h_n}} \mu_n(\sigma_{n-1} \vee \omega_n)\mathbf{1}(\sigma_{n-1} \vee \omega_n \in \Omega^a_n) = \mu_{n-1}(\sigma_{n-1}). \tag{3.12}$$

Here $\sigma_{n-1} \vee \omega_n$ is the concatenation of the configurations.

In this case there exists a unique measure μ on Ω^a such that, for all n and $\sigma_n \in \Omega^a_n$,

$$\mu(\{\sigma|_{L_n} = \sigma_n\}) = \mu_n(\sigma_n).$$

Such a measure is called a (splitting) *Gibbs measure* corresponding to the Hamiltonian (3.10) and functions $h_{i,l}, l \in L$.

The following statement describes conditions on $h_{i,l}$ guaranteeing compatibility of $\mu_n(\sigma_n)$.

Theorem 3.1. *Probability distributions* $\mu_n(\sigma_n)$, $n = 1, 2, \ldots$, *in* (3.11) *are compatible iff for any* $l \in L$ *the following equations hold*

$$z_{0,l} = \frac{1 + z_{l_0}}{\alpha + \alpha^{-1} z_{l_0}} \cdot \frac{1 + z_{l_1}}{\alpha + \alpha^{-1} z_{l_1}} \prod_{t \in S_0(l)} \frac{1 + z_{0,t} + z_{1,t}}{\alpha + z_{0,t} + \alpha^{-1} z_{1,t}}, \quad l \notin \mathbb{Z} - \text{path},$$

$$z_{1,l} = \frac{\alpha^{-1} + \alpha z_{l_0}}{\alpha + \alpha^{-1} z_{l_0}} \cdot \frac{\alpha^{-1} + \alpha z_{l_1}}{\alpha + \alpha^{-1} z_{l_1}} \prod_{t \in S_0(l)} \frac{\alpha^{-1} + z_{0,t} + \alpha z_{1,t}}{\alpha + z_{0,t} + \alpha^{-1} z_{1,t}}, \quad l \notin \mathbb{Z} - \text{path}, \tag{3.13}$$

$$z_l = \frac{\alpha^{-1} + \alpha z_{l_1}}{\alpha + \alpha^{-1} z_{l_1}} \prod_{t \in S_1(l)} \frac{\alpha^{-1} + z_{0,t} + \alpha z_{1,t}}{\alpha + z_{0,t} + \alpha^{-1} z_{1,t}}, \quad l \in \mathbb{Z} - \text{path}.$$

Here,

$$\alpha = \exp(J\beta);\ z_{i,l} = \exp\left(h_{i,l} - h_{-1,l}\right), i = 0, 1;\ z_l = \exp\left(h_{1,l} - h_{-1,l}\right). \tag{3.14}$$

Proof. Fix an edge $l_0 \in L$ as a "root" in L. Then there is one-to-one correspondence between elements of L and elements of V, which can be given as follows: let $l = \langle x, y\rangle$ be an arbitrary edge, without loss of generality we assume that x is closer (than y) to l_0. Then it is easy to see that the map $l \to y$ is one-to-one. Therefore in this proof for $l = \langle x, y\rangle$ the value $\sigma(l)$ is simply denoted by $\sigma(y)$. Then E_n is replaced by V_n.

Necessity. Suppose that (3.12) holds; we want to prove (3.13). Substituting (3.11) into (3.12), obtain that for any configurations $\sigma_{n-1}\colon x \in V_{n-1} \mapsto \sigma_{n-1}(x) \in \{-1, 0, 1\}$:

$$\frac{Z_{n-1}}{Z_n} \sum_{\omega_n \in \Omega_{W_n}} \exp\left(\sum_{x \in W_{n-1}} \sum_{y \in S(x)} (J\beta\sigma_{n-1}(x)\omega_n(y) + h_{\omega_n(y),y}) \right)$$
$$= \exp\left(\sum_{x \in W_{n-1}} h_{\sigma_{n-1}(x),x} \right), \tag{3.15}$$

where $\omega_n\colon x \in W_n \mapsto \omega_n(x)$ and $\sigma_{n-1} \vee \omega_n \in \Omega_n^a$.

From (3.15) we get:

$$\frac{Z_{n-1}}{Z_n} \sum_{\omega_n \in \Omega_{W_n}} \prod_{x \in W_{n-1}} \prod_{y \in S(x)} \exp\left(J\beta\sigma_{n-1}(x)\omega_n(y) + h_{\omega_n(y),y}\right)$$
$$= \prod_{x \in W_{n-1}} \exp\left(h_{\sigma_{n-1}(x),x}\right). \tag{3.16}$$

Fix a vertex (i.e., an endpoint of an edge) $x \in W_{n-1}$ and consider three configurations $\sigma_{n-1} = \overline{\sigma}_{n-1}$, $\sigma_{n-1} = \hat{\sigma}_{n-1}$ and $\sigma_{n-1} = \tilde{\sigma}_{n-1}$ on W_{n-1} which coincide on $W_{n-1} \setminus \{x\}$, and rewrite now the equality (3.16) for $\overline{\sigma}_{n-1}(x) = -1$, $\hat{\sigma}_{n-1}(x) = 0$ and $\tilde{\sigma}_{n-1}(x) = 1$, then dividing second and third equality to the first one (taking into account the relation of x with $\mathbb{Z}$-path and the admissibility of configuration) we get (3.13).

Sufficiency. This is similar to the corresponding part of the proof of Theorem 2.2. That is, assuming that (3.13) holds, one can rewrite it in the form (3.16) and then in the form (3.15). Consequently, one gets (3.12). □

From Theorem 3.1 it follows that for any set of vectors

$$\mathbf{z} = \{(z_{0,l}, z_{1,l}, z_t), l \notin \mathbb{Z} - \text{path}, \quad t \in \mathbb{Z}\},$$

satisfying the system of functional equations (3.13) there exists a unique Gibbs measure μ and vice versa. However, the analysis of solutions to (3.13) is not easy.

In the next section we shall give several solutions to (3.13). The number of the solutions of equation (3.13) depends on the temperature-parameter $\alpha = \exp(\frac{J}{T})$.

Remark 3.3. Note that if there is more than one solution to equation (3.13), then there is more than one Gibbs measure corresponding to these solutions. One says that a "phase" transition occurs for the model of DNAs, if equation (3.13) has more than one solution.

3.3 Translation-invariant Gibbs measures of the set of DNAs

In this section without loss of generality we take $J = 1$. Find solutions $\mathbf{z}_l$ to the system of functional equations (3.13), which do not depend on l, i.e.,

$$z_{0,l} = u, \quad z_{1,l} = v, \quad \forall l \notin \mathbb{Z} - \text{path}; \quad z_l = w, \quad \forall l \in \mathbb{Z} - \text{path}. \tag{3.17}$$

The Gibbs measure corresponding to such a solution is called *translation-invariant*.

For u, v, w from (3.13) we get

$$\begin{aligned} u &= \left(\frac{1+u+v}{\alpha+u+\alpha^{-1}v}\right)^{k-2}\left(\frac{1+w}{\alpha+\alpha^{-1}w}\right)^2, \\ v &= \left(\frac{1+\alpha u+\alpha^2 v}{\alpha^2+\alpha u+v}\right)^{k-2}\left(\frac{1+\alpha^2 w}{\alpha^2+w}\right)^2, \\ w &= \left(\frac{1+\alpha u+\alpha^2 v}{\alpha^2+\alpha u+v}\right)^{k-1}\left(\frac{1+\alpha^2 w}{\alpha^2+w}\right). \end{aligned} \tag{3.18}$$

Here $u, v, w > 0$. By assumption $J = 1$ we have $\alpha > 1$.

It is clear that $v = w = 1$ satisfies the system (3.18) for any $k \geq 2$ and $\alpha > 1$, then from the first equation of the system we obtain

$$u = \left(\frac{2+u}{\alpha+\alpha^{-1}+u}\right)^{k-2}\left(\frac{2}{\alpha+\alpha^{-1}}\right)^2. \tag{3.19}$$

Remark 3.4. For any $k \geq 2$ number of solutions to the equation (3.19) can be given depending on parameter α. Because this equation can be rewritten as the following well known equation (see, e.g., [220] or [271]):

$$ax = f(x), \qquad f(x) := \left(\frac{1+x}{b+x}\right)^k. \tag{3.20}$$

Proposition 3.1. *Equation (3.20) with $x \geq 0$, $k \geq 1$, $a, b > 0$ has a unique solution if either $k = 1$ or $b \leq (\frac{k+1}{k-1})^2$. If $k > 1$ and $b > (\frac{k+1}{k-1})^2$ then there exist $\nu_1(b,k)$, $\nu_2(b,k)$, with $0 < \nu_1(b,k) < \nu_2(b,k)$, such that the equation has three solutions if $\nu_1(b,k) < a < \nu_2(b,k)$ and has two if either $a = \nu_1(b,k)$ or $a = \nu_2(b,k)$. In fact:*

$$\nu_i(b,k) = \frac{1}{x_i}\left(\frac{1+x_i}{b+x_i}\right)^k,$$

where x_1, x_2 are the solutions of

$$x^2 + [2 - (b-1)(k-1)]x + b = 0.$$

But this proposition only gives the number of solutions. In this section for simplicity we consider the case $k = 2$ and will explicitly find the solutions.

For $k = 2$ from (3.19) we get

$$u = \left(\frac{2}{\alpha + \alpha^{-1}}\right)^2 = \cosh^{-2}(\beta).$$

Thus for $k = 2$ independently on α (i.e., $\beta = 1/T$) the vector $(\cosh^{-2}(\beta), 1, 1)$ is a solution to system (3.18). To find other solutions (for $k = 2$) we substitute u and v to the third equation of (3.18) and obtain

$$w = \frac{1+\alpha^2 w}{\alpha^2 + w}\left(\frac{(\alpha^2+w)^2 + \alpha^3(1+w)^2 + \alpha^2(1+\alpha^2 w)^2}{\alpha^2(\alpha^2+w)^2 + \alpha^3(1+w)^2 + (1+\alpha^2 w)^2}\right). \tag{3.21}$$

Define

$$\alpha_* = \frac{1}{3}\left(1 + (19+3\sqrt{33})^{1/3} + \frac{4}{(19+3\sqrt{33})^{1/3}}\right) \approx 1.839287.$$

Lemma 3.1. *The following statements hold*

1. *If a positive number w is a solution to (3.21) then the number $1/w$ is also solution to (3.21).*
2. *Positive solutions of (3.21) are $w_1 = 1$ and*

$$w_2 = w_2(\alpha) = \frac{\alpha^6 - \alpha^4 - 3\alpha^2 - 2\alpha - 1 - (\alpha^2-1)\sqrt{D}}{2(\alpha^2+\alpha+1)},$$

$$w_3 = w_3(\alpha) = \frac{\alpha^6 - \alpha^4 - 3\alpha^2 - 2\alpha - 1 + (\alpha^2-1)\sqrt{D}}{2(\alpha^2+\alpha+1)},$$

with $D = (\alpha^2+1)(\alpha^3+\alpha^2+\alpha+3)(\alpha^3-\alpha^2-\alpha-1)$. Moreover, if $\alpha \le \alpha_*$ then there is unique positive solution: $w_2 = w_3 = 1$. In case $\alpha > \alpha_*$, we have $w_2 = 1/w_3 < 1$.

Proof. Denote by $\varphi(w)$ the right side of (3.21). Then for any $w > 0$, it is easy to see that $\varphi(1/w) = 1/\varphi(w)$. From this equality it follows part 1.

It is easy, to check that $w = 1$ is solution of the equation. Rewrite the equation (3.21) as a polynomial equation $P_4(w) = 0$. By part 1 of this lemma, since P_4 is a polynomial of order 4 which has root 1, it follows that the polynomial has unique root bigger than 1 (denoted by w_3). Indeed, if it has another such solution: $\tilde{w}_3 > 1$, with $\tilde{w}_3 \ne w_3$, then by part 1, it has two solutions $1/w_3 < 1, 1/\tilde{w}_3 < 1$. So in total number of positive solutions

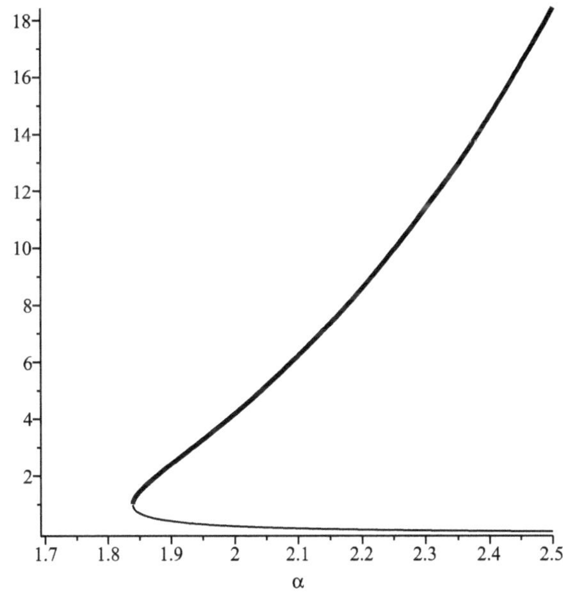

Fig. 3.6 The graph of the functions $w_2(\alpha)$ (thin curve) and $w_3(\alpha)$ (bold curve), for $\alpha \in [\alpha_*, \infty)$. The curves meet at α_* and $w_2(\alpha_*) = w_3(\alpha_*) = 1$.

is 5, since we already have solution 1 too. This is impossible, because P_4 may have up to 4 solutions.

Dividing the polynomial to $w - 1$ then solving the remaining equation one gets w_2 and $w_3 = 1/w_2$. The solutions w_2 (and w_3) exist iff $\alpha^3 - \alpha^2 - \alpha - 1 \geq 0$. It is easy to check that the last inequality is true for any $\alpha \geq \alpha_*$. As mentioned above, both w_2 and w_3 are positive and $w_2 w_3 = 1$, see Fig. 3.6 for graphs of the solutions as a function of $\alpha \in [\alpha_*, \infty)$.

This completes the proof. □

Thus, for $k = 2$ we have the following

Lemma 3.2. *The following hold*

- *If* $\alpha = \exp(\beta) < \alpha_*$ *then system (3.18) has unique solution*

$$\mathbf{z}_1 = (u_1, v_1, w_1) = (\cosh^{-2}(\beta), 1, 1);$$

- *If* $\alpha = \alpha_*$ *then system (3.18) has two solutions*

$$\mathbf{z}_1 = (\cosh^{-2}(\beta), 1, 1), \quad \mathbf{z}_2 = (u_2, v_2, w_2);$$

- *If $\alpha > \alpha_*$ then system (3.18) has three solutions*

$$\mathbf{z}_1 = (\cosh^{-2}(\beta), 1, 1), \quad \mathbf{z}_2 = (u_2, v_2, w_2), \quad \mathbf{z}_3 = (u_3, v_3, w_3),$$

where

$$u_i = \left(\frac{1 + w_i}{\alpha + \alpha^{-1} w_i}\right)^2, \quad v_i = \left(\frac{1 + \alpha^2 w_i}{\alpha^2 + w_i}\right)^2, \quad i = 2, 3.$$

Denote by μ_i the Gibbs measure which, by Theorem 3.1, corresponds to solution $\mathbf{z}_i$, $i = 1, 2, 3$.

Define critical temperature

$$T_c = \frac{1}{\ln \alpha_*}.$$

Summarizing above results we obtain the following

Theorem 3.2. *For the model (3.10) of DNAs on the Cayley tree of order $k = 2$ the following statements are true*

(1) If the temperature $T > T_c$ then there is unique translation-invariant Gibbs measure μ_1.
(2) If $T = T_c$ then there are 2 translation-invariant Gibbs measures μ_1, μ_2.
(2) If $T < T_c$ then there are 3 translation-invariant Gibbs measures μ_1, μ_2, μ_3.

Remark 3.5. Analysis of system (3.18) for the case $k \geq 3$ seems difficult. Thus remains open. The case $k = 2$ is already interesting enough to see biological interpretations of Theorem 3.2.

3.4 Biological interpretations: Holliday junction and branches of DNA

For marginals on the two-edge sets which consist of two neighbor edges l, t, considering a boundary law $\{(z_{0,l}, z_{1,l}, z_t), l \notin \mathbb{Z} - \text{path}, t \in \mathbb{Z}\}$ (i.e., the solutions of system (3.13)) we have[7]

$$\mu(\sigma(l) = a, \sigma(t) = b) = \frac{1}{Z} z_{a,l} \exp(\beta ab) z_{b,t}, \quad a, b = -1, 0, 1,$$

where Z is normalizing factor.

From this, the relation between the boundary law and the transition matrix for the associated tree-indexed Markov chain (Gibbs measure)

[7]This boundary law is normalized at -1, i.e., $z_{-1,l} = 1$, that is $h_{-1,l} = 0$.

is immediately obtained from the formula of the conditional probability. Indeed, once we have $h_{a,l}$ (given in formula (3.11)) which is independent on l we define tree-edge-indexed (non-homogenous) Markov chain with states $\{-1,0,1\}$ with transition matrix $\mathbb{P}^{[l,t]} = \left(P_{ij}^{[l,t]}\right)$, depending on $\langle l,t\rangle$ with

$$P_{ij}^{[l,t]} = \begin{cases} \frac{\exp(\beta ij+h_j)}{Z_1}, & \text{if } \ l,t \notin \mathbb{Z}-\text{path}, i,j=-1,0,1 \\ \frac{\exp(\beta ij+h_j)}{Z_2}, & \text{if } \ l \notin \mathbb{Z}-\text{path}, \ t \in \mathbb{Z}-\text{path}, i=-1,0,1; j=-1,1 \\ \frac{\exp(\beta ij+h_j)}{Z_1}, & \text{if } \ l \in \mathbb{Z}-\text{path}, t \notin \mathbb{Z}-\text{path}, i=-1,1; j=-1,0,1 \\ \frac{\exp(\beta ij+h_j)}{Z_2}, & \text{if } \ l,t \in \mathbb{Z}-\text{path}, i,j=-1,1. \end{cases}$$

where

$$Z_1 = \exp\left(-\beta i + h_{-1}\right) + \exp\left(h_0\right) + \exp\left(\beta i + h_1\right),$$

$$Z_2 = \exp\left(-\beta i + h_{-1}\right) + \exp\left(\beta i + h_1\right).$$

Here $P_{ij}^{[l,t]}$ is the probability to go from a state i at edge l to a state j at the neighbor edge t.

Using formulas (3.14) and (3.17) for solutions (u,v,w) to (3.18) we write the matrices $\mathbb{P}^{[l,t]} = \left(P_{ij}^{[l,t]}\right)$:

$$\mathbb{P}^{[l,t]} = \mathbb{P}^{[l,t]}_{(3\to3)} = \begin{pmatrix} \frac{\alpha}{\alpha+u+\alpha^{-1}v} & \frac{u}{\alpha+u+\alpha^{-1}v} & \frac{\alpha^{-1}v}{\alpha+u+\alpha^{-1}v} \\ \frac{1}{1+u+v} & \frac{u}{1+u+v} & \frac{v}{1+u+v} \\ \frac{\alpha^{-1}}{\alpha^{-1}+u+\alpha v} & \frac{u}{\alpha^{-1}+u+\alpha v} & \frac{\alpha v}{\alpha^{-1}+u+\alpha v} \end{pmatrix}, \quad \text{if } \ l,t \notin \mathbb{Z}-\text{path}.$$

$$\mathbb{P}^{[l,t]} = \mathbb{P}^{[l,t]}_{(3\to2)} = \begin{pmatrix} \frac{\alpha}{\alpha+\alpha^{-1}w} & 0 & \frac{\alpha^{-1}w}{\alpha+\alpha^{-1}w} \\ \frac{1}{1+w} & 0 & \frac{w}{1+w} \\ \frac{\alpha^{-1}}{\alpha^{-1}+\alpha w} & 0 & \frac{\alpha w}{\alpha^{-1}+\alpha w} \end{pmatrix}, \quad \text{if } \ l \notin \mathbb{Z}-\text{path}, t \in \mathbb{Z}-\text{path}.$$

$$\mathbb{P}^{[l,t]} = \mathbb{P}^{[l,t]}_{(2\to3)} = \begin{pmatrix} \frac{\alpha}{\alpha+u+\alpha^{-1}v} & \frac{u}{\alpha+u+\alpha^{-1}v} & \frac{\alpha^{-1}v}{\alpha+u+\alpha^{-1}v} \\ * & * & * \\ \frac{\alpha^{-1}}{\alpha^{-1}+u+\alpha v} & \frac{u}{\alpha^{-1}+u+\alpha v} & \frac{\alpha v}{\alpha^{-1}+u+\alpha v} \end{pmatrix}, l \in \mathbb{Z}-\text{path}, \ t \notin \mathbb{Z}-\text{path},$$

where $*$ means that $P_{0j}^{[l,t]}$ is not defined, because $\sigma(l) \neq 0$ for any $l \in \mathbb{Z}-$path.

$$\mathbb{P}^{[l,t]} = \mathbb{P}^{[l,t]}_{(2\to2)} = \begin{pmatrix} \frac{\alpha}{\alpha+\alpha^{-1}w} & \frac{\alpha^{-1}w}{\alpha+\alpha^{-1}w} \\ \frac{\alpha^{-1}}{\alpha^{-1}+\alpha w} & \frac{\alpha w}{\alpha^{-1}+\alpha w} \end{pmatrix}, \quad \text{if } \ l \in \mathbb{Z}-\text{path}, t \in \mathbb{Z}-\text{path}.$$

Remark 3.6. Matrices $\mathbb{P}^{[l,t]}_{(2\to 3)}$, $\mathbb{P}^{[l,t]}_{(3\to 2)}$ do not define a standard Markov chain. So we will not consider them. We are interested to see thermodynamics of Holliday junctions and branches of DNAs, for this reason it will be sufficient to study Markov chains generated by $\mathbb{P}^{[l,t]}_{(3\to 3)}$ (which gives a Markov chain on the subtree consisting edges which are not on a $\mathbb{Z}$-path). The case $k=2$ is special: in this case $\mathbb{P}^{[l,t]}_{(3\to 3)}$ is defined only for $l = t \notin \mathbb{Z}$-path. and $\mathbb{P}^{[l,t]}_{(2\to 2)}$ (which gives a Markov chain on the $\mathbb{Z}$-paths).

We note that each matrix $\mathbb{P}^{[l,t]}_{(n\to m)}$, $n,m=2,3$ is homogenous on the corresponding set of neighbor edges $\langle l,t\rangle$ where it is given, i.e., $\mathbb{P}^{[l,t]}_{(n\to m)}$ does not depend on $\langle l,t\rangle$ itself but only depends on its relation with $\mathbb{Z}-$path.

It is easy to find the following stationary distributions

$$\pi_{(n\to m)} = (\pi_{(n\to m),-1}, \pi_{(n\to m),0}, \pi_{(n\to m),1}),$$

of the matrix $\mathbb{P}^{[l,t]}_{(n\to m)}$, $n=m$.

$$\pi_{(3\to 3)} = \frac{1}{N}\begin{pmatrix}(\alpha^2+\alpha u+v)(u+\alpha v+\alpha)\alpha\\ (1+u+v)(\alpha^2+\alpha u+v)\alpha u\\ (\alpha u+\alpha^2 v+1)(1+\alpha u+v)v\end{pmatrix}^t,$$

where N the normalizing factor.

$$\pi_{(2\to 2)} = \left(\frac{1+\alpha^{-2}w}{w^2+2\alpha^{-2}w+1}, \frac{w(w+\alpha^{-2})}{w^2+2\alpha^{-2}w+1}\right).$$

In the case of non-uniqueness of Gibbs measure (and corresponding Markov chains) we have different stationary states for different measures. These depend on the temperature and on the fixed measure.

The following is known as ergodic theorem for positive stochastic matrices.

Theorem 3.3. *([98], p.55) Let $\mathbb{P}$ be a positive stochastic matrix and π the unique probability vector with $\pi\mathbb{P}=\pi$ (i.e., π is stationary distribution). Then*

$$\lim_{n\to\infty} x\mathbb{P}^n = \pi,$$

for all initial vector x.

As corollary of this theorem and above formulas of matrices and stationary distributions we obtain the following

Theorem 3.4. *In a stationary state of the set of DNAs, independently on $l \notin \mathbb{Z}$-path, a Holliday junction through l does not occur with the following probability (with respect to measure μ_i, $i = 1, 2, 3$)*

$$\pi_{(3\to 3),0} = \pi^{(i)}_{(3\to 3),0} = \frac{1}{N}(1 + u_i + v_i)(\alpha^2 + \alpha u_i + v_i)\alpha u_i.$$

(Consequently, a Holliday junction occurs with probability $1 - \pi^{(i)}_{(3\to 3),0}$)[8] *where (u_i, v_i) are defined in Lemma 3.2.*

Now we are interested to calculate the limit of stationary distribution vectors $\pi^{(i)}_{(3\to 3)}$, $\pi^{(i)}_{(2\to 2)}$ (which correspond to the Markov chain generated by the Gibbs measure μ_i) in case when temperature $T \to 0$ (i.e., $\beta \to \infty$ and $\alpha \to \infty$) and when temperature $T \to +\infty$ (i.e., $\beta \to 0$ and $\alpha \to 1$). To calculate the limit observe that values u_i, v_i, w_i, $i = 1, 2, 3$ vary with $T = 1/\beta$.

Lemma 3.3. *The following equalities hold*

- *The case of low temperature:*

$$\lim_{T\to 0} \pi^{(i)}_{(3\to 3)} = (1, 0, 0), \quad i = 1, 2; \quad \lim_{T\to 0} \pi^{(3)}_{(3\to 3)} = (0, 0, 1).$$

$$\lim_{T\to 0} \pi^{(1)}_{(2\to 2)} = (\frac{1}{2}, \frac{1}{2}), \quad \lim_{T\to 0} \pi^{(2)}_{(2\to 2)} = (1, 0), \quad \lim_{T\to 0} \pi^{(3)}_{(2\to 2)} = (0, 1).$$

- *The case of high temperature (recall that measures μ_2 and μ_3 do not exist for $T > T_c$):*

$$\lim_{T\to +\infty} \pi^{(1)}_{(3\to 3)} = (\frac{1}{3}, \frac{1}{3}, \frac{1}{3}).$$

$$\lim_{T\to T_c} \pi^{(i)}_{(3\to 3)} \approx (0.54246, 0.23555, 0.22199), i = 1, 2, 3.$$

$$\lim_{T\to +\infty} \pi^{(1)}_{(2\to 2)} = \lim_{T\to T_c} \pi^{(i)}_{(2\to 2)} = (\frac{1}{2}, \frac{1}{2}), i = 2, 3.$$

Proof. Since we know explicit formulas for u_i, v_i, w_i, $i = 1, 2, 3$ the proof consists simple calculations of limits. □

---

[8] One can see that $\pi^{(i)}_{(3\to 3),0}$ is a function of i and temperature only.

Structure of DNAs in low and high temperatures. Recall that $-1 = A+T$, moreover we assumed $A+T = T+A$. Similarly, $+1 = C+G = G+C$.

By Lemma 3.3 we have the following structures of the set of DNAs:

(i) In case $T \to 0$ the set of DNAs have the following stationary states (configurations):
Case μ_1: All neighboring DNAs connected to each other (Holliday junctions) with state $\sigma(l) = -1$ for any $l \notin \mathbb{Z}$-path. The sequence of ± 1s, in a DNA on the $\mathbb{Z}$-path, is free, i.e., can be any sequence, with iid and equiprobable ($= 1/2$), of -1 and $+1$.
Case μ_2: All neighboring DNAs connected to each other (Holliday junctions) with state $\sigma(l) = -1$ for any $l \notin \mathbb{Z}$-path. The sequence of ± 1s, in a DNA on the $\mathbb{Z}$-path, is rigid, i.e., $\sigma(l) = -1$ for all $l \in \mathbb{Z}$-path. Thus the system contains only one multiple (countable) branched DNA (which has tree structure)
Case μ_3: All neighboring DNAs connected to each other with state $\sigma(l) = +1$ for any $l \notin \mathbb{Z}$-path. The DNA on the $\mathbb{Z}$-path, is rigid, i.e., $\sigma(l) = +1$ for all $l \in \mathbb{Z}$-path. Thus this is similar to case of μ_2 but all -1s replaced by $+1$s.

(ii) In case $T = T_c$ the set of DNAs have the following stationary states: each neighboring DNAs have Holliday junction with probability 0.76445 (more precisely, a junction through state -1 with probability 0.54246 and a junction through state $+1$ with probability 0.22199) and no junction with probability 0.23555. The sequence of ± 1s, in a DNA on the $\mathbb{Z}$-path, is free, with iid and equiprobable ($= 1/2$), of -1 and $+1$s.

(iii) In case $T \to +\infty$ the set of DNAs have the following stationary states: each neighboring DNAs have Holliday junction with probability 2/3 (more precisely, a junction through state -1 or $+1$ with equiprobable 1/3) and no junction with probability 1/3. The sequence of ± 1s, in a DNA on the $\mathbb{Z}$-path, is free, with iid and equiprobable ($= 1/2$), of -1 and $+1$s.

Remark 3.7. Since each DNA has a countable set of neighbor DNAs, at the same temperature, it my have Holliday junctions with several of its neighbors. Such connected DNAs can be considered as a branched DNA. In case of coexistence of more than one Gibbs measures, branches of a DNA can consist different phases and different stationary states.

Chapter 4

Thermodynamics in a system of interacting DNAs

In this chapter we define a DNA as a sequence of $1, 2$'s and (as in previous chapter) embed DNAs on $\mathbb{Z}$-paths of the Cayley tree in such a way that each vertex of the Cayley tree belongs only to one of DNA and each DNA has its own countably many set of neighboring DNAs. The Hamiltonian of this set of DNAs is a 2-state model with nearest-neighboring interactions considered as DNA base pairs. It is derived a system of functional equations (the unknown functions defined on vertices of the tree), each solution of which defines a consistent family of finite-dimensional Gibbs distributions and guarantees existence of thermodynamic limit for such distributions.

It is shown that, depending on temperature, number of translation invariant Gibbs measures can be up to three. In the last subsection, by properties of Markov chains (corresponding to Gibbs measures), we study interaction nature of DNAs. In case of very high and very low temperatures we give stationary distributions and typical configurations of the model.

4.1 The model and finite dimensional distributions

The model. Let V be the set of vertices of a Cayley tree (see Subsection 2.1.1).

Consider function σ which assigns to each vertex $x \in V$, values $\sigma(x) \in \{1, 2\}$. Here $1 = A + T$ and $2 = C + G$ are DNA base pairs.

A configuration $\sigma = \{\sigma(x),\ x \in V\}$ on vertices of the Cayley tree is given by a function from V to $\{1, 2\}$. The set of all configurations in V is denoted by Ω. Configurations in V_n are defined analogously and the set of all configurations in V_n is denoted by Ω_n.

The restriction of a configuration on a $\mathbb{Z}$-path is called *a DNA* (see Section 3.1 for definitions).

We consider the following model of the energy of the configuration σ of a set of DNAs:

$$H(\sigma) = \sum_{\langle x,y\rangle \in L} f_{x,y}(\sigma(x), \sigma(y)), \tag{4.1}$$

where

$$f_{x,y}(\sigma(x), \sigma(y)) = \begin{cases} f(\sigma(x), \sigma(y)), & \text{if } \langle x,y\rangle \in \mathbb{Z}-\text{path} \\ J\delta_{\sigma(x),\sigma(y)}, & \text{if } \langle x,y\rangle \notin \mathbb{Z}-\text{path}. \end{cases} \tag{4.2}$$

$J > 0$ is a coupling constant between neighboring DNAs, δ is the Kronecker delta, $\sigma(x) \in \{1,2\}$ and $\langle x,y\rangle$ stands for nearest neighbor vertices. The function $f : \{1,2\}^2 \to \mathbb{R}$ gives interaction between DNA base pairs.

Remark 4.1. In case $f(\sigma(x), \sigma(y)) = J\delta_{\sigma(x),\sigma(y)}$ the Hamiltonian (4.1) coincides with the Potts model. It is natural to expect that the energy of edges of a DNA (i.e. on $\mathbb{Z}$-path) should be different from energies outside of DNA (i.e., on the edges outside of the $\mathbb{Z}$-paths) to control this difference we have introduced, in (4.2), the general function f.

System of functional equations for finite dimensional distributions.

Let Ω_n be the set of all configurations on V_n.

Define a finite-dimensional distribution of a probability measure μ on Ω_n as

$$\mu_n(\sigma_n) = Z_n^{-1} \exp\left\{\beta H_n(\sigma_n) + \sum_{y \in W_n} h_{\sigma(y),y}\right\}, \tag{4.3}$$

where $\beta = 1/T$, $T > 0$ is temperature, Z_n^{-1} is the normalizing factor,

$$\{h_{i,x} \in \mathbb{R}, i = 1, 2, \ x \in V\},$$

is a collection of real numbers and

$$H_n(\sigma_n) = \sum_{\langle x,y\rangle \in L_n} f_{x,y}(\sigma(x), \sigma(y)).$$

The probability distributions (4.3) are compatible if for all $n \geq 1$ and $\sigma_{n-1} \in \Omega_{n-1}$:

$$\sum_{\omega_n \in \Omega_{W_n}} \mu_n(\sigma_{n-1} \vee \omega_n) = \mu_{n-1}(\sigma_{n-1}). \tag{4.4}$$

Here $\sigma_{n-1} \vee \omega_n$ is the concatenation of the configurations.

In the formula (4.2) we take the function f such that

$$f(1,1)=f(2,2),\quad f(1,2)=f(2,1).$$

This assumption is natural in the structure of DNA.

For $x\in V_{n-1}$ denote

$$S(x)=\{t\in V_n:\langle x,t\rangle\}.$$

For $x\in V$ we denote by $x_\downarrow$ the unique point of the set $\{y\in V:\langle x,y\rangle\}\backslash S(x)$.

It is easy to see that

$$S(x)\cap\mathbb{Z}-\text{path}=\begin{cases}\{x_0,x_1\}\subset V, & \text{if }\ \langle x_\downarrow,x\rangle\notin\mathbb{Z}-\text{path},\\ \{x_1\}\subset V, & \text{if }\ \langle x_\downarrow,x\rangle\in\mathbb{Z}-\text{path}.\end{cases}$$

Denote

$$S_0(x)=S(x)\setminus\{x_0,x_1\},\quad \langle x_\downarrow,x\rangle\notin\mathbb{Z}-\text{path},$$

$$S_1(x)=S(x)\setminus\{x_1\},\quad \langle x_\downarrow,x\rangle\in\mathbb{Z}-\text{path}.$$

The following theorem can be proved as Theorem 3.1 (see also proof of Theorem 7.22 in [243]).

Theorem 4.1. *Probability distributions $\mu_n(\sigma_n)$, $n=1,2,\ldots$, in (4.3) are compatible iff for any $x\in V\setminus\{x^0\}$ the following equations hold*

$$z_x=\frac{\hat{z}_{x_0}+b}{b\hat{z}_{x_0}+1}\cdot\frac{\hat{z}_{x_1}+b}{b\hat{z}_{x_1}+1}\prod_{t\in S_0(x)}\frac{z_t+\theta}{\theta z_t+1},\quad \langle x_\downarrow,x\rangle\notin\mathbb{Z}-\text{path},\tag{4.5}$$

$$\hat{z}_x=\frac{\hat{z}_{x_1}+b}{b\hat{z}_{x_1}+1}\prod_{t\in S_1(x)}\frac{z_t+\theta}{\theta z_t+1},\quad \langle x_\downarrow,x\rangle\in\mathbb{Z}-\text{path}.\tag{4.6}$$

Here,

$$\begin{aligned}&\theta=\exp\{\beta(f(1,2)-f(1,1))\},\quad b=\exp(-J\beta),\\ &z_x=\exp\left(h_{1,x}-h_{2,x}\right),\quad \langle x_\downarrow,x\rangle\notin\mathbb{Z}-\text{path},\\ &\hat{z}_x=\exp\left(h_{1,x}-h_{2,x}\right),\quad \langle x_\downarrow,x\rangle\in\mathbb{Z}-\text{path}.\end{aligned}\tag{4.7}$$

From Theorem 4.1 it follows that for any set of vectors $\mathbf{z} = \{(z_x, \hat{z}_t)\}$ satisfying the system of functional equations (4.5) and (4.6) there exists a unique Gibbs measure μ and vice versa. However, the analysis of this system of nonlinear functional equations is not easy. In next section we shall give several solutions to the system.

Remark 4.2. The number of the solutions of equations (4.5) and (4.6) depends on the temperature and interaction parameters b, θ. Note that if there is more than one solution to the system (4.5) and (4.6), then there is more than one Gibbs measure (i.e. occurrence of "phase" transition for the model of DNAs).

4.2 TIGMs of the set of DNAs

In this subsection, we find solutions to the system of functional equations (4.5) and (4.6), which have the following form

$$z_x = u, \quad \forall\langle x_\downarrow, x\rangle \notin \mathbb{Z}-\text{path}; \quad \hat{z}_x = w, \quad \forall\langle x_\downarrow, x\rangle \in \mathbb{Z}-\text{path}. \tag{4.8}$$

The measure corresponding to such a solution is called *translation invariant Gibbs measure* (TIGM).

For u, v from (4.5) and (4.6) we get

$$\begin{aligned} u &= \left(\frac{v+b}{bv+1}\right)^2 \left(\frac{u+\theta}{\theta u+1}\right)^{k-2}, \\ v &= \left(\frac{v+b}{bv+1}\right) \left(\frac{u+\theta}{\theta u+1}\right)^{k-1}. \end{aligned} \tag{4.9}$$

Here $u, v > 0$.

It is clear that $u = v = 1$ satisfies the system (4.9) for any $k \geq 2$ and $b, \theta > 0$. In general, it is not easy to find all solution of this system. Therefore for simplicity we consider the case $k = 2$. Then from the system we get

$$u = \left(\frac{v+b}{bv+1}\right)^2,$$

and substituting it in the second equation (for $k = 2$) we obtain

$$(v^2-1)\big(b(b^2+\theta)(v^2+1) + (b^2(3+\theta)+\theta-1)v\big) = 0. \tag{4.10}$$

Since $v > 0$ we have three solutions of (4.10): $v_1 = 1$,

$$v_2 = \frac{-(b^2(3+\theta)+\theta-1) - |b^2-1|\sqrt{(\theta-1)^2-4b^2}}{2b(b^2+\theta)},$$

$$v_3 = \frac{-(b^2(3+\theta)+\theta-1)+|b^2-1|\sqrt{(\theta-1)^2-4b^2}}{2b(b^2+\theta)}.$$

The solutions v_2 and v_3 exist iff

$$\theta \in \begin{cases} (0, 1-2b] \cup [1+2b, +\infty) & \text{if } 0<b<\frac{1}{2} \\ (1+2b, +\infty) & \text{if } b \geq \frac{1}{2}. \end{cases}$$

Note that both v_2 and v_3 should be positive. This can be possible only if

$$0 < b < \frac{1}{2} \quad \text{and} \quad \theta \leq 1-2b. \tag{4.11}$$

Indeed, to see this rewrite equation

$$b(b^2+\theta)(v^2+1)+(b^2(3+\theta)+\theta-1)v=0$$

as

$$v+\frac{1}{v}=\frac{1-\theta-b^2(3+\theta)}{b(b^2+\theta)}.$$

In case $v>0$ we should have

$$\frac{1-\theta-b^2(3+\theta)}{b(b^2+\theta)} \geq 2.$$

which gives the condition (4.11). Note also that $v_2v_3=1$.

Introduce the following sets

$$U=(0,+\infty)\times(0,+\infty),$$
$$U_1=\{(b,\theta)\in U : b\leq 1/2, \quad \theta<1-2b\}.$$

Thus, for $k=2$ we proved the following

Lemma 4.1. *The following hold*

- *If* $(b,\theta)\in U\setminus U_1$ *then system (4.9) has unique solution*

$$\mathbf{z}_1=(u_1,v_1)=(1,1);$$

- *If* $(b,\theta)\in U_1$ *then system (4.9) has three solutions*

$$\mathbf{z}_1=(1,1), \quad \mathbf{z}_2=(u_2,v_2), \quad \mathbf{z}_3=(u_3,v_3),$$

where

$$u_i=\left(\frac{v_i+b}{bv_i+1}\right)^2, \quad i=2,3.$$

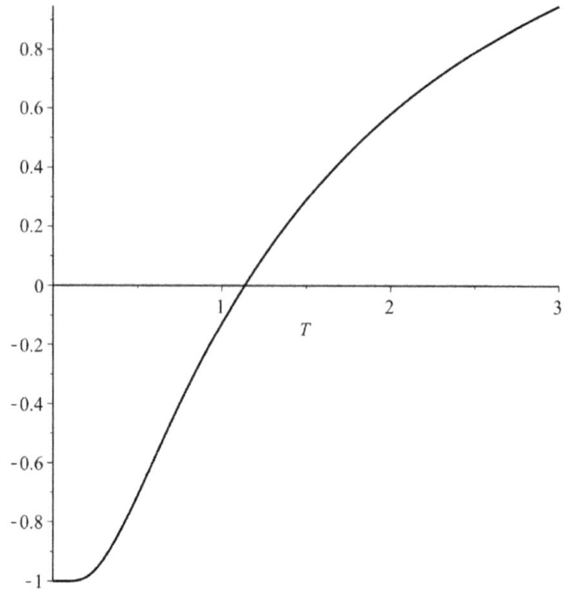

Fig. 4.1 The graph of $t(T)$ for $J = 1$, $f(1,2) - f(1,1) = -2$. In this case $T_c \approx 1.1346$.

Denote by μ_i the Gibbs measure which, by Theorem 4.1, corresponds to the solution $\mathbf{z}_i$, $i = 1, 2, 3$.

For given f and J the critical temperature T_c is defined by the equation $\theta + 2b = 1$, i.e., by formula (4.7) and $\beta = 1/T$ we have

$$\theta + 2b = \exp\left(\frac{1}{T}(f(1,2) - f(1,1))\right) + 2\exp\left(\frac{-J}{T}\right) = 1. \tag{4.12}$$

This equation has solution iff

$$J > 0, \quad f(1,2) - f(1,1) < 0. \tag{4.13}$$

Indeed, if $J \leq 0$ or $f(1,2) - f(1,1) \geq 0$ then LHS of the equation (4.12) is strongly large than 1. Moreover, in the case $J > 0$ and $f(1,2) - f(1,1) < 0$ the function $t(T) := \theta + 2b - 1$ is an increasing function of T with value -1 at $T = 0$ and value 2 at $T = +\infty$. Therefore, the equation has unique solution $T = T_c$ (see Fig. 4.1).

Summarizing above results we obtain the following

Theorem 4.2. *For the model (4.1) (given J and f with (4.13)) of DNAs on the Cayley tree of order $k = 2$ the following statements hold*

(1) If the temperature $T \geq T_c$ (where $T_c \equiv T_c(J, f)$ the unique solution of (4.12)) then there is unique translation-invariant Gibbs measure μ_1.
(2) If $T < T_c$ then there are 3 translation-invariant Gibbs measures μ_1, μ_2, μ_3.

Remark 4.3.

(1) For given J and f with the conditions of Theorem 4.2 are related to conditions of Lemma 4.1 as following
   (1) $T \geq T_c$ iff $(b, \theta) \in U \setminus U_1$.
   (2) $T < T_c$ iff $(b, \theta) \in U_1$.
(2) Analysis of system (4.9) for the case $k \geq 3$ seems difficult. The case $k = 2$ is already interesting enough to see biological interpretations of Theorem 4.2.

4.3 Markov chains of TIGM

Recall

$$z_x = \exp\left(h_{1,x} - h_{2,x}\right), \quad \langle x_\downarrow, x\rangle \notin \mathbb{Z} - \text{path},$$

$$\hat{z}_x = \exp\left(h_{1,x} - h_{2,x}\right), \quad \langle x_\downarrow, x\rangle \in \mathbb{Z} - \text{path}.$$

Using boundary law

$$\mathbf{z} = \{(z_x, \hat{z}_t): \quad \langle x_\downarrow, x\rangle \notin \mathbb{Z} - \text{path}; \quad \langle t_\downarrow, t\rangle \in \mathbb{Z} - \text{path}\},$$

i.e., the solutions of system (4.5), for marginals on the two-site volumes which consist of two adjacent sites x, y we have

$$\mu(\sigma(x) = s, \sigma(y) = r) = \frac{1}{Z}\exp(h_{s,x} + \beta f_{x,y}(s, r) + h_{r,y}), \quad s, r = 1, 2,$$

where Z is normalizing factor.

From this, the relation between the boundary law and the transition matrix for the associated tree-indexed Markov chain (Gibbs measure) is immediately obtained from the formula of the conditional probability. The transition matrix of this Markov chain is defined as follows

$$\mathbb{P}^{\langle x,y\rangle} = \left(P_{ij}^{\langle x,y\rangle}\right)_{i,j=1,2} = \begin{cases} \begin{pmatrix} \frac{u}{u+\theta} & \frac{\theta}{u+\theta} \\ \frac{\theta u}{\theta u+1} & \frac{1}{\theta u+1} \end{pmatrix}, & \text{if } \langle x, y\rangle \in \mathbb{Z} - \text{path} \\ \begin{pmatrix} \frac{v}{v+b} & \frac{b}{v+b} \\ \frac{bv}{bv+1} & \frac{1}{bv+1} \end{pmatrix}, & \text{if } \langle x, y\rangle \notin \mathbb{Z} - \text{path}, \end{cases}$$

where (u, v) a solution of system (4.9) (mentioned in Lemma 4.1).

We note that each matrix $\mathbb{P}^{\langle x,y\rangle}$ does not depend on $\langle x,y\rangle$ itself but only depends on its relation with $\mathbb{Z}-$path.

It is easy to find the following stationary distributions

$$\pi^{\langle x,y\rangle}=\begin{cases}\left(\frac{u^2+\theta u}{u^2+(\theta+1)u+1},\frac{u+1}{u^2+(\theta+1)u+1}\right), & \text{if } \langle x,y\rangle\in\mathbb{Z}-\text{path}\\ \left(\frac{v^2+bv}{v^2+(b+1)v+1},\frac{v+1}{v^2+(b+1)v+1}\right), & \text{if } \langle x,y\rangle\notin\mathbb{Z}-\text{path}.\end{cases}$$

In the case of non-uniqueness of Gibbs measure (and corresponding Markov chains) we have different stationary states for different measures. These depend on the temperature and on the fixed measure.

Recall that a DNA is a configuration $\sigma\in\{1,2\}^{\mathbb{Z}}$ on a $\mathbb{Z}$-path. According to the definition of the model only neighboring DNAs may interact. The interaction is through an edge $l=\langle x,y\rangle\notin\mathbb{Z}$-path connecting two DNAs when configuration on this endpoints of the edge satisfy $\sigma(x)=\sigma(y)$. The neighboring DNAs do not interact if $\sigma(x)\neq\sigma(y)$.

As a corollary of Theorem 3.3 and above formulas of matrices and stationary distributions we obtain the following:

Theorem 4.3. *In a stationary state of the set of DNAs we have*

1. *Two neighboring DNAs do not interact with the following probability, (denoted by $\mathbb{P}_i$ with respect to measure μ_i, $i=1,2,3$):*
$$\mathbb{P}_i=\frac{bv_i((b+1)v_i+2)}{(bv_i+1)(v_i^2+(b+1)v_i+1)},$$
(Consequently, they do interact with probability $1-\mathbb{P}_i$) where (u_i,v_i) are defined in Lemma 4.1.
2. *Two neighboring base pairs (laying on vertices of an edge $l=\langle x,y\rangle\in\mathbb{Z}$-path) in a DNA have distinct values (i.e., $\sigma(x)\neq\sigma(y)$) with probability (denoted by $\mathbb{Q}_i$ with respect to measure μ_i, $i=1,2,3$):*
$$\mathbb{Q}_i=\frac{\theta u_i((\theta+1)u_i+2)}{(\theta u_i+1)(u_i^2+(\theta+1)u_i+1)},$$
(Consequently, they have the same values with probability $1-\mathbb{Q}_i$).

Proof. Let $l=\langle x,y\rangle\notin\mathbb{Z}$-path be the edge connecting two neighboring DNAs. By the definition of $\mathbb{P}_i$ we have

$$\mathbb{P}_i=\mu_i(\sigma(x)=1,\sigma(y)=2)+\mu_i(\sigma(x)=2,\sigma(y)=1).$$

Since μ_i generates a Markov chain we have

$$\mu_i(\sigma(x) = a, \sigma(y) = c) = \pi_a^{\langle x,y\rangle} P_{ac}^{\langle x,y\rangle}.$$

This by using the above given formulas of $\pi^{\langle x,y\rangle}$ and $\mathbb{P}^{\langle x,y\rangle}$ completes the proof of part 1, the part 2 is similar. □

Remark 4.4. Since each DNA has a countable set of neighbor DNAs, at the same temperature, it my have interactions with several of its neighbors. In case when a DNA does not interact with its neighbors then it is an isolated one. Interacting DNAs can be considered as a branched DNA. In case of coexistence of more than one Gibbs measures, branches of a DNA can consist different phases and different stationary states.

Now we are interested to calculate the limit of stationary distribution vectors $\pi^{\langle x,y\rangle,i}$, (which correspond to the Markov chain generated by the Gibbs measure μ_i) in case when temperature $T \to 0$ and when temperature $T \to +\infty$. To calculate the limit observe that values u_i, v_i, $i = 1, 2, 3$ vary with $T = 1/\beta$.

Recall that measures μ_2 and μ_3 do not exist for $T > T_c$.

Proposition 4.1. *Independently on $\langle x, y\rangle$ the following equalities hold*

- *The case of low temperature:*

$$\lim_{T\to 0} \pi^{\langle x,y\rangle,1} = (\frac{1}{3}, \frac{2}{3}), \quad \lim_{T\to 0} \pi^{\langle x,y\rangle,2} = (0, 1), \quad \lim_{T\to 0} \pi^{\langle x,y\rangle,3} = (1, 0).$$

- *The case of high temperature:*

$$\lim_{T\to +\infty} \pi^{\langle x,y\rangle,1} = \lim_{T\to T_c} \pi^{\langle x,y\rangle,i} = (\frac{1}{2}, \frac{1}{2}), i = 1, 2, 3.$$

Proof. Using explicit formulas for u_i, v_i, $i = 1, 2, 3$ the proof consists simple calculations of limits. □

Remark 4.5. Using this proposition one can give the structure of DNAs in low and high temperatures. For example, in case $T \to 0$ the set of DNAs have the following stationary states (configurations):

Case μ_1: The base pairs $1 = A + T, 2 = C + G$s, in each point of a DNA can be seen with probability $1/3$ for 1 and $2/3$ for 2.

Case μ_2: All DNAs are rigid and interact, i.e. $\sigma(x) = 2$ for all $x \in \mathbb{Z}$-path.

Case μ_3: All DNAs are rigid and interact, with $\sigma(x) = 1$ for all $x \in \mathbb{Z}$-path.

In case $T \to +\infty$ the sequence of 1, 2s, in a DNA on the $\mathbb{Z}$-path, is free, with iid and equiprobable ($= 1/2$), of 1 and 2s.

Chapter 5

Holliday junctions for the Potts model of DNA

In this chapter a DNA is considered as a configuration of 4-state Potts model and it is embed on a path of a Cayley tree. Then we give a Hamiltonian of the set of DNAs by an analogue of Potts model with 5-spin values $0, \pm 1, \pm 2$ (considered as DNA base pairs and 0 means vacant) on a set of admissible configurations. To study thermodynamic properties of the model of DNAs we describe corresponding translation invariant Gibbs measures (TIGM) of the model on the Cayley tree of order two. Each such measure describes a phase of the set of DNAs. We show that there are two critical temperatures $T_{i,c}$, $i = 1, 2$ such that

(i) If the temperature $T > T_{1,c}$ then there is at last one (TIGM);
(ii) If $T = T_{1,c}$ then there are at least 4 TIGMs;
(iii) If $T_{2,c} < T < T_{1,c}$ then there are at least 7 TIGMs;
(iv) If $T = T_{2,c}$ then there are at least 10 TIGMs; and
(v) If $T < T_{2,c}$ then there are at least 13 TIGMs.

Moreover, we study distributions of Holliday junctions of DNAs. In case of very low temperatures we give stationary distributions and typical configurations of the Holliday junctions.

5.1 Definitions and system of functional equations

In this chapter we mainly follow [248]. Consider the set of DNAs as configurations of the 5-state Potts model on a Cayley tree.

The model
Consider a set of DNAs which 'live' on a Cayley tree.

Let l be an edge of this tree consider a function $\sigma(l)$ with five possible values $0, \pm 1, \pm 2$ (an analogue of spin values in physical systems), in case $\sigma(l) = 0$ the edge l does not belong to a DNA (i.e., a vacant).

Denote

$$-1 = A + T, \quad 1 = T + A, \quad -2 = C + G, \quad 2 = G + C. \tag{5.1}$$

If this l separates two DNA then the value $\sigma(l) \neq 0$ means that these two DNA have a Holliday junction (see Chapter 3).

Let L be the set of edges of a Cayley tree. Consider function σ which assigns to each edge $l \in L$, values $\sigma(l) \in \{-2, -1, 0, 1, 2\}$.

A configuration $\sigma = \{\sigma(l),\ l \in L\}$ on edges of the Cayley tree is given by a function from L to $\{-2, -1, 0, 1, 2\}$. The set of all configurations in L is denoted by Ω. Configurations in L_n are defined analogously and the set of all configurations in L_n is denoted by Ω_n.

A configuration $\sigma = \{\sigma(l),\ l \in L\}$ is called *admissible* if $\sigma(l) \neq 0$ for any $l \in \mathbb{Z}$-path.

The restriction of an admissible configuration on a $\mathbb{Z}$-path is called *a DNA*.

Note that for an admissible configuration σ we may have $\sigma(l) = 0$ only for edges $l = \langle x, y\rangle$ with $x \sim y$, i.e., $f_0(x) = f_0(y)$. Comparing this with Holliday junction one can see that DNAs corresponding to two $\mathbb{Z}$-paths can have a junction only through the edge $l = \langle x, y\rangle$ with $x \sim y$.

Consider the following Potts model (see [267]) (i.e., the energy of the configuration σ) of a set of DNAs:

$$H(\sigma) = J \sum_{\langle l,t\rangle \in L\times L} \delta_{\sigma(l)\sigma(t)}, \tag{5.2}$$

where $J > 0$ is a coupling constant, δ is the Kronecker delta, $\sigma(l) \in \{-2, -1, 0, 1, 2\}$ and $\langle l, t\rangle$ stands for nearest neighbor edges, i.e., edges which have a common endpoint.

Let Ω_n^a (resp. Ω^a) be the set of all admissible configurations on L_n (resp. L).

Denote

$$E_n = \{\langle x, y\rangle \in L : x \in W_{n-1},\ y \in W_n\},$$

$$\Omega_n^{ba} = \text{ the set of admissible configurations on } E_n.$$

For $l \in E_{n-1}$ denote

$$S(l) = \{t \in E_n : \langle l, t\rangle\}.$$

We have

$$S(l) \cap \mathbb{Z} - \text{path} = \begin{cases} \{l_0, l_1\} \subset L, & \text{if } l \notin \mathbb{Z} - \text{path}, \\ \{l_1\} \subset L, & \text{if } l \in \mathbb{Z} - \text{path}. \end{cases}$$

Denote

$$S_0(l) = S(l) \setminus \{l_0, l_1\}, \quad l \notin \mathbb{Z} - \text{path},$$

$$S_1(l) = S(l) \setminus \{l_1\}, \quad l \in \mathbb{Z} - \text{path}.$$

Define a finite-dimensional distribution of a probability measure μ on Ω_n^a as

$$\mu_n(\sigma_n) = Z_n^{-1} \exp\left\{\beta H_n(\sigma_n) + \sum_{l \in E_n} h_{\sigma(l),l}\right\}, \tag{5.3}$$

where $\beta = 1/T$, $T > 0$ is temperature, Z_n^{-1} is the normalizing factor, $\{h_{i,l} \in \mathbb{R}, i = -2, -1, 0, 1, 2, \, l \in L\}$ is a collection of real numbers and

$$H_n(\sigma_n) = J \sum_{\substack{l,t \in L_n: \\ \langle l,t \rangle}} \delta_{\sigma(l)\sigma(t)}.$$

The probability distributions (5.3) are compatible if for all $n \geq 1$ and $\sigma_{n-1} \in \Omega_{n-1}^a$:

$$\sum_{\omega_n \in \Omega_n^{ba}} \mu_n(\sigma_{n-1} \vee \omega_n) = \mu_{n-1}(\sigma_{n-1}). \tag{5.4}$$

Here $\sigma_{n-1} \vee \omega_n$ is the concatenation of the configurations. In this case there exists a unique measure μ on Ω^a such that, for all n and $\sigma_n \in \Omega_n^a$,

$$\mu(\{\sigma|_{L_n} = \sigma_n\}) = \mu_n(\sigma_n).$$

Such a measure is called a *Gibbs measure* corresponding to the Hamiltonian (5.2) and functions $h_{i,l}, l \in L$.

The following statement describes conditions on $h_{i,l}$ guaranteeing compatibility of $\mu_n(\sigma_n)$.

Theorem 5.1. *Probability distributions $\mu_n(\sigma_n)$, $n = 1, 2, \ldots$, in (5.3) are compatible iff for any $l \in L$ the following equations hold:*

if $l \notin \mathbb{Z} -$ path then

$$u_{-1,l} = \prod_{i=0}^{1} \frac{1 + \alpha v_{-1,l_i} + v_{1,l_i} + v_{2,l_i}}{\alpha + v_{-1,l_i} + v_{1,l_i} + v_{2,l_i}} \prod_{t \in S_0(l)} \frac{1 + \alpha u_{-1,t} + u_{0,t} + u_{1,t} + u_{2,t}}{\alpha + u_{-1,t} + u_{0,t} + u_{1,t} + u_{2,t}},$$

$$u_{0,l} = \prod_{i=0}^{1} \frac{1 + v_{-1,l_i} + v_{1,l_i} + v_{2,l_i}}{\alpha + v_{-1,l_i} + v_{1,l_i} + v_{2,l_i}} \prod_{t \in S_0(l)} \frac{1 + u_{-1,t} + \alpha u_{0,t} + u_{1,t} + u_{2,t}}{\alpha + u_{-1,t} + u_{0,t} + u_{1,t} + u_{2,t}},$$

$$u_{1,l}=\prod_{i=0}^{1}\frac{1+v_{-1,l_i}+\alpha v_{1,l_i}+v_{2,l_i}}{\alpha+v_{-1,l_i}+v_{1,l_i}+v_{2,l_i}}\prod_{t\in S_0(l)}\frac{1+u_{-1,t}+u_{0,t}+\alpha u_{1,t}+u_{2,t}}{\alpha+u_{-1,t}+u_{0,t}+u_{1,t}+u_{2,t}},$$

$$u_{2,l}=\prod_{i=0}^{1}\frac{1+v_{-1,l_i}+v_{1,l_i}+\alpha v_{2,l_i}}{\alpha+v_{-1,l_i}+v_{1,l_i}+v_{2,l_i}}\prod_{t\in S_0(l)}\frac{1+u_{-1,t}+u_{0,t}+u_{1,t}+\alpha u_{2,t}}{\alpha+u_{-1,t}+u_{0,t}+u_{1,t}+u_{2,t}},\tag{5.5}$$

if $l\in\mathbb{Z}-$ path *then*

$$v_{-1,l}=\frac{1+\alpha v_{-1,l_1}+v_{1,l_1}+v_{2,l_1}}{\alpha+v_{-1,l_1}+v_{1,l_1}+v_{2,l_1}}\prod_{t\in S_1(l)}\frac{1+\alpha u_{-1,t}+u_{0,t}+u_{1,t}+u_{2,t}}{\alpha+u_{-1,t}+u_{0,t}+u_{1,t}+u_{2,t}},$$

$$v_{1,l}=\frac{1+v_{-1,l_1}+\alpha v_{1,l_1}+v_{2,l_1}}{\alpha+v_{-1,l_1}+v_{1,l_1}+v_{2,l_1}}\prod_{t\in S_1(l)}\frac{1+u_{-1,t}+u_{0,t}+\alpha u_{1,t}+u_{2,t}}{\alpha+u_{-1,t}+u_{0,t}+u_{1,t}+u_{2,t}},$$

$$v_{2,l}=\frac{1+v_{-1,l_1}+v_{1,l_1}+\alpha v_{2,l_1}}{\alpha+v_{-1,l_1}+v_{1,l_1}+v_{2,l_1}}\prod_{t\in S_1(l)}\frac{1+u_{-1,t}+u_{0,t}+u_{1,t}+\alpha u_{2,t}}{\alpha+u_{-1,t}+u_{0,t}+u_{1,t}+u_{2,t}},$$

Here, $\alpha=\exp(J\beta)$,

$$\begin{aligned}u_{i,l}&=\exp\left(h_{i,l}-h_{-2,l}\right),i=-1,0,1,2,\ l\notin\mathbb{Z}-\text{path};\\ v_{j,l}&=\exp\left(h_{j,l}-h_{-2,l}\right),j=-1,1,2,\ l\in\mathbb{Z}-\text{path}.\end{aligned}\tag{5.6}$$

Proof. Similar to the proof of Theorem 5.1 in [243], see also the proof of Theorem 1 in [247]. □

From Theorem 5.1 it follows that for any set of vectors

$$\mathbf{z}=\{(u_{i,l},i=-1,0,1,2,\ l\notin\mathbb{Z}-\text{path}),(v_{j,t},j=-1,1,2,\ t\in\mathbb{Z}-\text{path})\},$$

satisfying the system of functional equations (5.5) there exists a unique Gibbs measure μ and vice versa. However, the analysis of solutions to (5.5) is not easy.

In next section we shall give several solutions to (5.5).

One says that a "phase" transition occurs for the model of DNAs, if equation (5.5) has more than one solution [98].

5.2 TIGMs of the set of DNAs

The number of the solutions of equation (5.5) depends on the temperature-parameter $\alpha=\exp(\frac{J}{T})$. Here, without loss of generality we take $J=1$

and study Gibbs measures corresponding to solutions $\mathbf{z}_l$ of the system of functional equations (5.5), which do not depend on l, i.e.,

$$u_{i,l} = u_i, \quad \forall l \notin \mathbb{Z} - \text{path}; \quad v_{j,l} = v_j, \; \forall l \in \mathbb{Z} - \text{path}. \tag{5.7}$$

For u_i, v_j from (5.5) we get

$$\begin{aligned}
u_{-1} &= \left(\frac{1+\alpha v_{-1}+v_1+v_2}{\alpha+v_{-1}+v_1+v_2}\right)^2 \left(\frac{1+\alpha u_{-1}+u_0+u_1+u_2}{\alpha+u_{-1}+u_0+u_1+u_2}\right)^{k-2},\\
u_0 &= \left(\frac{1+v_{-1}+v_1+v_2}{\alpha+v_{-1}+v_1+v_2}\right)^2 \left(\frac{1+u_{-1}+\alpha u_0+u_1+u_2}{\alpha+u_{-1}+u_0+u_1+u_2}\right)^{k-2},\\
u_1 &= \left(\frac{1+v_{-1}+\alpha v_1+v_2}{\alpha+v_{-1}+v_1+v_2}\right)^2 \left(\frac{1+u_{-1}+u_0+\alpha u_1+u_2}{\alpha+u_{-1}+u_0+u_1+u_2}\right)^{k-2},\\
u_2 &= \left(\frac{1+v_{-1}+v_1+\alpha v_2}{\alpha+v_{-1}+v_1+v_2}\right)^2 \left(\frac{1+u_{-1}+u_0+u_1+\alpha u_2}{\alpha+u_{-1}+u_0+u_1+u_2}\right)^{k-2}, \qquad (5.8)\\
v_{-1} &= \frac{1+\alpha v_{-1}+v_1+v_2}{\alpha+v_{-1}+v_1+v_2} \left(\frac{1+\alpha u_{-1}+u_0+u_1+u_2}{\alpha+u_{-1}+u_0+u_1+u_2}\right)^{k-1},\\
v_1 &= \frac{1+v_{-1}+\alpha v_1+v_2}{\alpha+v_{-1}+v_1+v_2} \left(\frac{1+u_{-1}+u_0+\alpha u_1+u_2}{\alpha+u_{-1}+u_0+u_1+u_2}\right)^{k-1},\\
v_2 &= \frac{1+v_{-1}+v_1+\alpha v_2}{\alpha+v_{-1}+v_1+v_2} \left(\frac{1+u_{-1}+u_0+u_1+\alpha u_2}{\alpha+u_{-1}+u_0+u_1+u_2}\right)^{k-1},
\end{aligned}$$

where $u_i, v_j > 0$. By assumption $J = 1$ we have $\alpha > 1$.

Remark 5.1. The system (5.8) has one parameter α and 7 unknowns. For general $k \geq 2$ there is no any result about its solutions. Therefore, in this chapter we give results related to the case $k = 2$ only. The case $k \geq 3$ remaines open (as in Chapter 3 and 4).

For simplicity we consider the case $k = 2$. Then from (5.8) denoting $x = v_{-1}$, $y = v_1$, $z = v_2$ we get

$$\begin{cases} x = f(x, y, z, \alpha) \\ y = f(y, x, z, \alpha) \\ z = f(z, y, x, \alpha) \end{cases}, \tag{5.9}$$

where

$$f(x,y,z,\alpha) \equiv \frac{1+\alpha x+y+z}{\alpha+x+y+z} \times \frac{A}{B}$$

with

$$\begin{aligned} A &= (\alpha+x+y+z)^2 + \alpha(1+\alpha x+y+z)^2 \\ &\quad + (1+x+y+z)^2 + (1+x+\alpha y+z)^2 + (1+x+y+\alpha z)^2, \\ B &= \alpha(\alpha+x+y+z)^2 + (1+\alpha x+y+z)^2 \\ &\quad + (1+x+y+z)^2 + (1+x+\alpha y+z)^2 + (1+x+y+\alpha z)^2. \end{aligned}$$

Lemma 5.1. *We have*

1. *If* $\mathbf{v} = (x, y, z)$ *is a solution to (5.9) then any vector obtained by permutations of coordinates of* $\mathbf{v}$ *is a solution to (5.9).*
2. $x = y = z = 1$ *is a solution to (5.9).*

Proof. It is easy to see that the function f has the following properties:

(1) $f(x,y,z,\alpha) = f(x,z,y,\alpha)$, for any $x,y,z > 0$, $\alpha > 1$.
(2) $f(1,y,z,\alpha) = 1$, for any $y,z > 0$, $\alpha > 1$.

Using these properties one completes the proof. □

Now we are going to find special solutions of (5.9):

Case: $x \neq 1$, $y = z = 1$. Then from the first equation of (5.9) we obtain

$$x = f(x,1,1,\alpha) = \frac{3+\alpha x}{\alpha+x+2} \cdot \frac{\alpha(3+\alpha x)^2 + (3+x)^2 + 3(2+x+\alpha)^2}{(3+\alpha x)^2 + (3+x)^2 + (2+\alpha)(2+x+\alpha)^2}.$$

This equation can be written as $P_4(x,\alpha) = 0$ with a polynomial P_4 of degree 4. It has root $x = 1$, therefore we reduce our equation to

$$\begin{aligned} \frac{P_4(x,\alpha)}{x-1} &= (\alpha^2+\alpha+3)x^3 - (\alpha^4 - \alpha^3 - 6\alpha^2 - 16\alpha - 23)x^2 \\ &\quad - (\alpha^4 + 5\alpha^3 - 24\alpha^2 - 52\alpha - 65)x + 9(\alpha^2 + 7\alpha + 7) = 0. \quad (5.10) \end{aligned}$$

Using solutions of cubic equations one obtains the following

Lemma 5.2. *There exists* $\alpha^* \approx 4.538695$ *such that*

(i) If $\alpha \in [1, \alpha^*)$ *then equation (5.10) has not positive solutions.*
(ii) If $\alpha = \alpha^*$ *then equation (5.10) has unique positive solution,* x_1^*.
(iii) If $\alpha \in (\alpha^*, +\infty)$ *then equation (5.10) has two positive solutions,* x_1^*, x_2^*, *with* $x_1^* < x_2^*$.

Case: $x = y \neq 1$, $z = 1$. Then from the first equation of (5.9) we obtain

$$x = f(x, x, 1, \alpha) = \frac{2 + (1+\alpha)x}{\alpha + 2x + 1} \times \frac{2(1+\alpha+2x)^2 + (1+\alpha)(2+(1+\alpha)x)^2 + 4(1+x)^2}{(1+\alpha)(1+\alpha+2x)^2 + 2(2+(1+\alpha)x)^2 + 4(1+x)^2}.$$

Since $x \neq 1$ the last equation can be reduced to the following cubic equation

$$4(\alpha^2+4\alpha+5)(x^3+1) - (\alpha^4+2\alpha^3-16\alpha^2-50\alpha-57)x(x+1) = 0. \quad (5.11)$$

Solving this equation we get two *positive* solutions:

$$x_3^* = \frac{\alpha^4 - 12\alpha^2 + 2\alpha^3 - 34\alpha - 37 - (\alpha - 1)\sqrt{D}}{8(5 + 4\alpha + \alpha^2)},$$

$$x_4^* = \frac{\alpha^4 - 12\alpha^2 + 2\alpha^3 - 34\alpha - 37 + (\alpha - 1)\sqrt{D}}{8(5 + 4\alpha + \alpha^2)},$$

where

$$D = (\alpha+3)(1+\alpha)(\alpha^4 + 2\alpha^3 - 20\alpha^2 - 66\alpha - 77).$$

Note that solutions x_3^* and x_4^* exist iff

$$\alpha^4 + 2\alpha^3 - 20\alpha^2 - 66\alpha - 77 \geq 0. \quad (5.12)$$

It is well known (see [221]) that the number of positive roots of a polynomial does not exceed the number of sign changes of its coefficients. Using this fact for the polynomial $p_4(\alpha) := \alpha^4 + 2\alpha^3 - 20\alpha^2 - 66\alpha - 77$ (i.e., LHS of (5.12)) we note that it has up to one positive root. Moreover, since $p_4(0) = -77 < 0$ and $p_4(6) > 0$, it has unique positive root $\alpha^{**} \approx 5.0806$. Therefore, the inequality (5.12) is true for any $\alpha \geq \alpha^{**}$.

Thus, we proved the following

Lemma 5.3. *The following hold*

- *If* $\alpha < \alpha^{**}$ *then equation (5.11) has not positive solution.*
- *If* $\alpha = \alpha^{**}$ *then (5.11) has one solution,* x_3^*.
- *If* $\alpha > \alpha^{**}$ *then (5.11) has two solutions,* $x_3^* < x_4^*$.

Case: $x = y = z \neq 1$. From the first equation of (5.9) we obtain

$$x = f(x, x, x, \alpha) = \frac{1 + (2+\alpha)x}{\alpha + 3x} \times \frac{(\alpha+3x)^2 + (2+\alpha)(1+(2+\alpha)x)^2 + (1+3x)^2}{\alpha(\alpha+3x)^2 + 3(1+(2+\alpha)x)^2 + (1+3x)^2}.$$

This equation can be reduced to the following

$$9(\alpha^2+7\alpha+7)x^3-(\alpha^4+5\alpha^3-24\alpha^2-52\alpha-65)x^2$$
$$-(\alpha^4-\alpha^3-6\alpha^2-16\alpha-23)x+\alpha^2+\alpha+3=0. \qquad (5.13)$$

Comparing (5.10) and (5.13) one can see that if x is a solution for one of these equations then $1/x$ will be solution to the second one. Therefore, by Corollary 2 of [138] (that is Corollary 8.1 in the Chapter 9 below) we conclude that Gibbs measures corresponding to solutions of equation (5.10) coincide with Gibbs measures corresponding to solutions of (5.13).

By above results we get the following list of solutions (v_{-1}, v_1, v_2) to (5.9):

$$\mathbf{v}_{10}=(1,1,1),\ \mathbf{v}_{11}=(x_1^*,1,1),\ \mathbf{v}_{12}=(1,x_1^*,1),\ \mathbf{v}_{13}=(1,1,x_1^*);$$

$$\mathbf{v}_{21}=(x_2^*,1,1),\ \mathbf{v}_{22}=(1,x_2^*,1),\ \mathbf{v}_{23}=(1,1,x_2^*);$$

$$\mathbf{v}_{31}=(1,x_3^*,x_3^*),\ \mathbf{v}_{32}=(x_3^*,1,x_3^*),\ \mathbf{v}_{33}=(x_3^*,x_3^*,1);$$

$$\mathbf{v}_{41}=(1,x_4^*,x_4^*),\ \mathbf{v}_{42}=(x_4^*,1,x_4^*),\ \mathbf{v}_{43}=(x_4^*,x_4^*,1).$$

Denote by μ_{ij} the Gibbs measure which, by Theorem 5.1, corresponds to the solution $\mathbf{v}_{ij}$.

Define critical temperatures

$$T_{1,\mathrm{c}}=\frac{1}{\ln\alpha^*},\quad T_{2,\mathrm{c}}=\frac{1}{\ln\alpha^{**}}.$$

Note that $T_{1,\mathrm{c}}>T_{2,\mathrm{c}}$.

Summarizing above results we obtain the following

Theorem 5.2. *For the model (5.2) of DNAs on the Cayley tree of order $k=2$ the following statements are true*

(1) If the temperature $T>T_{1,\mathrm{c}}$ then there is at last one translation-invariant Gibbs measure (TIGM), μ_{10}.
(2) If $T=T_{1,\mathrm{c}}$ then there are at least 4 TIGMs μ_{1j}, $j=0,1,2,3$.
(3) If $T_{2,\mathrm{c}}<T<T_{1,\mathrm{c}}$ then there are at least 7 TIGMs μ_{1j}, $j=0,1,2,3$, μ_{2j}, $j=1,2,3$.
(4) If $T=T_{2,\mathrm{c}}$ then there are at least 10 TIGMs μ_{1j}, $j=0,1,2,3$, μ_{ij}, $i=2,3$, $j=1,2,3$.
(5) If $T<T_{2,\mathrm{c}}$ then there are at least 13 TIGMs μ_{1j}, $j=0,1,2,3$, μ_{ij}, $i=2,3,4$, $j=1,2,3$.

5.3 Markov chains of TIGMs and Holliday junction of DNA

We note that the quantities $u_{i,l}$ and $v_{i,l}$ of (5.5) define a boundary law of the biological system of DNAs.

For marginals on the two-edge sets which consist of two neighbor edges l, t, considering a boundary law

$$\{(u_{i,l}, i = -1, 0, 1, 2;\ v_{j,t}, j = -1, 1, 2), l \notin \mathbb{Z} - \text{path},\ t \in \mathbb{Z} - \text{path}\},$$

i.e., the solutions of system (5.5) (this boundary law is normalized at -2, i.e., $h_{-2,l} = 0$), we have in the case of the Potts model that

$$\mu(\sigma(l) = a, \sigma(t) = b) = \frac{1}{Z} z_{a,l} \exp(\beta\delta_{ab}) z_{b,t}, \quad a, b = -2, -1, 0, 1, 2,$$

where Z is normalizing factor and $z_{a,l}$ takes values $u_{a,l}$ or $v_{a,l}$ depending on the relation of l to $\mathbb{Z}$ – path.

From this, the relation between the boundary law and the transition matrix for the associated tree-indexed Markov chain (Gibbs measure) is obtained from the formula of the conditional probability.

By Theorem 5.2 we have up to 13 TIGMs, and to each of them we can associate a Markov chain.

Now we are interested to study thermodynamics of Holliday junctions of DNAs, for this reason it will be sufficient to study Markov chains on the subtree consisting edges which are not on a $\mathbb{Z}$-path and Markov chains on the $\mathbb{Z}$-paths. Here we define these two Markov chains:

- Define tree-edge-indexed Markov chain with states $\{-2, -1, 0, 1, 2\}$ with transition matrix $\mathbf{P}^{[l,t]} = \left(P_{ij}^{[l,t]}\right)$, where $P_{ij}^{[l,t]}$ is the probability to go from a state i at edge l to a state j at the neighbor edge t. Using solutions (u_i, v_j) to (5.8) we write the matrices $\mathbf{P}^{[l,t]} = \left(P_{ij}^{[l,t]}\right)$:

$$\mathbf{P}^{[l,t]} = \mathbf{P}^{[l,t]}_{(5\to5)} = \begin{pmatrix} \frac{\alpha}{Z_1} & \frac{u_{-1}}{Z_1} & \frac{u_0}{Z_1} & \frac{u_1}{Z_1} & \frac{u_2}{Z_1} \\ \frac{1}{Z_2} & \frac{\alpha u_{-1}}{Z_2} & \frac{u_0}{Z_2} & \frac{u_1}{Z_2} & \frac{u_2}{Z_2} \\ \frac{1}{Z_3} & \frac{u_{-1}}{Z_3} & \frac{\alpha u_0}{Z_3} & \frac{u_1}{Z_3} & \frac{u_2}{Z_3} \\ \frac{1}{Z_4} & \frac{u_{-1}}{Z_4} & \frac{u_0}{Z_4} & \frac{\alpha u_1}{Z_4} & \frac{u_2}{Z_4} \\ \frac{1}{Z_5} & \frac{u_{-1}}{Z_5} & \frac{u_0}{Z_5} & \frac{u_1}{Z_5} & \frac{\alpha u_2}{Z_1} \end{pmatrix}, \quad \text{if } l, t \notin \mathbb{Z} - \text{path}.$$

  Here $Z_1 = \alpha + u_{-1} + u_0 + u_1 + u_2$, $Z_2 = 1 + \alpha u_{-1} + u_0 + u_1 + u_2$, $Z_3 = 1 + u_{-1} + \alpha u_0 + u_1 + u_2$, $Z_4 = 1 + u_{-1} + u_0 + \alpha u_1 + u_2$, $Z_5 = 1 + u_{-1} + u_0 + u_1 + \alpha u_2$.

- Define tree-edge-indexed Markov chain with states $\{-2,-1,1,2\}$

$$\mathbf{P}^{[l,t]}=\mathbf{P}^{[l,t]}_{(4\to4)}=\begin{pmatrix}\frac{\alpha}{Y_1} & \frac{v_{-1}}{Y_1} & \frac{v_1}{Y_1} & \frac{v_2}{Y_1}\\ \frac{1}{Y_2} & \frac{\alpha v_{-1}}{Y_2} & \frac{v_1}{Y_2} & \frac{v_2}{Y_2}\\ \frac{1}{Y_3} & \frac{v_{-1}}{Y_3} & \frac{\alpha v_1}{Y_3} & \frac{v_2}{Y_3}\\ \frac{1}{Y_4} & \frac{v_{-1}}{Y_4} & \frac{v_1}{Y_4} & \frac{\alpha v_2}{Y_4}\end{pmatrix}, \quad \text{if } \ l,t\notin\mathbb{Z}-\text{path}.$$

Here $Y_1=\alpha+v_{-1}+v_1+v_2$, $Y_2=1+\alpha v_{-1}+v_1+v_2$, $Y_3=1+v_{-1}+\alpha v_1+v_2$, $Y_4=1+v_{-1}+v_1+\alpha v_2$.

We note that each matrix $\mathbf{P}^{[l,t]}_{(n\to n)}$, $n=4,5$ is homogenous on the corresponding set of neighbor edges $\langle l,t\rangle$ where it is given, i.e., $\mathbf{P}^{[l,t]}_{(n\to m)}$ does not depend on $\langle l,t\rangle$ itself but only depends on its relation with $\mathbb{Z}$−path.

It is easy to find the following stationary distributions

$$\pi_{(5\to5)}=(\pi_{(5\to5),-2},\pi_{(5\to5),-1},\pi_{(5\to5),0},\pi_{(5\to5),1},\pi_{(5\to5),2}),$$

of the matrix $\mathbf{P}^{[l,t]}_{(5\to5)}$:

$$\pi_{(5\to5)}=\frac{1}{N}\begin{pmatrix}\alpha+u_{-1}+u_0+u_1+u_2\\ (1+\alpha u_{-1}+u_0+u_1+u_2)u_{-1}\\ (1+u_{-1}+\alpha u_0+u_1+u_2)u_0\\ (1+u_{-1}+u_0+\alpha u_1+u_2)u_1\\ (1+u_{-1}+u_0+u_1+\alpha u_2)u_2\end{pmatrix}^t,$$

where N the normalizing factor.

$$\pi_{(4\to4)}=\frac{1}{M}\begin{pmatrix}\alpha+v_{-1}+v_1+v_2\\ (1+\alpha v_{-1}+v_1+v_2)v_{-1}\\ (1+v_{-1}+\alpha v_1+v_2)v_1\\ (1+v_{-1}+v_1+\alpha v_2)v_2\end{pmatrix}^t,$$

where M the normalizing factor.

In the case of non-uniqueness of Gibbs measure (and corresponding Markov chains) we have different stationary states for different measures. These depend on the temperature and on the fixed measure.

As a corollary of Theorem 3.3 and above formulas of matrices and stationary distributions we obtain the following

Theorem 5.3. *In a stationary state of the set of DNAs, independently on $l\notin\mathbb{Z}$-path, a Holliday junction through l does not occur with the following probability (with respect to a measure μ_{ij} corresponding to a solution (u_{-1},u_0,u_1,u_2) of system (5.8))*

$$\pi_{(5\to5),0}=\frac{1}{N}(1+u_{-1}+\alpha u_0+u_1+u_2)u_0.$$

(Consequently, a Holliday junction occurs with probability $1-\pi_{(5\to5),0}$).

One can see that $\pi_{(5\to5),0}$ is a function of solution and temperature only.

Now we are interested to calculate the limit of stationary distribution vectors $\pi_{(5\to5)}$, $\pi_{(4\to4)}$ in case when temperature $T\to 0$ (i.e., $\beta\to\infty$ and $\alpha\to\infty$).

We calculate the limits for measures μ_{3j} and μ_{4j} mentioned in Theorem 5.2, because for these measures we have explicit formulas of solutions (see x_3^* and x_4^*), i.e., for these measures we have explicit formulas of solution u_i, $i=-1,0,1,2$; v_j, $j=-1,1,2$ which vary with $T=1/\beta$. By symmetry of measures μ_{3j} and μ_{4j} it will be sufficient to consider only μ_{31} and μ_{41}.

Recall that measures μ_{31} and μ_{41} do not exist for $T>T_{2,c}$.

For $k=2$, by above-mentioned solutions $\mathbf{v}_{31}$, $\mathbf{v}_{41}$ of (5.9) from (5.8), we obtain

$$u_{-1}=1,\quad u_0=4\left(\frac{1+x_i^*}{1+\alpha+2x_i^*}\right)^2,\quad u_1=u_2=\left(\frac{2+(1+\alpha)x_i^*}{1+\alpha+2x_i^*}\right)^2,\tag{5.14}$$

where $i=3,4$.

Lemma 5.4. *The following equalities hold*

$$\lim_{T\to 0}\pi^{(10)}_{(5\to5)}=(\frac{1}{5},\frac{1}{5},\frac{1}{5},\frac{1}{5},\frac{1}{5}).$$

$$\lim_{T\to 0}\pi^{(31)}_{(5\to5)}=(\frac{1}{2},\frac{1}{2},0,0,0),\quad \lim_{T\to 0}\pi^{(41)}_{(5\to5)}=(0,0,0,\frac{1}{2},\frac{1}{2}).$$

$$\lim_{T\to T_{2,c}}\pi^{(31)}_{(5\to5)}$$
$$=(0.2433190909,0.2433190909,0.03715503259,0.2381033927,0.2381033927).$$

$$\lim_{T\to 0}\pi^{(31)}_{(4\to4)}=(\frac{1}{2},\frac{1}{2},0,0),\quad \lim_{T\to 0}\pi^{(41)}_{(4\to4)}=(0,0,\frac{1}{2},\frac{1}{2}).$$

Proof. For measure μ_{10} we have $u_{-2}=u_{-1}=u_0=u_1=u_2=1$ this by formula $\pi_{(5\to5)}$ gives the first limit. Since we know explicit formulas for u_i,v_i (see (5.14) and x_3^* and x_4^*) the remaining limits follow by simple calculations. □

Structure of DNAs in low temperatures

See (5.1) for the relation of our spin values and the base pairs of a DNA.

By Lemma 5.4, in case $T\to 0$, the set of DNAs have the following stationary states (configurations):

Case μ_{10}: All neighboring DNAs connected to each other (Holliday junctions) with probability $\frac{4}{5}$ with state one of $\sigma(l) \in \{-2, -1, 1, 2\}$ for any $l \notin \mathbb{Z}$-path. The sequence of $\pm 1, \pm 2$s, in a DNA on the $\mathbb{Z}$-path, is free, i.e. can be any sequence, with iid and equiprobable $(= 1/4)$.

Case μ_{31}: All neighboring DNAs connected to each other (Holliday junctions) with state $\sigma(l) = -2$ or $\sigma(l) = -1$ for any $l \notin \mathbb{Z}$-path. A DNA on a $\mathbb{Z}$-path is constructed with sequence of -2 and -1s and iid and equiprobable $(= 1/2)$. Thus the system contains only multiple (countable) branched DNAs (which has tree structure).

Case μ_{41}: All neighboring DNAs connected to each other (Holliday junctions) with state $\sigma(l) = 1$ or $\sigma(l) = 2$ for any $l \notin \mathbb{Z}$-path. A DNA on a $\mathbb{Z}$-path is constructed with sequence of 1 and 2s and iid and equiprobable $(= 1/2)$.

In case $T = T_{2,c}$ the set of DNAs have the following stationary states with respect to measure $\mu_{31} = \mu_{41}$: each neighboring DNAs have Holliday junction with probability 0.9628449675 (more precisely, a junction through state -2 (and -1) with probability 0.2433190909 and a junction through state $+1$ (and $+2$) with probability 0.2381033927) and no junction with probability 0.03715503259.

Chapter 6

Thermodynamics of DNA-RNA renaturation

This chapter is devoted to a new model which consists of a DNA together with a RNA. The DNA may be from a mammal or bird but RNA can be from a virus. We use Gibbs measures to describe phases (states) of the DNA-RNA system. Using a Markov chain (corresponding to Gibbs measure) we give conditions (on temperature) of DNA-RNA renaturation.

6.1 Introduction and definitions

Recall that each molecule of DNA is a double helix formed by two complementary strands of nucleotides held together by hydrogen bonds between $G+C$ and $A+T$ base pairs, where C = cytosine, G = guanine, A = adenine, and T = thymine. Duplication of the genetic information occurs by the use of one DNA strand as a template for formation of a complementary strand. The genetic information stored in an organism's DNA contains the instructions for all the proteins the organism will ever synthesize. It is known that (see, for example, [6]) genetic information is carried in the linear sequence of nucleotides in DNA.

RNA[1] is a polymeric molecule essential in various biological roles in coding, decoding, regulation and expression of genes. RNA is assembled as a chain of nucleotides, but unlike DNA, RNA is found in nature as a single strand folded onto itself, rather than a paired double strand. Cellular organisms use messenger RNA to convey genetic information (using the nitrogenous bases of C, G, A and U = uracil) that directs synthesis of specific proteins.

[1] https://en.wikipedia.org/wiki/RNA.

All viruses contain[2] nucleic acid, either DNA or RNA (but not both), and a protein coat, which encases the nucleic acid. Coronaviruses are a group of related RNA viruses that cause diseases in mammals and birds. In humans, these viruses cause respiratory tract infections that can range from mild to lethal.

This chapter is devoted to thermodynamic properties of a model which consists a DNA (from a mammal or bird) together with a RNA (from a virus).

Studying DNA's thermodynamics one wants to know how temperature affects the nucleic acid structure of double-stranded DNA [166]. There are few models of thermodynamics of DNAs ([43], [213], [282]). In previous chapters we gave Ising and Potts models of DNAs and studied their thermodynamics. Here we shall use the arguments of these chapters to study thermodynamic behavior of a system consisting a DNA and an RNA.

The structure of DNA, at the microscopic level, can be described using ideas from statistical physics (see [277], [287]), where by a single DNA strand is modeled as a stochastic system of interacting bases with long-range correlations. This approach makes an important connection between the structure of DNA sequence and *temperature*; e.g., phase transitions in such a system may be interpreted as a conformational (topological) restructuring.

In this section, following [253], we consider a new model which consists a DNA together with an RNA. The bases in nucleic acids can interact via hydrogen bonds. Base pairing stabilizes the native three-dimensional structures of DNA and RNA. Our interpretation of this system is that RNA tries to denature the DNA and renature a new DNA by adding its own nitrogenous bases (as analogue of corona virus's RNA).

A DNA denaturation process is the breaking of the hydrogen bonds connecting the two stands under treatment by heat [61], [287]. The process consists of the splitting of DNA base pairs, or nucleotides, resulting in the separation of two complementary DNA strands.[3]

The area of partially separated DNA is known[4] as the denaturation bubble, which can be more specifically defined as the opening of a DNA double helix through the coordinated separation of base pairs. In 1966 the first model (called the Poland-Scheraga model) that attempted to describe the thermodynamics of the denaturation bubble was introduced. This model

---

[2]https://micro.magnet.fsu.edu/cells/virus.html.
[3]https://www.ncbi.nlm.nih.gov/books/NBK21514/.
[4]https://en.wikipedia.org/wiki/Denaturation_(biochemistry).

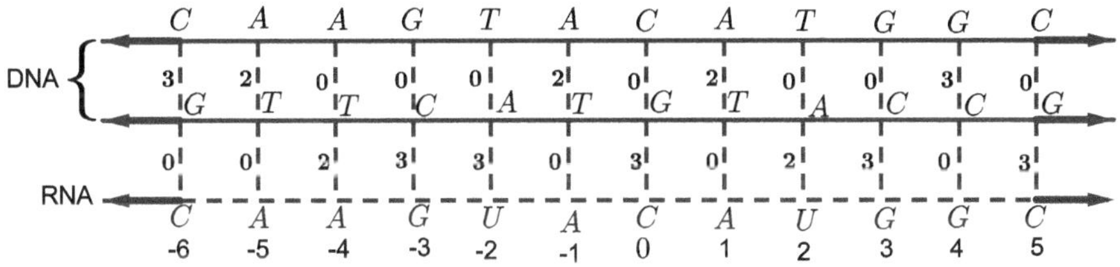

Fig. 6.1 The common picture of DNA and RNA (the double-ladder). Configuration consisting $0, 2, 3$ is the state of DNA-RNA denaturation-renaturation at a given temperature T. The value 0 means that the corresponding pair is broken (melted).

describes the denaturation of DNA strands as a function of temperature. As the temperature increases, the hydrogen bonds between base pairs are increasingly disturbed and "denatured loops" begin to form. Thus at a given temperature, the two strands may either be bound or unbound at a given site.

Biophysical and biochemical researchers are trying to fully describe the thermodynamic details of the denaturation bubble [8]. In the past decades DNA denaturation attracted the interest of various researchers, which introduced and studied statistical and dynamical models of this fundamental biological process (see [137], recent paper [151] and the references therein).

It is known that in a DNA each $A+T$ pair connected by two hydrogen bonds, while each $C+G$ pair connected by three hydrogen bonds. Therefore in this section we model them as (spin value) $2 = A + T$, $3 = C + G$. A melted (broken) under treatment by heat hydrogen bond assigned to (spin value) 0. The base pairs $A+T$ (in DNA), and $A+U$ (in RNA) considered as identical in process of renaturation of DNA from the RNA (of the virus).

Then a DNA can be considered as a ladder shown in Fig. 6.1.

An RNA is a one-dimensional line (also showed in Fig. 6.1). Thus the (spin) system is a double-ladder levels of which denoted by integer numbers $n \in \mathbb{Z}$. Assume a base pair is either broken or intact.

To each base pair of level $i \in \mathbb{Z}$ assign two parameters d_i (to base pair of DNA) and r_i (to the base pair of renatured DNA, i.e., between old DNA and RNA). These parameters are defined as

$$d_i = \begin{cases} 0, & \text{if the } i\text{th base pair of DNA is broken} \\ 2, & \text{if the } i\text{th base pair of DNA is intact and at state } A+T \\ 3, & \text{if the } i\text{th base pair of DNA is intact and at state } C+G, \end{cases}$$

$$r_i = \begin{cases} 0, & \text{if the } i\text{th BPB is broken} \\ 2, & \text{if the } i\text{th BPB is intact and at state } A+T \\ 3, & \text{if the } i\text{th BPB is intact and at state } C+G, \end{cases}$$

where BPB means “base pair between DNA and RNA”.

Since RNA (as coronavirus) will break base pair of DNA and puts its own pair, we have condition

$$d_i r_i = 0, \quad \text{for all level } i \in \mathbb{Z}. \tag{6.1}$$

Thus the configuration space Ω of our system is build by configurations

$$d = \{d_i \in \{0,2,3\} : i \in \mathbb{Z}\}, \quad r = \{r_i \in \{0,2,3\} : i \in \mathbb{Z}\},$$

as

$$\Omega = \left\{\sigma = (d,r) \in \{0,2,3\}^{\mathbb{Z}} \times \{0,2,3\}^{\mathbb{Z}} : d_i r_i = 0, \forall i \in \mathbb{Z}\right\}.$$

For each $\sigma \in \Omega$ define its energy (Hamiltonian) by

$$H(\sigma) = H(d,r) = -J \sum_{i=-\infty}^{+\infty} \left(\delta_{d_i d_{i+1}} + \delta_{r_i r_{i+1}}\right) - \alpha \sum_{i=-\infty}^{+\infty} (d_i + r_i), \tag{6.2}$$

where $J \in \mathbb{R}$ is coupling constant between base pairs, $\alpha \in \mathbb{R}$ is external field and δ is Kronecker delta.

6.2 Equations describing DNA-RNA renaturation

Denote by σ_n the restriction of the configuration $\sigma \in \Omega$ on $\mathbb{Z}_n = \{-n, -n+1, \dots, n-1, n\}$ and by Ω_n the set of all such configurations. In general, for a subset $A \subset \mathbb{Z}$ denote by Ω_A the set of all configurations restricted on A.

Define a finite-dimensional distribution of a probability measure μ on Ω_n as

$$\mu_n(\sigma_n) = Z_n^{-1} \exp\left\{-\beta H_n(\sigma_n) + \sum_{m \in \{-n,n\}} h_{m,d_m,r_m}\right\}, \tag{6.3}$$

where $\beta = 1/T$, $T > 0$ is temperature, Z_n^{-1} is the normalizing factor,

$$h_{m,i,j} \in \mathbb{R}, \quad i,j = 0,2,3, \quad m = -n, n \tag{6.4}$$

are real numbers and

$$H_n(\sigma_n) = -J \sum_{i=-n}^{n} \left(\delta_{d_i d_{i+1}} + \delta_{r_i r_{i+1}}\right) - \alpha \sum_{i=-n}^{n} (d_i + r_i).$$

The probability distributions (6.3) are compatible if for all $n \geq 1$ and $\sigma_{n-1} \in \Omega_{n-1}$:

$$\sum_{\omega_n \in \Omega_{\{-n,n\}}} \mu_n(\sigma_{n-1} \vee \omega_n) = \mu_{n-1}(\sigma_{n-1}). \tag{6.5}$$

Here $\sigma_{n-1} \vee \omega_n$ is the concatenation of the configurations. In this case there exists a unique measure μ on Ω such that, for all n and $\sigma_n \in \Omega_n$,

$$\mu(\{\sigma|_{\mathbb{Z}_n} = \sigma_n\}) = \mu_n(\sigma_n).$$

Such a measure is called a *Gibbs measure* corresponding to the Hamiltonian (6.2) and values (6.4).

For simplicity assume that

$$h_{-n,i,j} = h_{n,i,j}, \quad i,j = 0,2,3. \tag{6.6}$$

Under this condition the following statement describes conditions on $h_{n,i,j}$ guaranteeing compatibility of $\mu_n(\sigma_n)$.

Theorem 6.1. *Probability distributions $\mu_n(\sigma_n)$, $n = 1, 2, \ldots$, in (6.3) are compatible iff for any $n \geq 1$ the following hold*

$$\begin{aligned}
x_{n-1} &= \frac{\theta + \eta^2(\theta^2 x_n + y_n) + \eta^3(\theta u_n + v_n)}{\theta^2 + \theta\eta^2(x_n + y_n) + \theta\eta^3(u_n + v_n)} \\
y_{n-1} &= \frac{\theta + \eta^2(x_n + \theta^2 y_n) + \eta^3(u_n + \theta v_n)}{\theta^2 + \theta\eta^2(x_n + y_n) + \theta\eta^3(u_n + v_n)} \\
u_{n-1} &= \frac{\theta + \eta^2(\theta x_n + y_n) + \eta^3(\theta^2 u_n + v_n)}{\theta^2 + \theta\eta^2(x_n + y_n) + \theta\eta^3(u_n + v_n)} \\
v_{n-1} &= \frac{\theta + \eta^2(x_n + \theta y_n) + \eta^3(u_n + \theta^2 v_n)}{\theta^2 + \theta\eta^2(x_n + y_n) + \theta\eta^3(u_n + v_n)}.
\end{aligned} \tag{6.7}$$

Here,

$$\begin{aligned}
&\theta = \exp(J\beta), \quad \eta = \exp(\alpha\beta), \\
&x_n = \exp\left(h_{n,0,2} - h_{n,0,0}\right), \quad y_n = \exp\left(h_{n,2,0} - h_{n,0,0}\right), \\
&u_n = \exp\left(h_{n,0,3} - h_{n,0,0}\right), \quad v_n = \exp\left(h_{n,3,0} - h_{n,0,0}\right).
\end{aligned} \tag{6.8}$$

Proof. The proof is similar to the proof of Theorem 2.1 of [243]. □

It is difficult to find general solutions to (6.7).

Remark 6.1. For $\theta = 1$ (i.e., $J = 0$) the system (6.7) has unique solution $x_n = y_n = v_n = u_n = 1$. Therefore below we consider the case $\theta \neq 1$.

6.3 Translation-invariant solutions

The case when unknowns x_n, y_n, u_n, v_n depend on n is not studied yet. Here for simplicity, we assume that the unknowns do not depend on n, i.e., the value of each unknown is translation invariant.

Therefore denote

$$x=\theta\eta^2x_n,\quad y=\theta\eta^2y_n,\quad u=\theta\eta^3u_n,\quad v=\theta\eta^3v_n.$$

Define mapping

$$F:(x,y,u,v)\in\mathbb{R}^4_+\to(x',y',u',v')\in\mathbb{R}^4_+$$

as

$$\begin{aligned}
x'&=\eta^2\cdot\tfrac{\theta^2+\theta^2x+y+\theta u+v}{\theta^2+x+y+u+v}\\
y'&=\eta^2\cdot\tfrac{\theta^2+x+\theta^2y+u+\theta v}{\theta^2+x+y+u+v}\\
u'&=\eta^3\cdot\tfrac{\theta^2+\theta x+y+\theta^2u+v}{\theta^2+x+y+u+v}\\
v'&=\eta^3\cdot\tfrac{\theta^2+x+\theta y+u+\theta^2v}{\theta^2+x+y+u+v}.
\end{aligned}\tag{6.9}$$

Then the system (6.7) is reduced to the finding of fixed points of the mapping F, i.e., to solving of system $(x,y,u,z)=F(x,y,u,v)$.

Denote

$$M=\{(x,y,u,v)\in\mathbb{R}^4_+:x=y,u=v\}.$$

It is easy to see that $F(M)\subset M$, i.e. M is invariant with respect to F.

6.3.1 *Solutions in the set M*

Restricting F on M the fixed point problem reduced to the following system

$$\begin{aligned}
x&=\eta^2\cdot\tfrac{\theta^2+(\theta^2+1)x+(\theta+1)u}{\theta^2+2x+2u}\\
u&=\eta^3\cdot\tfrac{\theta^2+(\theta+1)x+(\theta^2+1)u}{\theta^2+2x+2u}.
\end{aligned}\tag{6.10}$$

From the first equation of this system we get

$$(\eta^2(1+\theta)-2x)u=2x^2+(\theta^2-\eta^2(1+\theta^2))x-\eta^2\theta^2.\tag{6.11}$$

(1) *Case:* $\eta^2(1+\theta)-2x=0$. In this case we get $x=\frac{\eta^2(1+\theta)}{2}$ and substituting this in the RHS of (6.11) we get $\eta=\sqrt{\frac{\theta}{1+\theta}}$. Consequently,

$x = x_0 = \frac{\theta}{2}$. For this value of η and x one can explicitly find unique positive value of $u = u_0(\theta)$.

Thus for any $\theta > 0$ and $\eta = \sqrt{\frac{\theta}{1+\theta}}$ there exists unique solution (x_0, u_0) for (6.10).

(2) *Case:* $\eta^2(1+\theta) - 2x \neq 0$.

$$u = u(x) := \frac{2x^2 + (\theta^2 - \eta^2(1+\theta^2))x - \eta^2\theta^2}{\eta^2(1+\theta) - 2x}. \tag{6.12}$$

Substituting this in the second equation of (6.10) we get

$$Ax^3 + Bx^2 + Cx + D = 0, \tag{6.13}$$

where

$A = 4(1-\eta)$,

$B = 2[(\eta^3 - \eta^2 - \eta + 1)\theta^2 + (2\eta^3 - \eta - 1)\theta + \eta^2(3\eta - 1)]$,

$C = (-\eta^5 + \eta^3 + \eta^2 - 1)\theta^3 - (2\eta^5 - 4\eta^3 + 2\eta^2)\theta^2 - (3\eta^5 - \eta^3 - \eta^2)\theta - 2\eta^5$,

$D = \eta^2\theta^2(\theta - \eta^3(1+\theta))$.

All of the roots of the cubic equation can be found[5]. We are interested in positive solutions, $x_i = x_i(\theta, \eta)$ of the cubic equation. Moreover, the corresponding $u(x_i)$ defined in (6.12) should be positive too. Thus condition for parameters $(\theta, \eta) \in \mathbb{R}^2_+$ of the existence of positive solutions can be explicitly written $x_i(\theta, \eta) > 0, \quad u(x_i) > 0$. But the explicit solutions of the cubic equation have some bulky formulas, therefore we do not present the solution here. Instead we consider some concrete cases:

(3) In the above-mentioned case 1) we solved the first equation of the system (6.10) with respect to u. Doing similar argument starting from the second equation of (6.10) and solving it with respect to x one gets $u = u_1 = \frac{\theta}{2}$, if $\eta = \sqrt[3]{\frac{\theta}{1+\theta}}$. Corresponding to u_1 one can explicitly find unique positive value of $x = x_1(\theta)$. Thus for any $\theta > 0$ and $\eta = \sqrt[3]{\frac{\theta}{1+\theta}}$ we can explicitly give unique solution (x_1, u_1) of (6.10).

(4) *Case*: $\eta = 1$. In this case the cubic equation is reduced to quadratic equation, which has unique positive solution:

$$x_2 = x_2(\theta) := \frac{1}{8}(\theta + 2 + \sqrt{17\theta^2 + 4\theta + 4}).$$

Corresponding $u_2 = u(x_2(\theta))$ is also positive. Note that $x_2(\theta)$ and $u_2(\theta)$ have value $+\infty$ (resp. $1/2$) if $\theta \to \infty$ (resp. $\theta \to 0$).

[5] https://en.wikipedia.org/wiki/Cubic_equation.

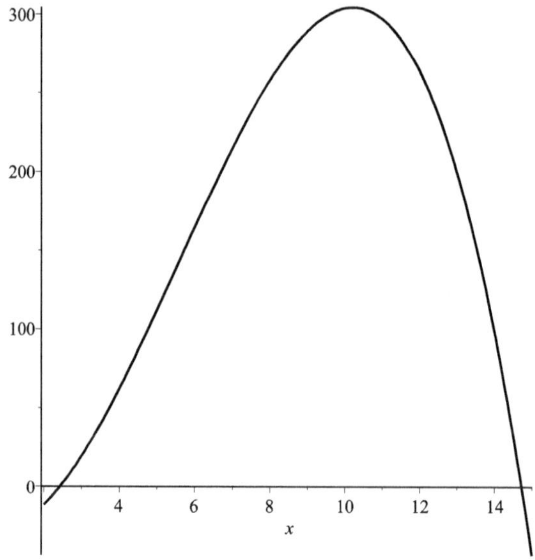

Fig. 6.2 The graph of cubic polynomial (6.13) for $\theta = 2$, $\eta = 1.2$. In this case there are two positive roots of the polynomial, which approximately: $x_1 = 2.43547$, $x_2 = 14.72382$ (the negative solution: -0.63929).

Thus in the case $\eta = 1$ the system (6.10) has unique solution (x_2, u_2).

(5) Several numerical analysis show that for $\eta \neq 1$ again we have unique solution (see Figs. 6.2 and 6.3).

6.3.2 *Solutions in the set* $\mathbb{R}^2_+ \setminus M$

Recall that $F(x, y, u, v) = (x, y, u, v)$ has the form

$$\begin{aligned} x &= \eta^2 \cdot \frac{\theta^2+\theta^2 x+y+\theta u+v}{\theta^2+x+y+u+v} \\ y &= \eta^2 \cdot \frac{\theta^2+x+\theta^2 y+u+\theta v}{\theta^2+x+y+u+v} \\ u &= \eta^3 \cdot \frac{\theta^2+\theta x+y+\theta^2 u+v}{\theta^2+x+y+u+v} \\ v &= \eta^3 \cdot \frac{\theta^2+x+\theta y+u+\theta^2 v}{\theta^2+x+y+u+v}. \end{aligned} \tag{6.14}$$

Subtracting from the first equation of this system the second one (resp. from the third equation of the last one) we get

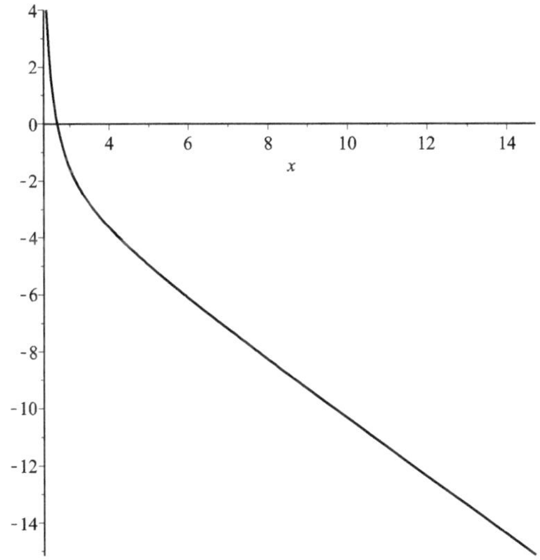

Fig. 6.3 The graph of function u defined in (6.12) for $\theta = 2$, $\eta = 1.2$ on $[2.4, 14.73]$, which contains both positive roots shown in Fig. 6.2. Thus only one $u_1 = u(x_1)$ is positive.

$$\begin{aligned} x - y &= L \cdot [(\theta - \theta^{-1})(x - y) + (1 - \theta^{-1})(u - v)] \\ u - v &= \eta L \cdot [(1 - \theta^{-1})(x - y) + (\theta - \theta^{-1})(u - v)], \end{aligned} \tag{6.15}$$

where

$$L \equiv L(x, y, u, v) = \eta^2 \cdot \frac{\theta}{\theta^2 + x + y + u + v}.$$

Recall that $\theta \neq 1$.

Lemma 6.1. *If (x, y, u, v) is a solution to system (6.14) then $x = y$ iff $u = v$.*

Proof. Since $L > 0$, $\theta \neq 1$, if $x = y$ then from the first equation of (6.15) we get $u = v$. If $u = v$ then from the second equation of (6.15) we get $x = y$. □

Assume $x \neq y$. Then find $u - v$ from the first equation of (6.15) and substituting it in the second equation we obtain

$$(\theta - 1)^2(\theta + 2)\eta L^2 - (1 + \eta)(\theta^2 - 1)L + \theta = 0. \tag{6.16}$$

Theorem 6.2. *The system (6.14) does not have any solution in $\mathbb{R}^2_+ \setminus M$.*

Proof. From Lemma 6.1 it follows that in $\mathbb{R}^2_+ \setminus M$ may only exist solutions with $x \neq y$ and $u \neq v$. Therefore we denote $t = \frac{u-v}{x-y}$.

Case 1: $t > 0$. Assuming $t > 0$ from (6.15) we get

$$t = \eta \cdot \frac{1+(1+\theta)t}{1+\theta+t} \quad \Leftrightarrow \quad t^2 + (1-\eta)(1+\theta)t - \eta = 0. \tag{6.17}$$

The last equation has unique positive root:

$$t_1 = \frac{1}{2}((\eta-1)(1+\theta) + \sqrt{[(\eta-1)(1+\theta)]^2 + 4\eta}\,).$$

Thus $u - v = t_1 \cdot (x - y)$. Using this from the first equation of (6.15) we get

$$1 = L \cdot [(\theta - \theta^{-1}) + (1 - \theta^{-1})t_1] \quad \Leftrightarrow$$

$$L = \frac{\eta^2\theta}{\theta^2 + x + y + u + v} = \frac{\theta}{(\theta-1)(1+\theta+t_1)}. \tag{6.18}$$

One can see that L satisfies (6.16).

By the last formula we get

$$\begin{aligned} x &= B \cdot [\theta + \theta x + \theta^{-1} y + u + \theta^{-1} v] \\ y &= B \cdot [\theta + \theta^{-1} x + \theta y + \theta^{-1} u + v] \\ u &= \eta B \cdot [\theta + x + \theta^{-1} y + \theta u + \theta^{-1} v] \\ v &= \eta B \cdot [\theta + \theta^{-1} x + y + \theta^{-1} u + \theta v], \end{aligned} \tag{6.19}$$

where the constant B is

$$B = B(\theta, \eta) = \frac{\theta}{(\theta-1)(1+\theta+t_1)}.$$

Case 1.1.: $\theta < 1$. It is clear that (6.19) does not have any positive solution if $\theta < 1$ (because in this case $B < 0$).

Case 1.2.: $\theta > 1$. In this case the system is a linear system of equation of the form $\mathbf{Mv} = \mathbf{b}$, where

$$\mathbf{M} = \begin{pmatrix} A & 1 & \theta & 1 \\ 1 & A & 1 & \theta \\ \theta & 1 & C & 1 \\ 1 & \theta & 1 & C \end{pmatrix}, \quad A = \theta^2 - \frac{\theta}{B}, \quad C = \theta^2 - \frac{\theta}{\eta B},$$

$$\mathbf{v} = \begin{pmatrix} x \\ y \\ u \\ v \end{pmatrix}, \quad \mathbf{b} = -\theta^2 \begin{pmatrix} 1 \\ 1 \\ 1 \\ 1 \end{pmatrix}.$$

We are interested in positive solutions of the system. By this system of linear equations we get

$$t_1 = \frac{u-v}{x-y} = \frac{1-A}{\theta-1} = \frac{\theta-1}{1-C} \quad \Leftrightarrow \quad AC = A + C + \theta^2 - 2\theta. \tag{6.20}$$

Using formula of A, C and t_1 one can see that (6.20) is satisfied. Moreover, we have

$$\det(\mathbf{M}) = (AC + A + C - \theta^2 - 2\theta)(AC - A - C - \theta^2 + 2\theta). \tag{6.21}$$

In case $\det(\mathbf{M}) \neq 0$ the system $\mathbf{Mv} = \mathbf{b}$ has unique solution with $x = y$ and $u = v$. To have its other solutions (with the condition $x \neq y$ and $u \neq v$) we need to the condition $\det(\mathbf{M}) = 0$ which by (6.20) is satisfied and $\operatorname{rank}(\mathbf{M}) = 3$. Solving the linear system $\mathbf{Mv} = \mathbf{b}$, under condition (6.20), we explicitly obtain infinitely many solutions:

$$x = -\frac{(\theta^2+v)t_1 + v}{t_1(1+t_1)}, \quad y = \frac{v}{t_1}, \quad u = -\frac{(\theta^2+v)t_1+v}{1+t_1}, \quad v > 0. \tag{6.22}$$

Thus $x < 0$ and $u < 0$, i.e., there is no positive solution $x \neq y$ and $u \neq v$.

Case 2: $t < 0$. In this case from (6.17) we get unique negative root:

$$t = t_2 = \frac{1}{2}\left((\eta-1)(1+\theta) - \sqrt{[(\eta-1)(1+\theta)]^2 + 4\eta}\,\right).$$

Thus $u - v = t_2 \cdot (x - y)$. Using this from the first equation of (6.15) we get

$$1 = L \cdot [(\theta - \theta^{-1}) + (1 - \theta^{-1})t_2] \quad \Leftrightarrow \quad \theta = L \cdot (\theta - 1)(1 + \theta + t_2).$$

In the last equality, since $\theta > 0$ and $L > 0$, it is necessary that $(\theta-1)(1+\theta+t_2) > 0$. It is easy to see that $t_2 > -1-\theta$. Consequently, the system may have solution only for $\theta > 1$. Therefore we have

$$\frac{\eta^2\theta}{\theta^2 + x + y + u + v} = \frac{\theta}{(\theta-1)(1+\theta+t_2)}. \tag{6.23}$$

By the last formula we get

$$\begin{aligned} x &= B_2 \cdot [\theta + \theta x + \theta^{-1}y + u + \theta^{-1}v] \\ y &= B_2 \cdot [\theta + \theta^{-1}x + \theta y + \theta^{-1}u + v] \\ u &= \eta B_2 \cdot [\theta + x + \theta^{-1}y + \theta u + \theta^{-1}v] \\ v &= \eta B_2 \cdot [\theta + \theta^{-1}x + y + \theta^{-1}u + \theta v], \end{aligned} \tag{6.24}$$

where the constant B_2 is

$$B_2 = B_2(\theta, \eta) = \frac{\theta}{(\theta - 1)(1 + \theta + t_2)}.$$

Thus for $\theta > 1$ the system is a linear system of equation of the form $\mathbf{N}\mathbf{v} = \mathbf{b}$ where

$$A_2 = \theta^2 - \frac{\theta}{B_2},\ C_2 = \theta^2 - \frac{\theta}{\eta B_2},$$

$$\mathbf{N} = \begin{pmatrix} A_2 & 1 & \theta & 1 \\ 1 & A_2 & 1 & \theta \\ \theta & 1 & C_2 & 1 \\ 1 & \theta & 1 & C_2 \end{pmatrix}, \quad \mathbf{v} = \begin{pmatrix} x \\ y \\ u \\ v \end{pmatrix}, \quad \mathbf{b} = -\theta^2 \begin{pmatrix} 1 \\ 1 \\ 1 \\ 1 \end{pmatrix}. \tag{6.25}$$

To have a positive solution of this system it is necessary that

$$A_2 = \theta^2 - \frac{\theta}{B_2} < 0,\ C_2 = \theta^2 - \frac{\theta}{\eta B_2} < 0.$$

But we have

$$A_2 = \theta^2 - \frac{\theta}{B_2} < 0 \quad \Leftrightarrow \quad \theta - \frac{\theta - 1}{\theta}(1 + \theta + t_2) < 0 \quad \Leftrightarrow \quad 1 < (\theta - 1)t_2.$$

The last equality does not hold because $\theta > 1$ and $t_2 < 0$. Thus in the case $t < 0$ the system (6.24) does *not* have any positive solution. This completes the proof of theorem. □

6.4 Gibbs measures: Conditions of DNA-RNA renaturation

Let μ be the Gibbs measure corresponding to a solution (x, y, u, v) of the system (6.14).

The measure μ defines joint distribution

$$\begin{aligned} &\mu\Big(\sigma(n) = (d_n, r_n), \sigma(n+1) = (d_{n+1}, r_{n+1})\Big) \\ &= \frac{1}{Z} \exp\Big(J\beta\left(\delta_{d_n d_{n+1}} + \delta_{r_n r_{n+1}}\right) \\ &\quad + \sum_{m \in \{n, n+1\}} \left[\alpha\beta\left(d_m + r_m\right) + h_{m, d_m, r_m}\right]\Big), \end{aligned}$$

where Z is normalizing factor.

From this, the relation between the solutions (x, y, u, v) and the transition matrix for the associated Markov chain is obtained from the formula of the conditional probability. The transition matrix of this Markov chain is defined as follows

$$\mathbb{P} = (P_{ij})_{i,j=\hat{1},\hat{2},\dots,\hat{5}} = \begin{pmatrix} \frac{\theta^2}{Z_1} & \frac{x}{Z_1} & \frac{y}{Z_1} & \frac{u}{Z_1} & \frac{v}{Z_1} \\ \frac{\theta}{Z_2} & \frac{\theta x}{Z_2} & \frac{\theta^{-1}y}{Z_2} & \frac{u}{Z_2} & \frac{\theta^{-1}v}{Z_2} \\ \frac{\theta}{Z_3} & \frac{\theta^{-1}x}{Z_3} & \frac{\theta y}{Z_3} & \frac{\theta^{-1}u}{Z_3} & \frac{v}{Z_3} \\ \frac{\theta}{Z_4} & \frac{x}{Z_4} & \frac{\theta^{-1}y}{Z_4} & \frac{\theta u}{Z_4} & \frac{\theta^{-1}v}{Z_4} \\ \frac{\theta}{Z_5} & \frac{\theta^{-1}x}{Z_5} & \frac{y}{Z_5} & \frac{\theta^{-1}u}{Z_5} & \frac{\theta v}{Z_5} \end{pmatrix}, \tag{6.26}$$

where

$$\hat{1} = (0,0), \quad \hat{2} = (0,2), \quad \hat{3} = (2,0), \quad \hat{4} = (0,3), \quad \hat{5} = (3,0); \tag{6.27}$$

$$Z_1 = \theta^2 + x + y + u + v, \quad Z_2 = \theta + \theta x + \theta^{-1}y + u + \theta^{-1}v,$$

$$Z_3 = \theta + \theta^{-1}x + \theta y + \theta^{-1}u + v, \quad Z_4 = \theta + x + \theta^{-1}y + \theta u + \theta^{-1}v,$$

$$Z_5 = \theta + \theta^{-1}x + y + \theta^{-1}u + \theta v,$$

and (x, y, u, v) a solution of system (6.14) (which depends on both parameters θ and η).

Since $\mathbb{P}$ is a positive stochastic matrix there exists unique probability vector $\pi = (\pi_1, \dots, \pi_5)$ which satisfies the system of linear equations $\pi\mathbb{P} = \pi$ (i.e., π is stationary distribution). Note that this linear system can be explicitly solved. Its solution π depends on both parameters θ and η subject to the constraint that elements must sum to 1. But coordinates of the vector π has a bulky form. Therefore we do not present it here.

Recall $r_i = 0$ means that RNA does not renature DNA at level $i \in \mathbb{Z}$.

Thus for a given DNA and RNA we say that the RNA (virus) do not destroy the DNA if for any $\epsilon > 0$ there exists $N \geq 1$ such that for any $n \geq N$ the following inequality holds

$$\mu(\Omega_n^0) > 1 - \epsilon,$$

where μ is Gibbs measure corresponding to a solution of system (6.7) and

$$\Omega_n^0 = \{\sigma_n = \{(d_i, r_i)\} \in \Omega_n : r_i = 0, \forall i \in \mathbb{Z}_n\}.$$

Note that each element of Ω_n^0 defines a DNA, which has thermodynamic behavior.

For the measure corresponding to a solution (x, y, u, v) of system (6.14) we have (Markov measure):

$$\mu(\Omega_n^0) = \sum_{\sigma_n \in \Omega_n^0} \pi_{\sigma(-n)} \prod_{i=-n}^{n} P_{\sigma(i),\sigma(i+1)},$$

where $\pi_{\sigma(-n)} \in \{\pi_1, \ldots, \pi_5\}$ (a coordinate of the stationary distribution) and $P_{\sigma(i),\sigma(i+1)} \in \{P_{\hat{i},\hat{j}} : \text{ with odd } i, j\}$ see (6.27).

For a given solution (x, y, u, v) corresponding to it value $\mu(\Omega_n^0)$ depends only on parameters θ, η and n, i.e. $\mu_n(\theta, \eta) := \mu(\Omega_n^0)$. Using explicit formula of solution (x, y, u, v) one can explicitly calculate $\mu_n(\theta, \eta)$. But it will have a bulky form.

For fixed parameters J and α of the model (6.2) the parameters θ and η are functions of temperature $T = \frac{1}{\beta}$ (see (6.8)). Therefore for fixed parameters of the model, the quantity $\mu_n(T) := \mu_n(\theta, \eta)$ is the function of temperature T and n only. By Theorem 3.3 and above formulas of matrices we have the following:

Corollary 6.1. *For given parameters J and α of the model (6.2) RNA of the virus do not destroy DNA (with respect to measure μ) if **temperature** T satisfies the following condition: $\forall \epsilon > 0$ there exists $N \geq 1$ such that $\forall n \geq N$ the following inequality holds*

$$\mu_n(T) > 1 - \epsilon.$$

Since $\mu_n(T)$ has a bulky form to check this condition one can use numerical analysis.

Remark 6.2. For a configuration $\sigma \in \Omega$ we have condition $d_i r_i = 0$, moreover the condition $d_i + r_i \neq 0$ is needed to have a base pair at each level $i \in \mathbb{Z}$. Therefore introduce

$$\Omega^+ = \left\{\sigma = (d, r) \in \{0, 2, 3\}^{\mathbb{Z}} \times \{0, 2, 3\}^{\mathbb{Z}} : d_i r_i = 0, \quad d_i + r_i \neq 0, \quad \forall i \in \mathbb{Z}\right\}.$$

Take configuration $\hat{\sigma} \in \Omega^+$ defined by $d_i = 2$ and $r_i = 0$ for all $i \in \mathbb{Z}$. Then, for $J > 0$ and $\alpha < 0$, this configuration $\hat{\sigma}$ is a ground state of Hamiltonian (6.2), i.e.,

$$H(\hat{\sigma}_A) \leq H(\sigma_A), \quad \text{for any connected set } A \text{ and any } \sigma_A \in \Omega_A^+.$$

It is known that a system at absolute zero temperature ($T = 0$) exists in its ground state. In language of Gibbs measure μ, which depends on the temperature $T \geq 0$, through solutions (x, y, u, v), the last fact is formulated as

$$\mu\{\sigma : \sigma_n = \hat{\sigma}_n\} \to 1, \quad \text{as} \quad T \to 0.$$

Since $\hat{\sigma} \in \Omega_n^0 = \{\sigma_n = \{(d_i, r_i)\} \in \Omega_n : r_i = 0, \forall i \in \mathbb{Z}_n\}$ for any $\epsilon > 0$, fixed n and sufficiently small T we have $\mu_n(T) = \mu(\Omega_n^0) > 1 - \epsilon$.

Chapter 7

On quadratic stochastic operators of Gibbs measures

In this chapter, we give another approach to a natural introduction of thermodynamics in studying some models of heredity. This approach is related to the discrete-time dynamical systems generated by quadratic stochastic operators. We construct quadratic stochastic operators which act to the set of all probability measures on some measurable space. This construction depends on a probability measure μ and cardinality of a set of cells (configurations) which can be finite or continual. The behavior of trajectories of such operators is studied when the probability measure μ is a Gibbs measure. For the continual case we compare the quadratic operators which correspond to well-known Gibbs measures of the Potts model on $\mathbb{Z}^d$. It is shown that any trajectory of the quadratic stochastic operator generated by a Gibbs measure μ of the Potts model converges to this measure.

7.1 Definitions

As it was mentioned in Chapter 1 a well-known phenomenon among reptile species is temperature-dependent sex determination in which the temperature of egg incubation determines the sex of the hatchlings. Many researchers (see for example, [199] and the references therein) have developed a delay differential equation model describing the nesting habits of crocodilian; the delay accounts for some of the dependence of birth and death rates on the age of the population members. The differential equations are continuous-time dynamical systems.

In this chapter, we consider discrete-time dynamical systems related to quadratic stochastic operators. Quadratic stochastic operator frequently arise in many models of mathematical genetics [21], [69], [82], [83], [85], [116], [161], [162], [251], [289].

The approach to population genetics, posed within that scheme the problem of an explicit description of evolutionary operators of free populations.

Quadratic stochastic operator of a free population [162] is a mapping of the simplex

$$S^{n-1} = \{x = (x_1, ..., x_n) \in \mathbb{R}^n : x_i \geq 0, \sum_{i=1}^{n} x_i = 1\},$$

into itself, of the form

$$V : x'_k = \sum_{i,j=1}^{n} p_{ij,k} x_i x_j, \quad (k = 1, ..., n), \tag{7.1}$$

where $p_{ij,k}$ are coefficients of heredity and

$$p_{ij,k} \geq 0, \quad \sum_{k=1}^{n} p_{ij,k} = 1, \quad (i, j, k = 1, ..., n). \tag{7.2}$$

Note that each element $x \in S^{n-1}$ is a probability distribution on $E = \{1, ..., n\}$.

The population evolves by starting from an arbitrary state (probability distribution on E) $x \in S^{n-1}$ then passing to the state Vx (in the next "generation"), then to the state V^2x, and so on.

In [21], [69], [82], [83], [85], [116], [161], [162], [289] the theory of quadratic operators for $E = \{1, ..., n\}$ is developed (see also [84], [191] and [251] for more details).

Note that each quadratic operator V can be uniquely defined by a cubic matrix $\mathbf{P} \equiv \mathbf{P}(V) = (p_{ij,k})_{i,j,k=1}^{n}$ with condition (7.2).

In this chapter, we mainly follow [87] and give a constructive description of $\mathbf{P}$ (i.e., quadratic stochastic operator). This construction depends on cardinality of E, namely we separately consider two cases: (i) E is finite; and (ii) E is a continual set.

In [85] some particular cases of this construction were considered . Note that for the second case one of the key problem is to determine the set of coefficients of heredity which is already infinite dimensional; the second problem is to investigate the quadratic operator which corresponds to this set of coefficients. By the construction the operator V depends on a probability measure μ on a measurable space $(E, \mathcal{F})$.

The next goal is to study the quadratic operators which corresponds to Gibbs measures of the Potts model on $\mathbb{Z}^d$ [267]. It is well known that for the

Potts model with $q \geq 2$ spin values, there exists a critical temperature $T_{\rm cr}$ such that for $T < T_{\rm cr}$ there are q distinct Gibbs measures μ_i, $i = 1, \ldots, q$.

Here, following [85] and [87], we shall study and compare properties of quadratic operators generated by $\mu_i, i = 1, ..., q$. These investigations allows a natural introduction of thermodynamics in studying some models of heredity.

Let $(E, \mathcal{F})$ be a measurable space and $S(E, \mathcal{F})$ be the set of all probability measures on $(E, \mathcal{F})$.

Definition 7.1. A mapping $V : S(E, \mathcal{F}) \to S(E, \mathcal{F})$ is called a quadratic stochastic operator (QSO) if, for an arbitrary measure $\lambda \in S(E, \mathcal{F})$, the measure $\lambda' = V\lambda$ is defined as

$$\lambda'(A) = \int_E \int_E P(x, y, A) d\lambda(x) d\lambda(y), \quad \forall A \in \mathcal{F},$$

where $P(x, y, A)$ satisfies the following conditions:

(i) $P(x, y, \cdot) \in S(E, \mathcal{F})$ for any fixed $x, y \in E$;
(ii) $P(x, y, A)$, regarded as a function of two variables x and y, is measurable on $(E \times E, \mathcal{F} \otimes \mathcal{F})$ for any fixed $A \in \mathcal{F}$.

When E is finite, a QSO on $S(E, \mathcal{F}) = S^{n-1}$ is defined as in (7.1) with $p_{ij,k} = P(i, j, k)$.

The following model described by QSOs prompted the construction of QSOs suggested here. In [69], the transmission of a trait from parents to their offspring is described by the following three probabilities of transmission:

$p_{AA,A}$: from a parent with a genotype AA, the child receives the allele A;
$p_{Aa,A}$: from a parent with a genotype Aa, the child receives the allele A;
$p_{aa,A}$: from a parent with a genotype aa, the child receives the allele A;

then $p_{..,a} = 1 - p_{..,A}$.

7.2 Construction of QSO for finite E

Let $G = (\Lambda, L)$ be a finite graph without loops and multiple edges, where Λ is the set of vertexes and L is the set of edges of the graph.

A finite set Φ is called the set of alleles. Recall that in problems of statistical mechanics, Φ is called the set of spin values.

The function $\sigma : \Lambda \to \Phi$ is called a cell.[1]

Denote by Ω the set of all cells, this set corresponds to E in (7.1).

Let $S(\Lambda, \Phi)$ be the set of all probability measures defined on the finite set Ω.

Let $\{\Lambda_i, i = 1, ..., m\}$ be the set of maximal connected subgraphs (components) of the graph G.

For a configuration $\sigma \in \Omega$ denote by $\sigma(M)$ its "projection" (or "restriction") to $M \subset \Lambda : \sigma(M) = \{\sigma(x)\}_{x\in M}$.

Fix two cells $\sigma_1, \sigma_2 \in \Omega$, and put

$$\Omega(G, \sigma_1, \sigma_2) \\ = \{\sigma \in \Omega : \sigma(\Lambda_i) = \sigma_1(\Lambda_i) \text{ or } \sigma(\Lambda_i) = \sigma_2(\Lambda_i) \text{ for all } i = 1, ..., m\}.$$

Let $\mu \in S(\Lambda, \Phi)$ be a probability measure defined on Ω such that $\mu(\sigma) > 0$ for any cell $\sigma \in \Omega$; i.e., μ is a Gibbs measure with some potential [220].

The heredity coefficients $p_{\sigma_1\sigma_2,\sigma}$ are defined as

$$p_{\sigma_1\sigma_2,\sigma} = \begin{cases} \frac{\mu(\sigma)}{\mu(\Omega(G,\sigma_1,\sigma_2))}, & \text{if } \sigma \in \Omega(G, \sigma_1, \sigma_2), \\ 0 & \text{otherwise.} \end{cases} \tag{7.3}$$

Obviously, $p_{\sigma_1\sigma_2,\sigma} \geq 0$, $p_{\sigma_1\sigma_2,\sigma} = p_{\sigma_2\sigma_1,\sigma}$ and $\sum_{\sigma\in\Omega} p_{\sigma_1\sigma_2,\sigma} = 1$ for all $\sigma_1, \sigma_2 \in \Omega$.

The QSO $V \equiv V_\mu$ acting on the simplex $S(\Lambda, \Phi)$ and determined by coefficients (7.3) is defined as follows: for an arbitrary measure $\lambda \in S(\Lambda, \Phi)$, the measure $V(\lambda) = \lambda' \in S(\Lambda, \Phi)$ is defined by the equality

$$\lambda'(\sigma) = \sum_{\sigma_1,\sigma_2\in\Omega} p_{\sigma_1\sigma_2,\sigma}\lambda(\sigma_1)\lambda(\sigma_2) \tag{7.4}$$

for any cell $\sigma \in \Omega$.

Thus Gibbs measures arise in a natural way when QSOs are constructed.

The QSO constructed from the potential of the nearest neighbor or, equivalently, from a Markov random field $\mu \in S(\Lambda, \Phi)$ (see [220]) is called a Markov quadratic operator.

The QSO construction is also closely related to the graph structure on the set Λ.

A quadratic operator (7.1) is called Volterra [116] if the coefficients $p_{ij,k}$ may be nonzero only when $i = k$ or $j = k$ and vanish in all the remaining cases. The biological treatment of such models is rather clear: the offspring repeats the genotype of one of its parents.

[1] In mechanics it is called a configuration.

Theorem 7.1. *The QSO (7.3) is Volterra if and only if the graph G is connected.*

Proof. Let G be a connected graph then we have $\Omega(G, \sigma_1, \sigma_2) = \{\sigma_1, \sigma_2\}$ for any $\sigma_1, \sigma_2 \in \Omega$. Therefore, by (7.3) it follows that the corresponding QSO is Volterra. Conversely, if the coefficients (7.3) satisfy that $p_{\sigma_1\sigma_2,\sigma} = 0$ for any $\sigma \notin \{\sigma_1, \sigma_2\}$ then by condition $\mu > 0$ it follows that G is a connected graph. □

From Theorem 7.1 and Theorem 1.1 stated in [220] we get the following

Corollary 7.1. *If a connected graph G is a simplex (i.e. any of its two vertices are neighboring see [220]), then any QSO (7.4) is simultaneously Markov and Volterra.*

7.3 Construction of QSO for continual case

Now let $G = (\Lambda, L)$ be a countable graph. For a finite set Φ denote by Ω the set of all functions $\sigma : \Lambda \to \Phi$.

Let $S(\Omega, \Phi)$ be the set of all probability measures defined on $(\Omega, \mathcal{F})$, where $\mathcal{F}$ is the standard $\sigma-$algebra generated by the finite-dimensional cylindrical set.

Let μ be a measure on $(\Omega, \mathcal{F})$ such that $\mu(B) > 0$ for any finite-dimensional cylindrical set $B \in \mathcal{F}$. Note that only Gibbs measure have this property [220].

Fix a finite connected subset $M \subset \Lambda$. Elements $\sigma \in \Omega$ and $\varphi \in \Omega$ are equivalent iff $\sigma(x) = \varphi(x)$ for any $x \in M$, i.e. $\sigma(M) = \varphi(M)$.

Let $\xi = \{\Omega_i, i = 1, 2, ..., |\Phi|^{|M|}\}$, be the partition of Ω generated by this equivalent relation, where $|\cdot|$ denotes the cardinality of the set and Ω_i contains all equivalent elements.

Denote

$$b_{ij,k} = \begin{cases} 1 \;\text{ if }\; i = j = k \\ \frac{\mu(\Omega_k)}{\mu(\Omega_i)+\mu(\Omega_j)}, \;\text{ if }\; k = i \;\text{ or }\; k = j, i \neq j \\ 0 \;\text{ otherwise}, \end{cases} \tag{7.5}$$

where $i, j, k \in \{1, 2, ..., |\Phi|^{|M|}\}$.

Assume

$$p_{\sigma_1\sigma_2,\sigma} = b_{ij,k} \;\text{ if }\; \sigma_1 \in \Omega_i, \sigma_2 \in \Omega_j, \sigma \in \Omega_k. \tag{7.6}$$

Then the heredity coefficients $P(\sigma_1,\sigma_2,A)$ $(\sigma_1,\sigma_2\in\Omega,\quad A\in\mathcal{F})$ are defined as

$$P(\sigma_1,\sigma_2,A)=Z(\sigma_1,\sigma_2)\int_A p_{\sigma_1\sigma_2,\sigma}d\mu(\sigma)$$

$$=Z(\sigma_1,\sigma_2)\cdot\begin{cases}\mu(\Omega_i)\mu(A\cap\Omega_i)+\mu(\Omega_j)\mu(A\cap\Omega_j), & \text{if}\quad \sigma_1\in\Omega_i,\sigma_2\in\Omega_j,i\neq j\\ \mu(A\cap\Omega_i)\ \text{if}\quad \sigma_1,\sigma_2\in\Omega_i.\end{cases}\tag{7.7}$$

where

$$Z(\sigma_1,\sigma_2)=\begin{cases}\frac{1}{\mu^2(\Omega_i)+\mu^2(\Omega_j)}, & \text{if}\quad \sigma_1\in\Omega_i,\sigma_2\in\Omega_j,i\neq j\\ \frac{1}{\mu(\Omega_i)} & \text{if}\quad \sigma_1,\sigma_2\in\Omega_i.\end{cases}\tag{7.8}$$

The QSO V acting on the set $S(\Omega,\Phi)$ and determined by coefficients (7.7) is defined as follows: for an arbitrary measure $\lambda\in S(\Omega,\Phi)$, the measure $\lambda'=V\lambda$ is

$$\lambda'(A)=\int_\Omega\int_\Omega P(\sigma_1,\sigma_2,A)d\lambda(\sigma_1)d\lambda(\sigma_2)=\sum_{i=1}^{|\Phi|^{|M|}}\frac{\mu(A\cap\Omega_i)}{\mu(\Omega_i)}\lambda^2(\Omega_i)$$
$$+2\sum_{1\leq i<j\leq|\Phi|^{|M|}}\frac{\mu(\Omega_i)\mu(A\cap\Omega_i)+\mu(\Omega_j)\mu(A\cap\Omega_j)}{\mu^2(\Omega_i)+\mu^2(\Omega_j)}\lambda(\Omega_i)\lambda(\Omega_j).\tag{7.9}$$

For a given measure $\lambda\in S(\Lambda,\Phi)$ the trajectory $\{\lambda^{(n)}\},n=1,2,...$ of the operator (7.9) is defined by $\lambda^{(n+1)}(A)=V(\lambda^{(n)})(A)$, where $n=0,1,2,...$ and $\lambda^{(0)}=\lambda$, $A\in\mathcal{F}$.

Now we shall investigate the behavior of trajectories of the operator (7.9).

By (7.9) we have

$$\lambda'(\Omega_i)=\lambda^2(\Omega_i)+2\sum_{\substack{j=1\\ j\neq i}}^{m}\frac{\mu^2(\Omega_i)}{\mu^2(\Omega_i)+\mu^2(\Omega_j)}\lambda(\Omega_i)\lambda(\Omega_j),\tag{7.10}$$

where $m=|\Phi|^{|M|}$.

Denoting $\lambda_i=\lambda(\Omega_i)$, $a_{ij}=\frac{\mu^2(\Omega_i)-\mu^2(\Omega_j)}{\mu^2(\Omega_i)+\mu^2(\Omega_j)}$ from (7.10) we get

$$\lambda_i'=\lambda_i\left(1+\sum_{\substack{j=1\\ j\neq i}}^{m}a_{ij}\lambda_j\right),\quad i=1,...,m.\tag{7.11}$$

It is easy to see that $a_{ij}=-a_{ji}$ and $|a_{ij}|<1$.

Note that the $n-$th iteration, i.e., $\lambda^{(n)} = V^{(n)}\lambda^{(0)}$ of the operator (7.9) can be written as

$$\lambda^{(n)}(A) = \sum_{i=1}^{m} \frac{\mu(A\cap\Omega_i)}{\mu(\Omega_i)}(\lambda_i^{(n-1)})^2$$

$$+2\sum_{1\leq i<j\leq m} \frac{\mu(\Omega_i)\mu(A\cap\Omega_i)+\mu(\Omega_j)\mu(A\cap\Omega_j)}{\mu^2(\Omega_i)+\mu^2(\Omega_j)}\lambda_i^{(n-1)}\lambda_j^{(n-1)}, \qquad (7.12)$$

where $\lambda_j^{(n)}$, $j = 1,...,m$ are coordinates of the trajectory of operator (7.11) for the given λ.

Thus in order to study the trajectory of operator (7.9) it suffices to know the behavior of trajectories of the operator (7.11).

After a suitable renumbering of the classes Ω_i we can assume that

$$\mu(\Omega_1) = \mu(\Omega_2) = ... = \mu(\Omega_r) > \mu(\Omega_{r+1}) \geq \mu(\Omega_{r+1}) \geq ... \geq \mu(\Omega_m) > 0, \qquad (7.13)$$

for some integer $r \geq 1$.

Theorem 7.2. *Assume for a given measure μ (7.13) holds then for any $\lambda = (\lambda_1, ..., \lambda_m) \in S^{m-1}$ we have*

$$\lim_{n\to\infty} \lambda^{(n)} = (\lambda_1^*, \lambda_2^*, ..., \lambda_r^*, 0, 0, ..., 0), \qquad (7.14)$$

where $\lambda_i^* = \frac{\lambda_i}{\sum_{j=1}^{r}\lambda_j}$.

Proof. By (7.13) we have

$$\begin{aligned} &a_{ij} = 0 \text{ if } i,j = 1,...,r \text{ and} \\ &a_{ij} = a_{1j} > 0 \text{ if } i = 1,...,r, j = r+1,...,m. \\ &a_{ij} \geq 0 \text{ if } i < j \text{ and} \\ &a_{ij} \leq 0 \text{ if } i > j. \end{aligned} \qquad (7.15)$$

Using (7.15) the operator (7.11) can be rewritten as

$$\begin{cases} \lambda_i' = \lambda_i^{(0)}\left(1 + \sum_{j=r+1}^{m} a_{1j}\lambda_j^{(0)}\right), i = 1,...,r \\ \lambda_k' = \lambda_k^{(0)}\left(1 - \sum_{j=1}^{k-1} a_{jk}\lambda_j^{(0)} + \sum_{j=k+1}^{m} a_{kj}\lambda_j^{(0)}\right), k = r+1,...,m. \end{cases} \qquad (7.16)$$

For arbitrary $\lambda^{(0)} = (\lambda_1^{(0)}, ..., \lambda_m^{(0)})$ we consider two cases.

Case 1: Assume $\prod_{i=1}^{m}\lambda_i^{(0)}\neq 0$, then since $a_{j1}>0$, $j=r+1,...,m$ we have that $\lambda'_i\geq\lambda_i^{(0)}, i=1,...,r$. Consequently $\lambda_i^{(n)}\geq\lambda_i^{(n-1)}, i=1,...,r$ and $\lambda_j^{(n)}\leq\lambda_j^{(n-1)}, j=r+1,...,m$. Thus, there exists the following limit

$$\lambda=\lim_{n\to\infty}\lambda^{(n)}=\lim_{n\to\infty}(\lambda_1^{(n)},...,\lambda_m^{(n)}).$$

Obviously, λ is a fixed point to (7.16). Note that any fixed point of (7.16) has the form $(\lambda_1,\lambda_2,...,\lambda_r,0,0,...0)$. In the other hand we have $\lambda_i^{(n)}/\lambda_j^{(n)}=\lambda_i^{(0)}/\lambda_j^{(0)}$, for any $i,j=1,...,r, n=1,2,...$ Consequently $\lambda_i/\lambda_j=\lambda_i^{(0)}/\lambda_j^{(0)}$, which for $j=1$ implies $\lambda_i=(\lambda_i^{(0)}/\lambda_1^{(0)})\lambda_1$. Using $\sum_{j=1}^{r}\lambda_i=1$ we obtain $\lambda_i=(\lambda_i^{(0)}/\sum_{j=1}^{r}\lambda_j^{(0)}), i=1,...,r$.

Case 2: Assume $\prod_{i=1}^{m}\lambda_i^{(0)}=0$ which implies $\lambda_{i_1}^{(0)}=\lambda_{i_2}^{(0)}=...=\lambda_{i_s}^{(0)}=0$ for some $s\in\{1,...,m-1\}$. In this case operator (7.11) can be reduced to the similar operator which acts on the simplex S^{m-s-1} with $\prod_{j\neq i_1,...,i_s}\lambda_j^{(0)}\neq 0$. Thus case 2 reduced to the case 1. Theorem is proved. □

Remark 7.1. 1. In [83] Theorem 7.2 was proven for $a_{ij}\neq 0$ using theory of tournaments. In (7.16) some a_{ij} can be zero, but a_{ij} has more specific form than in [83].

2. If $r=m$, then $V^2=V$, i.e., V is stationary QSO [161], [162].

Using (7.14) from (7.12) we get the following:

Theorem 7.3. *For any $\lambda\in S(\Omega,\Phi)$ the trajectory $\{\lambda^{(n)}(A)\}$ of operator (7.9) has the following limit*

$$\lambda(A)=\lim_{n\to\infty}\lambda^{(n)}(A)=\sum_{i=1}^{r}\frac{\mu(A\cap\Omega_i)}{\mu(\Omega_i)}(\lambda_i^*)^2$$
$$+2\sum_{1\leq i<j\leq m}\frac{\mu(\Omega_i)\mu(A\cap\Omega_i)+\mu(\Omega_j)\mu(A\cap\Omega_j)}{\mu^2(\Omega_i)+\mu^2(\Omega_j)}\lambda_i^*\lambda_j^*, \quad (7.17)$$

where λ^* defined in Theorem 7.2 and $\lambda_i=\lambda(\Omega_i)$.

Remark 7.2. The QSO V satisfies the ergodic theorem (see [289]) if the limit

$$\lim_{n\to\infty}\frac{1}{n}\sum_{k=0}^{n-1}V^{(k)}\lambda,$$

exists for any $\lambda\in S^{m-1}$. On the basis of numerical calculations Ulam conjectured ([289]) the ergodic theorem holds for any QSO. In [308] it was proven that this conjecture is false in general. From Theorem 7.3 follows that the ergodic theorem holds for any QSO determined by (7.9).

7.4 Measure of the Potts model

In this section we shall apply Theorem 7.3 to the Potts model on $\mathbb{Z}^d$. In this case Ω is the set of all configurations (cells) $\sigma : \mathbb{Z}^d \to \Phi = \{1, ..., q\}$.

The (formal) Hamiltonian of the Potts model is

$$H(\sigma) = -J \sum_{\substack{x,y\in\mathbb{Z}^d \\ \|x-y\|=1}} \delta_{\sigma(x)\sigma(y)},$$

where $J \in \mathbb{R}$.

It is well-known (see [267], Theorem 2.3) that for the Potts model with $q \geq 2$ there exist critical temperature T_{cr} such that for any $T < T_{\text{cr}}$ there are q distinct extreme Gibbs measures μ_i, $i = 1, ..., q$. For T low enough each measure μ_i is a small deviation of the constant configuration $\sigma^{(i)} \equiv i, i = 1, ..., q$. Thus condition (7.13) is satisfied for any measure μ_i with $r = 1$, i.e., $\mu_i(\Omega_i) > \mu_i(\Omega_j)$ for any $j \neq i$, here Ω_i is the set of all configurations $\sigma \in \Omega$ such that $\sigma(x) = i$ for any $x \in M$, where M is a fixed (as above) subset of $\mathbb{Z}^d$.

As corollary of Theorem 7.3 we get

Theorem 7.4. *Any trajectory of QSO (7.9) corresponding to measure μ_i of the Potts model has the following limit*

$$\lim_{n\to\infty} \lambda^{(n)}(A) = \frac{\mu_i(A \cap \Omega_i)}{\mu_i(\Omega_i)}, \quad i = 1, ..., q \tag{7.18}$$

Remark 7.3. By the construction the heredity coefficients (7.7) depend on fixed M. Consider an increasing sequence of connected, finite sets $M_1 \subset M_2 \subset ... \subset M_n \subset ...$ such that $\cup_n M_n = \mathbb{Z}^d$. Fix two configuration σ_1 and σ_2 and consider sequence of boundary conditions: $\sigma_1(\mathbb{Z}^d \backslash M_k)$ or $\sigma_2(\mathbb{Z}^d \backslash M_k), k = 1, 2, \ldots$ These sequences of subsets M_k and boundary conditions (for the Potts model) define a sequence of conditional Gibbs distributions μ_k on Ω (see [220], [267]).

For any measurable set A we denote by $P^{(n)}(\sigma_1, \sigma_2, A)$ the heredity coefficients (7.7) which is constructed by M_n and μ_n. An interesting problem is to describe the set of all configurations σ_1, σ_2 such that the following limit exist

$$P(\sigma_1, \sigma_2, A) = \lim_{n\to\infty} P^{(n)}(\sigma_1, \sigma_2, A). \tag{7.19}$$

Note that if σ_1 and σ_2 are equal almost sure (i.e., the set $\{x \in \mathbb{Z}^d : \sigma_1(x) \neq \sigma_2(x)\}$ is finite) then the limit (7.19) exists. It is easy to see that the limit does not exist if $d = 1$, and

$$\sigma_1 = \{..., 1, 1, 2, 2, \sigma_1(-1) = 1, \sigma_1(0) = 1, \sigma_1(1) = 1, 2, 2, 1, 1, ...\};$$

$$\sigma_2 = \{..., 2, 2, 1, 1, 2, \sigma_2(-1) = 2, \sigma_2(0) = 1, \sigma_2(1) = 2, 2, 1, 1, 2, ...\}.$$

This example can be easily extended to case $d > 1$.

Conjecture. The set of configurations (σ_1, σ_2) such that limit (7.19) does not exist has $\mu \times \mu$ measure zero.

Remark 7.4. The RHS of (7.18) is the conditional probability $\mu_i(A|\Omega_i)$. Note (see [267]) that $\mu_i(\Omega_i) \to 1$ if $T \to 0$. Thus for T low enough, from Theorem 7.4 follows that trajectory of any measure $\lambda \in S(\Lambda, \Phi)$ with respect to QSO constructed by μ_i tends to μ_i. In other words, any trajectory of the quadratic stochastic operator generated by a Gibbs measure μ of the Potts model converges to this measure. These investigations allows a natural introduction of thermodynamics in studying some models of heredity.

7.5 Comments and references for Part II

This part of the book is mainly based on results of recently written papers [87], [242], [246], [248], [251]-[253].

Since the theory of Gibbs measures for the Hamiltonians (models) of DNA systems is not developed yet, we were only able to obtain some results about translation-invariant Gibbs measures of the DNA models and Holliday junctions. Therefore, investigations of such models can be continued in many distinct directions. For example, one can define periodic Gibbs measures for such models and study periodic solutions of the equation giving compatibility of finite dimensional Gibbs measures. Moreover, these models can be investigated on other lattices.

To choose another lattice one can follow [145], which reviews creating a genetic code for crystal engineering. These codes are useful to mathematical modeling of the lattice on which we want to study DNA crystals. Because, rather than directing biological processes, these codes use synthetic forms of DNA to programme the assembly of nano-particles and micro-particles into crystalline architectures, in which almost every aspect of the resultant structures can be systematically controlled. The structural and functional advances are described in an effort to define the present level of sophistication and to predict future directions. In particular, these advances include exotic structures with programmable lattice symmetries and well-defined crystal habits, responsive materials that leverage the intrinsic properties of nucleic acids to manipulate structures on demand.

I hope the methods and results used in this part of the book will open a new direction in the study of systems of interacting DNAs and RNAs.

In a fundamental work [67] it was suggested that the transient opening of adjacent base pairs of a DNA, associated with torsional oscillations around the helix axis, could generate coherent thermally activated soliton excitations propagating along the DNA backbone [309]. These observations have fostered a rich line of research on the nonlinear dynamics of DNA which continues nowadays (see [57], [309] and references therein).

For further reading see [10], [40], [41], [43], [53], [63], [66], [127], [128], [135], [151], [169], [181], [183], [190], [216], [234], [242], [262], [282], [284], [288] and the references therein.

PART III

Thermodynamics in physics: The Potts model

Chapter 8

Full description of translation-invariant Gibbs measures for the Potts model on trees

In this chapter, for the q-state Potts model on a Cayley tree of order $k \geq 2$ we give *all* translation-invariant Gibbs measures. Such measures are called translation-invariant splitting Gibbs measures (TISGMs). In particular, it is shown that at sufficiently low temperatures their number is $2^q - 1$. It is proved that there are $[q/2]$ (where $[a]$ is the integer part of a) critical temperatures at which the number of TISGMs changes and the exact number of TISGMs for each intermediate temperature is calculated. In case $k = 2$ explicit formulae for the critical temperatures and the possible TISGMs are given. While we show that these measures are never convex combinations of each other, the question which of these measures are extremals in the set of all Gibbs measures will be treated in the next chapter. In the last section of this chapter for each TISGM μ we construct the set of boundary conditions such that limiting Gibbs measures with respect to these boundary conditions coincide with μ.

8.1 Translation-invariant SGMs

Consider a Cayley tree $\Gamma^k = (V, L)$ (see Subsection 2.1.1).

Recall the basic Hamiltonian of Potts model:

$$H(\sigma) = -J \sum_{\langle x,y\rangle \in L} \delta_{\sigma(x)\sigma(y)}, \tag{8.1}$$

where $J \in \mathbb{R}$, $\sigma(x) \in \Phi = \{1, \ldots, q\}$ and $\langle x, y\rangle$ stands for nearest neighbor vertices.

For the ferromagnetic Potts model in [94], [95] it was proved that for some temperatures the model possesses at least $q + 1$ TISGMs and an uncountable number of extremal non-TISGM. Here following [138] we show that the maximal number of TISGMs is $2^q - 1$. For other results related to the Potts model on Cayley trees see [4], [89], [109], [110], [293], [300] and the references therein.

Let S_q be the group of permutations on Φ.

Take $\sigma \in \Phi^V$ and define an action of $\pi = (\pi_1, \dots, \pi_q) \in S_q$ on σ (denoted by $\pi\sigma$) as

$$(\pi\sigma)(x) = \pi_{\sigma(x)}, \qquad \text{for all } \ x \in V.$$

The following lemma gives the permutation symmetry of Hamiltonian (8.1).

Lemma 8.1. *For any $\sigma \in \Phi^V$ and any $\pi \in S_q$ we have $H(\pi\sigma) = H(\sigma)$.*

Proof. Since π is one-to-one map on Φ, we have $\delta_{ij} = \delta_{\pi_i \pi_j}$ for any i, j. This completes the proof. □

By Theorem 2.2 to each splitting Gibbs measure (SGM) of the Hamiltonian (8.1) there is a vector-valued function h_x, such that for any $x \in V \setminus \{x^0\}$ the following equation holds:

$$h_x = \sum_{y \in S(x)} F(h_y, \theta), \tag{8.2}$$

where $F : h = (h_1, \dots, h_{q-1}) \in \mathbb{R}^{q-1} \to F(h, \theta) = (F_1, \dots, F_{q-1}) \in \mathbb{R}^{q-1}$ is defined as

$$F_i = \ln \left(\frac{(\theta - 1)e^{h_i} + \sum_{j=1}^{q-1} e^{h_j} + 1}{\theta + \sum_{j=1}^{q-1} e^{h_j}} \right),$$

$\theta = \exp(J\beta)$, $S(x)$ is the set of direct successors of x and $h_x = (h_{1,x}, \dots, h_{q-1,x})$ with

$$h_{i,x} = \tilde{h}_{i,x} - \tilde{h}_{q,x}, \quad i = 1, \dots, q-1. \tag{8.3}$$

Moreover, for any $h = \{h_x, \quad x \in V\}$ satisfying (8.2) there exists a unique SGM μ for the Potts model.

In this section, we follow [138] and consider SGMs which are translation-invariant, i.e., assume $h_x = h = (h_1, \dots, h_{q-1}) \in \mathbb{R}^{q-1}$ for all $x \in V$. Then from equation (8.2) we get $h = kF(h, \theta)$, i.e.,

$$h_i = k \ln \left(\frac{(\theta - 1)e^{h_i} + \sum_{j=1}^{q-1} e^{h_j} + 1}{\theta + \sum_{j=1}^{q-1} e^{h_j}} \right), \quad i = 1, \dots, q-1. \tag{8.4}$$

Denoting $z_i = \exp(h_i), i = 1, \dots, q-1$, from (8.4) we get

$$z_i = \left(\frac{(\theta - 1)z_i + \sum_{j=1}^{q-1} z_j + 1}{\theta + \sum_{j=1}^{q-1} z_j} \right)^k, \quad i = 1, \dots, q-1. \tag{8.5}$$

Remark 8.1. The permutation symmetry of Hamiltonian (8.1) (see Lemma 8.1) consequents such symmetry to solutions of (8.5): if $z = (z_1, \dots, z_{q-1})$ is a solution to (8.5) then by any permutation of its coordinates we get a solution of (8.5) too.

The following theorem characterizes all solutions of (8.5).

Theorem 8.1. *For any solution $z = (z_1, \dots, z_{q-1})$ of the system of equations (8.5) there exists $M \subset \{1, \dots, q-1\}$ and $z^* > 0$ such that*

$$z_i = \begin{cases} 1, & \text{if } i \notin M \\ z^*, & \text{if } i \in M. \end{cases}$$

Proof. It is easy to see that $z_i = 1$ is a solution of ith equation of the system (8.5) for each $i = 1, 2, \dots, q-1$. Thus for a given $M \subset \{1, \dots, q-1\}$ one can take $z_i = 1$ for any $i \notin M$. Let $\emptyset \neq M \subset \{1, \dots, q-1\}$, without loss of generality (see Remark 8.1) we can take $M = \{1, 2, \dots, m\}$, $m \leq q-1$, i.e., $z_i = 1$, $i = m+1, \dots, q$. Now we shall prove that $z_1 = z_2 = \dots = z_m$.

Denote $\sqrt[k]{z_i} = x_i$, $i = 1, \dots, m$. Then from (8.5) we have

$$x_i = \frac{(\theta - 1)x_i^k + \sum_{j=1}^m x_j^k + q - m}{\sum_{j=1}^m x_j^k + q - m - 1 + \theta}, \quad i = 1, \dots, m. \tag{8.6}$$

By assumption $x_i \neq 1, i = 1, 2, ..., m$, since $x_i > 0$, from (8.6) we get

$$\theta - 1 = \frac{(x_i - 1)\left(\sum_{j=1}^m x_j^k + q - m\right)}{x_i^k - x_i}$$
$$= \frac{\sum_{j=1}^m x_j^k + q - m}{x_i(x_i^{k-2} + x_i^{k-3} + \dots + 1)}, \quad i = 1, \dots, m.$$

From these equations we get

$$x_i(x_i^{k-2} + x_i^{k-3} + \dots + 1) = x_j(x_j^{k-2} + x_j^{k-3} + \dots + 1),$$

which gives $x_i = x_j$ for any $i, j \in \{1, \dots, m\}$, because

$$p(x) = x(x^{k-2} + x^{k-3} + \dots + 1),$$

is an increasing function of $x > 0$. □

By this theorem we have that any TISGM of the Potts model corresponds to a solution of the following equation

$$z = f_m(z) \equiv \left(\frac{(\theta + m - 1)z + q - m}{mz + q - m - 1 + \theta}\right)^k, \tag{8.7}$$

for some $m = 1, \dots, q-1$.

Lemma 8.2. *If $z(m_1)$ is a solution to (8.7) with $m = m_1$ then $z^{-1}(m_1)$ is a solution to (8.7) with $m = q - m_1$.*

Proof. We have

$$f_m(z) = \left(\frac{(\theta+m-1)z+q-m}{mz+q-m-1+\theta}\right)^k = \left(\frac{(\theta+m-1)+(q-m)z^{-1}}{m+(q-m-1+\theta)z^{-1}}\right)^k$$
$$= \left(\frac{m+(q-m-1+\theta)z^{-1}}{(\theta+m-1)+(q-m)z^{-1}}\right)^{-k} = \frac{1}{f_{q-m}(z^{-1})}.$$

Therefore, from $z = f_m(z)$ it follows that

$$\frac{1}{z} = \frac{1}{f_m(z)} = f_{q-m}(\frac{1}{z}).$$

This completes the proof. □

Let $M \subset \{1, \ldots, q-1\}$, with $|M| = m$. Then we denote the corresponding solution of (8.7) by $z(M) = \exp(h(M))$. It is clear that $h(M) = h(m)$, i.e., it only depends on the cardinality of M. Put

$$\mathbf{1}_M = (e_1, \ldots, e_q), \quad \text{with } \ e_i = 1 \ \text{ if } \ i \in M, \quad e_i = 0 \ \text{ if } \ i \notin M.$$

Denote by $\mu_{h(M)\mathbf{1}_M}$ the TISGM corresponding to the solution $h(M)$.

Remark 8.2. By formula (8.3) one has

$$\tilde{h}_i(M) = \ln(\tilde{z}_i(M)) = \begin{cases} h(M) + \tilde{h}_q(M), & \text{if } \ i \in M \\ \tilde{h}_q(M), \ \text{ if } \ i \notin M \end{cases},$$

i.e.,

$$\tilde{h}(M)\mathbf{1}_M = h(M)\mathbf{1}_M + \tilde{h}_q(M)\mathbf{1}_{\{1,\ldots,q\}}.$$

Hence for a given M, $|M| = m$ and a solution $h(M)$ the number of vectors $\tilde{h}(M)\mathbf{1}_M$ is equal to $\binom{q}{m}$.

The following proposition is useful.

Proposition 8.1. *For any finite $\Lambda \subset V$ and any $\sigma_\Lambda \in \{1, \ldots, q\}^\Lambda$ we have*

$$\mu_{h(M)\mathbf{1}_M}(\sigma_\Lambda) = \mu_{h(M^c)\mathbf{1}_{M^c}}(\sigma_\Lambda), \tag{8.8}$$

where $M^c = \{1, \ldots, q\} \setminus M$ and $h(M^c) = -h(M)$.

Proof. Let $\mathbb{I}$ be the indicator function. Then we have $\mathbb{I}(x \in M) + \mathbb{I}(x \in M^c) = 1$. By Lemma 8.2 we get

$$1/z(M) = z(M^c), \quad \text{i.e.,} \quad -h(M) = h(M^c).$$

Using these formulas and Remark 8.2 for $\partial\Lambda = \{x \in V \setminus \Lambda : \exists y \in \Lambda, \langle x, y\rangle\}$ we get

$$\begin{aligned}
\mu_{h(M)\mathbf{1}_M}(\sigma_\Lambda) &= \frac{\exp\left(-\beta H(\sigma_\Lambda)\right)(\tilde z_q(M))^{|\partial\Lambda|}(z(M))^{\sum_{x\in\partial\Lambda}\mathbb{I}(\sigma(x)\in M)}}{\sum_{\varphi_\Lambda}\exp\left(-\beta H(\varphi_\Lambda)\right)(\tilde z_q(M))^{|\partial\Lambda|}(z(M))^{\sum_{x\in\partial\Lambda}\mathbb{I}(\varphi(x)\in M)}}\\
&= \frac{\exp\left(-\beta H(\sigma_\Lambda)\right)(z(M))^{|\partial\Lambda|-\sum_{x\in\partial\Lambda}\mathbb{I}(\sigma(x)\in M^c)}}{\sum_{\varphi_\Lambda}\exp\left(-\beta H(\varphi_\Lambda)\right)(z(M))^{|\partial\Lambda|-\sum_{x\in\partial\Lambda}\mathbb{I}(\varphi(x)\in M^c)}}\\
&= \frac{\exp\left(-\beta H(\sigma_\Lambda)\right)(1/z(M))^{\sum_{x\in\partial\Lambda}\mathbb{I}(\sigma(x)\in M^c)}}{\sum_{\varphi_\Lambda}\exp\left(-\beta H(\varphi_\Lambda)\right)(1/z(M))^{\sum_{x\in\partial\Lambda}\mathbb{I}(\varphi(x)\in M^c)}}\\
&= \frac{\exp\left(-\beta H(\sigma_\Lambda)\right)(\tilde z_q(M^c))^{|\partial\Lambda|}(z(M^c))^{\sum_{x\in\partial\Lambda}\mathbb{I}(\sigma(x)\in M^c)}}{\sum_{\varphi_\Lambda}\exp\left(-\beta H(\varphi_\Lambda)\right)(\tilde z_q(M^c))^{|\partial\Lambda|}(z(M^c))^{\sum_{x\in\partial\Lambda}\mathbb{I}(\varphi(x)\in M^c)}}\\
&= \mu_{h(M^c)\mathbf{1}_{M^c}}(\sigma_\Lambda).
\end{aligned} \tag{8.9}$$

□

The following is a corollary of Theorem 8.1 and Proposition 8.1.

Corollary 8.1. *Each TISGM corresponds to a solution of (8.7) with some $m \le [q/2]$, where $[a]$ is the integer part of a. Moreover, for a given $m \le [q/2]$, a fixed solution to (8.7) generates $\binom{q}{m}$ vectors $\tilde h$ giving $\binom{q}{m}$ TISGMs.*

Case: $k = 2$. In this case we find explicit solutions of (8.7). Denoting $x = \sqrt{z}$ from (8.7) we get

$$mx^3 - (\theta + m - 1)x^2 + (\theta + q - m - 1)x - q + m = 0. \tag{8.10}$$

Note that $x = 1$ is a solution to (8.10). Dividing the left hand side of this equation by $x - 1$ we obtain

$$mx^2 - (\theta - 1)x + q - m = 0. \tag{8.11}$$

This equation has solutions

$$\begin{aligned}
x_1(m) &= \frac{\theta - 1 - \sqrt{(\theta-1)^2 - 4m(q-m)}}{2m},\\
x_2(m) &= \frac{\theta - 1 + \sqrt{(\theta-1)^2 - 4m(q-m)}}{2m},
\end{aligned} \tag{8.12}$$

if $(\theta - 1)^2 - 4m(q - m) \ge 0$, i.e.,

$$\theta \ge \theta_m = 1 + 2\sqrt{m(q-m)}, \quad m = 1, \dots, q-1. \tag{8.13}$$

Lemma 8.3. *If $x > 0$ is a common solution to equations (8.11) for different values of m, then $x = 1$.*

Proof. Let $x > 0$ be a common solution to equations (8.11) for different values of m, say m_1 and m_2. Then for this x we should have

$$m_1x^2 - (\theta - 1)x + q - m_1 = 0, \quad m_2x^2 - (\theta - 1)x + q - m_2 = 0.$$

Subtracting from the first equation the second one we obtain

$$(m_1 - m_2)(x^2 - 1) = 0.$$

Hence, if $m_1 \neq m_2$ then we have only $x = 1$. □

Remark 8.3. We note that $x_1(m) \neq 1$ and $x_2(m) \neq 1$ if $\theta \neq q + 1$. Moreover, if $\theta = q + 1$ then

$$x_1(m) = \begin{cases} 1, & \text{if } q \geq 2m \\ \frac{q}{m} - 1, & \text{if } q < 2m; \end{cases} \qquad x_2(m) = \begin{cases} 1, & \text{if } q \leq 2m \\ \frac{q}{m} - 1, & \text{if } q > 2m. \end{cases} \tag{8.14}$$

In particular, $x_1(m) = x_2(m) = 1$ if $q = 2m$.

It is easy to see that

$$\theta_m = \theta_{q-m} \text{ and } \theta_1 < \theta_2 < \cdots < \theta_{[\frac{q}{2}]-1} < \theta_{[\frac{q}{2}]} \leq q + 1. \tag{8.15}$$

Thus we have the following

Proposition 8.2. *Let $k = 2$, $J > 0$.*

1. *If $\theta < \theta_1$ then the system of equations (8.4) has a unique solution $h_0 = (0, 0, \ldots, 0)$;*
2. *If $\theta_m < \theta < \theta_{m+1}$ for some $m = 1, \ldots, [\frac{q}{2}] - 1$ then the system of equations (8.4) has solutions*

$$h_0 = (0, 0, \ldots, 0), \quad h_{1i}(s), \quad h_{2i}(s), \quad i = 1, \ldots, \binom{q-1}{s},$$

$$h'_{1i}(q - s), \quad h'_{2i}(q - s), \quad i = 1, \ldots, \binom{q-1}{q-s}, \quad s = 1, 2, \ldots, m,$$

where $h_{ji}(s)$, (resp. $h'_{ji}(q - s)$) $j = 1, 2$ is a vector with s (resp. $q - s$) coordinates equal to $2 \ln x_j(s)$ (resp. $2 \ln x_j(q - s)$) and the remaining $q-s-1$ (resp. $s-1$) coordinates equal to 0. The number of such solutions is equal to

$$1 + 2\sum_{s=1}^{m} \binom{q}{s};$$

3. *If $\theta_{[\frac{q}{2}]} < \theta \neq q + 1$ then there are $2^q - 1$ solutions to (8.4);*

4. *If $\theta = q+1$ then the number of solutions is as follows*

$$\begin{cases} 2^{q-1}, & \text{if } q \text{ is odd} \\ 2^{q-1} - \binom{q-1}{q/2}, & \text{if } q \text{ is even;} \end{cases}$$

5. *If $\theta = \theta_m$, $m = 1, \dots, [\frac{q}{2}]$, $(\theta_{[\frac{q}{2}]} \neq q+1)$ then $h_{1i}(m) = h_{2i}(m)$. The number of solutions is equal to*

$$1 + \binom{q}{m} + 2\sum_{s=1}^{m-1} \binom{q}{s}.$$

Proof. 1. Straightforward.

2. Fix $m \in \{1, \dots, q-1\}$. Then for each $M \subset \{1, \dots, q-1\}$ with cardinality $q-m-1$, we can have two solutions (8.12). Note that m is the number of different from 0 coordinates of a solution to the system (8.4). By (8.12)–(8.15) the number of such solutions (vectors) is equal to two times of

$$\binom{q-1}{m} + \binom{q-1}{q-m} = \binom{q}{m}.$$

3. If $\theta > \theta_{[q/2]}$ and $\theta \neq q+1$ then the number of solutions depending on the parity of q is as follows:

$$\begin{cases} 1 + 2\sum_{s=1}^{[q/2]} \binom{q}{s}, & \text{if } q \text{ is odd} \\ 1 + \binom{q}{q/2} + 2\sum_{s=1}^{q/2-1} \binom{q}{s}, & \text{if } q \text{ is even} \end{cases} = 2^q - 1, \tag{8.16}$$

where we used the following formulas

$$\binom{q}{s} = \binom{q}{q-s}, \quad \sum_{s=1}^{q} \binom{q}{s} = 2^q - 1.$$

4. For $\theta = q+1$ we use (8.14), Lemma 8.3, Remark 8.3, (8.15) and (8.16) to get the following formulas for the number of solutions:

q is odd:

$$1 + \sum_{s=1}^{[q/2]} \binom{q}{s} = 2^{q-1};$$

q is even:

$$1 + \sum_{s=1}^{q/2-1} \binom{q}{s} = 2^{q-1} - \binom{q-1}{q/2},$$

in the last formula we used $\frac{1}{2}\binom{q}{q/2} = \binom{q-1}{q/2}$.

5. On critical point θ_m the equation (8.7) has a unique solution. So omitting a factor 2 in front of $\binom{q}{m}$ we get the formula in a similar way as in case 2. □

To illustrate Proposition 8.2 we consider $q = 4$, then the system of equations (8.4) has the following solutions:

(a) If $\theta < \theta_1$ then $h_0 = (0, 0, 0)$ is the unique solution.
(b) If $\theta = \theta_1$ then there are 5 solutions

$$h_0, \quad h_{11}(1) = (2\ln x_1(1), 0, 0), \quad h_{12}(1) = (0, 2\ln x_1(1), 0),$$

$$h_{13}(1) = (0, 0, 2\ln x_1(1)).$$

$$h_{11}(3) = (2\ln x_1(3), 2\ln x_1(3), 2\ln x_1(3)).$$

(c) If $\theta_1 < \theta < \theta_2$ then there are 9 solutions

$$h_0, \quad h_{11}(1), \quad h_{12}(1), \quad h_{13}(1), \quad h_{11}(3),$$

$$h_{21}(3) = (2\ln x_2(3), 2\ln x_2(3), 2\ln x_2(3)),$$

$$h_{21}(1) = (2\ln x_2(1), 0, 0), \quad h_{22}(1) = (0, 2\ln x_2(1), 0),$$

$$h_{23}(1) = (0, 0, 2\ln x_2(1)).$$

(d) If $\theta_1 < \theta = \theta_2$ then there are $2^{q-1} - \binom{q-1}{q/2} = 5$ solutions

$$h_0, \quad h_{21}(1), \quad h_{22}(1), \quad h_{23}(1), \quad h_{11}(3).$$

(e) If $\theta_2 < \theta$ then there are 15 solutions

$$h_0, \quad h_{11}(1), \quad h_{12}(1), \quad h_{13}(1), \quad h_{21}(1), \quad h_{22}(1),$$

$$h_{23}(1), \quad h_{11}(3), \quad h_{21}(3), \quad h_{11}(2) = (2\ln x_1(2), 2\ln x_1(2), 0),$$

$$h_{12}(2) = (2\ln x_1(2), 0, 2\ln x_1(2)), \quad h_{13}(2) = (0, 2\ln x_1(2), 2\ln x_1(2)).$$

$$h_{21}(2) = (2\ln x_2(2), 2\ln x_2(2), 0),$$

$$h_{22}(2) = (2\ln x_2(2), 0, 2\ln x_2(2)), \quad h_{23}(2) = (0, 2\ln x_2(2), 2\ln x_2(2)).$$

Remark 8.4. We note that

(1) By Proposition 8.2 for $k = 2$ and $J > 0$ we have the full description of solutions to the system of equations (8.4). Consequently, this gives the full description of TISGMs. Moreover, by Remark 8.2 and Corollary 8.1, depending on parameter θ the maximal number of such measures can be $2^q - 1$.

(2) The well-known critical value of θ which implies non-uniqueness of Gibbs measure of the Potts model is $\theta_c = \frac{k+q-1}{k-1}$, $k \geq 2$, $q \geq 2$ (see [94], [95], [243]). For $k = 2$ by (8.13) we have

$$\begin{cases} \theta_c > \theta_m, & \text{for all } \ m \in \{1, \dots, q-1\} \setminus \{q/2\} \\ \theta_c = \theta_m, & \text{for } \ m = q/2. \end{cases}$$

Case $k \geq 2$.

In the following theorems we consider any $k \geq 2$.

Theorem 8.2. *For the q-state anti-ferromagnetic ($J < 0$) Potts model on the Cayley tree of order $k \geq 2$ there is unique TISGM.*

Proof. This theorem is known (see [236] or Theorem 5.2 of [243]). Here we give a new proof. In case $J < 0$, i.e., $\theta < 1$, it suffices to show that for each m the equation (8.7) has unique solution $z = 1$. This equation can be rewritten in the following form

$$(x-1)\varphi_m(x,\theta) = 0, \quad \text{with } \ x = \sqrt[k]{z}, \tag{8.17}$$

where

$$\varphi_m(x,\theta) = mx^k - (\theta - 1)(x^{k-1} + x^{k-2} + \cdots + x) + q - m. \tag{8.18}$$

Indeed, denoting $x = \sqrt[k]{z}$ from (8.7) we get

$$x = \frac{(\theta + m - 1)x^k + q - m}{mx^k + q - m - 1 + \theta} \quad \Leftrightarrow \quad x - 1 = \frac{(\theta - 1)(x^k - 1)}{mx^k + q - m - 1 + \theta}$$

$$\Leftrightarrow \quad (x-1)(mx^k + q - m - 1 + \theta) = (\theta - 1)(x-1)(x^{k-1} + \cdots + 1)$$

$$\Leftrightarrow \quad (x-1)\varphi_m(x,\theta) = 0.$$

For $\theta < 1$ we have $\varphi_m(x,\theta) > 0$, hence the equation (8.17) has unique solution $x = 1$. Therefore, $z = x^k = 1$ is unique solution to (8.7), for any m and any $\theta < 1$. □

Put

$$T_{cr} = \frac{J}{\ln\left(1 + \frac{q}{k-1}\right)}.$$

Theorem 8.3. *For the q-state ferromagnetic ($J > 0$) Potts model on the Cayley tree of order $k \geq 2$ there are critical temperatures $T_{c,m} \equiv T_{c,m}(k,q)$, $m = 1, \dots, [q/2]$ such that the following statements hold*

1.

$$T_{c,1} > T_{c,2} > \cdots > T_{c,[\frac{q}{2}]-1} > T_{c,[\frac{q}{2}]} \geq T_{cr};$$

2. If $T > T_{c,1}$ then there exists a unique TISGM;

3. If $T_{c,m+1} < T < T_{c,m}$ for some $m = 1, \ldots, [\frac{q}{2}] - 1$ then there are $1 + 2\sum_{s=1}^{m} \binom{q}{s}$ TISGMs.

4. If $T_{cr} \neq T < T_{c,[\frac{q}{2}]}$ then there are $2^q - 1$ TISGMs.

5. If $T = T_{cr}$ then the number of TISGMs is as follows

$$\begin{cases} 2^{q-1}, & \text{if } q-\text{odd} \\ 2^{q-1} - \binom{q-1}{q/2}, & \text{if } q-\text{even}. \end{cases};$$

6. If $T = T_{c,m}$, $m = 1, \ldots, [\frac{q}{2}]$, $(T_{c,[q/2]} \neq T_{cr})$ then there are $1 + \binom{q}{m} + 2\sum_{s=1}^{m-1} \binom{q}{s}$ TISGMs.

Proof. We divide the proof to several steps:

Step 1. First we shall show that for each m the equation (8.7) has up to three solutions. This equation can be rewritten as (8.17). It is well known (see [221]) that the number of positive roots of a polynomial does not exceed the number of sign changes of its coefficients. Using this for φ_m (defined in (8.18)) we see that the equation

$$\varphi_m(x, \theta) = 0 \tag{8.19}$$

has up to two positive solutions.

Step 2. In this step we shall check that Lemma 8.3 is also true for $k \geq 3$, i.e., if $x > 0$ is a common solution to equation (8.19) for different values of m then $x = 1$. Indeed, if $x > 0$ is a solution to (8.19) for $m = m_1$ and $m = m_2$ with $m_1 \neq m_2$. Then subtracting one equation from the second one we get $(m_1 - m_2)(x^k - 1) = 0$, i.e $x = 1$.

Moreover, it is easy to see that $x = 1$ is a solution to (8.19) iff

$$\theta = \theta_c = \frac{k + q - 1}{k - 1}.$$

As mentioned above the equation has up to two solutions, assume now one of the solutions is $x = 1$ then dividing the polynomial $\varphi_m(x, \theta_c)$ by $x - 1$ we get

$$\psi_m(x) = \frac{\varphi_m(x, \theta_c)}{x - 1} = m\sum_{i=0}^{k-1} x^i - \frac{q}{k-1}\sum_{i=0}^{k-2}(k - i - 1)x^i.$$

Thus the second solution of (8.19) will be 1 if $\psi_m(1) = 0$ which holds iff $q = 2m$.

Step 3. Now we shall prove that for each m there exists a unique critical temperature $T_{c,m}$. A critical value θ_m of the parameter θ is the value such that the equation $\varphi_m(x,\theta_m)=0$ has a unique solution x. It is easy to see that this critical value satisfies the following system

$$\begin{cases} \varphi_m(x,\theta_m)=0 \\ \varphi_m'(x,\theta_m)=mkx^{k-1}-(\theta_m-1)((k-1)x^{k-2}+(k-2)x^{k-3}+\cdots+1)=0. \end{cases} \tag{8.20}$$

From the first equation of the system (8.20) we get

$$\theta_m-1=\frac{mx^k+q-m}{x^{k-1}+x^{k-2}+\cdots+x}. \tag{8.21}$$

Substituting this to the second equation of (8.20) we get

$$m\sum_{i=1}^{k-1} ix^{2k-i-1}-(q-m)\sum_{i=1}^{k-1} ix^{i-1}=0. \tag{8.22}$$

By the above-mentioned property of polynomials we see that the equation (8.22) has a unique positive solution, say $x=x_*=x_*(m)$. Indeed, the number of sign changes of coefficients in (8.22) is equal to one. Moreover, for $x=0$ the LHS of (8.22) is negative and for x large enough it is positive. Therefore, (8.22) has unique positive solution.

Then for each m the unique critical value θ_m is given by (8.21), i.e.,

$$\theta_m=1+\frac{mx_*^k+q-m}{x_*^{k-1}+x_*^{k-2}+\cdots+x_*}. \tag{8.23}$$

The critical temperature is $T_{c,m}=J\ln^{-1}(\theta_m)$. The following equality holds

$$\varphi_m(x,\theta)=x^k\varphi_{q-m}(x^{-1},\theta),$$

from which we get $T_{c,m}=T_{c,q-m}$. Hence it remains critical values for $m=1,\dots,[q/2]$.

Step 4. Now we shall show that the critical temperature $T_{c,m}$ is a decreasing function of $m=1,\dots,[q/2]$, so they are distinct.

In order to show that x_* is a decreasing function of m, we find m from (8.22) as a function of x_*, i.e.

$$m=\xi(x_*),\quad \text{with}\quad \xi(x)=q\cdot\frac{\sum_{i=1}^{k-1} ix^{i-1}}{\sum_{i=1}^{k-1} i\left(x^{i-1}+x^{2k-i-1}\right)}.$$

We have

$$\xi'(x)=-q\cdot\frac{\sum_{i=1}^{k-1}\sum_{j=1}^{k-1} ij(2k-i-j)x^{2k+i-j-3}}{\left(\sum_{i=1}^{k-1} i\left(x^{i-1}+x^{2k-i-1}\right)\right)^2}<0.$$

Hence m is a decreasing function of x_*. Consequently, $x_* = \xi^{-1}(m)$ is a decreasing function of m.

Now let us write θ_m as function of x_*, i.e. substitute $m = \xi(x_*)$ in (8.23):

$$\theta_m = \eta(x_*) = 1 + \frac{\xi(x_*)(x_*^k - 1) + q}{\sum_{i=1}^{k-1} x_*^i}. \tag{8.24}$$

We have

$$\eta'(x) = \left(\sum_{i=1}^{k-1} x^i\right)^{-2} \times \left\{\xi'(x)(x^k-1)\sum_{i=1}^{k-1} x^i + \xi(x)\sum_{i=1}^{k-1}\left((k-i)x^{k+i-1} + ix^{i-1}\right) - q\sum_{i=1}^{k-1} ix^{i-1}\right\}. \tag{8.25}$$

Substituting the formulas for $\xi(x)$ and $\xi'(x)$ in (8.25) we get

$$\eta'(x) = -\frac{q}{\left(\sum_{i=1}^{k-1} x^i\right)^2 \left(\sum_{i=1}^{k-1} i\left(x^{i-1} + x^{2k-i-1}\right)\right)^2} \times \left[(x^k-1)\left(\sum_{i=1}^{k-1} x^i\right)\left(\sum_{i=1}^{k-1}\sum_{j=1}^{k-1} ij(2k-i-j)x^{2k+i-j-3}\right) + \left(\sum_{i=1}^{k-1} ix^{i-1}\right)\left(\sum_{i=1}^{k-1} i\left(x^{i-1}+x^{2k-i-1}\right)\right) \times \left(\sum_{i=1}^{k-1}\left(ix^{2k-i-1} - (k-i)x^{k+i-1}\right)\right)\right]. \tag{8.26}$$

We have

$$\sum_{i=1}^{k-1}\left(ix^{2k-i-1} - (k-i)x^{k+i-1}\right) = \sum_{i=1}^{k-1} ix^{2k-i-1} - \sum_{i'=1}^{k-1} i'x^{2k-i'-1} = 0,$$

where $i' = k - i$. Hence from (8.26) we obtain

$$\eta'(x) = -q \cdot \frac{(x^k-1)\left(\sum_{i=1}^{k-1}\sum_{j=1}^{k-1} ij(2k-i-j)x^{2k+i-j-3}\right)}{\left(\sum_{i=1}^{k-1} x^i\right)\left(\sum_{i=1}^{k-1} i\left(x^{i-1}+x^{2k-i-1}\right)\right)^2}. \tag{8.27}$$

Note that $\xi(1) = q/2$, thus for $m \in \{1, \dots, [q/2]\}$ we have $x_*(m) > 1$. Thus for $x > 1$ by (8.27) we have $\eta'(x) < 0$. Consequently, θ_m is a decreasing function of x_*. Since x_* is a decreasing function of m we conclude that θ_m is an increasing function of $m \in \{1, \dots, [q/2]\}$. Moreover,

$$\max_{x \geq 1} \eta(x) = \eta(1) = \theta_c = 1 + \frac{q}{k-1}.$$

Hence using formula $\theta_m = \exp(J/T_{c,m})$ we get

$$T_{c,1} > T_{c,2} > \cdots > T_{c,[\frac{q}{2}]} \geq T_{cr}.$$

Step 5. In this last step we calculate the number of TISGMs. In above steps we showed that if $T > T_{c,m}$ then the equation (8.7) has the unique solution $x = 1$, if $T = T_{c,m}$ then this equation has two solutions $x = 1$ and $x = x'$. If $T < T_{c,m}$ then the equation (8.7) has three solutions $x = 1$, $x = x'$ and $x = x''$. Consequently, the number of solutions of the equation (8.7) is at most three, independently of $k \geq 2$. Thus the number of solutions is the same as in the case of $k = 2$ mentioned in Proposition 8.2, hence by using Corollary 8.1 and Remark 8.2 we get exact numbers of TISGMs. This completes the proof. □

Remark 8.5. Let us make some useful remarks

1. The critical temperature $T_{cr} = J/\ln\left(1 + \frac{q}{k-1}\right)$ is explicit for any $k \geq 2$ and any $q \geq 2$. For other critical values we know only existence and explicit formulas for $k = 2$:

$$T_{c,m} \equiv T_{c,m}(2, q) = \frac{J}{\ln\left(1 + 2\sqrt{m(q-m)}\right)}, \quad m = 1, 2, \dots [\frac{q}{2}]. \quad (8.28)$$

2. For the case $k = 2$, $q = 3$ we have two critical temperatures

$$T_{cr} = \frac{J}{\ln 4}, \quad T_{c,1} = \frac{J}{\ln(1 + 2\sqrt{2})},$$

and up to $2^3 - 1 = 7$ TISGMs, denoted by μ_i, $i = 0, \dots, 6$. Here, μ_0 is free measure, corresponding to solution 1. Each TISGM defines its own measure of the cylinder $\{\sigma \in \Omega : \sigma(x) = j\}$, $x \in V$, $j = 1, 2, 3$. Moreover, since these measures are translation invariant, the measure of this cylinder does not depend on $x \in V$:

$$\mu_i(\{\sigma \in \Omega : \sigma(x) = j\}) = p_{ij}.$$

Depending on the value of the solution z of (8.7) one can calculate p_{ij}. In case $z = 1$, $p_{0j} = 1/3$ (since $q = 3$). If μ_i corresponds to a solution

$z > 1$ then for sufficiently low temperatures we have $p_{i1} > 2/3$, and for $z < 1$ we have $p_{i1} < 1/3$. Consequently, we can see typical configurations (of the seven TISGMs, i.e., phases) of the two-dimensional 3-state Potts model, for different values of temperature as shown in Fig. 8.1.

8.2 Non-uniqueness of extreme Gibbs measure

The following theorem improves the critical value (see Remark 8.4) of phase transitions for the Potts model, i.e., for the q-state Potts model ($q \geq 3$) on the Cayley tree of order two, the Gibbs measure is not unique starting from $\theta = \theta_1 = 1 + 2\sqrt{q-1} < \theta_c = q + 1$.

Theorem 8.4. *If $k = 2$ then for any $q \geq 3$ and $\theta_1 \leq \theta$ there are at least two extreme (not-necessarily TI) Gibbs measures.*

Proof. By the exact solutions (8.12) we know that if $\theta < \theta_1$ then there is unique solution $h = (0, \ldots, 0)$. For any $\theta < \theta_c$ the Gibbs measure μ_0 corresponding to $h = (0, \ldots, 0)$ is extreme (see [243]).

For $\theta \geq \theta_1$ we have measure μ_0 and at least q new measures mentioned in Proposition 8.2. If we assume that all the new measures are not extreme then there remains only one known extreme measure μ_0. But in this case the non-extreme measures can not be decomposed only into the unique measure μ_0. Consequently, at least one of the new measures must be extreme or decomposable into other extreme measures. □

From Theorem 8.3 similarly as Theorem 8.4 we get the following

Corollary 8.2. *If $T \leq T_{c,1}$ then there are at least two extreme Gibbs measures for the q-state Potts model on the Cayley tree of order $k \geq 2$.*

Let $\mathcal{G}_{ti}$ be the set of all TISGMs. The following gives relations between TISGMs.

Theorem 8.5. *Any measure $\mu \in \mathcal{G}_{ti}$ can not be non-trivial convex mixture of measures from $\mathcal{G}_{ti} \setminus \{\mu\}$.*

Proof. Let $\alpha_i \geq 0$ be such that $\sum_{i=1}^{|\mathcal{G}_{ti}|-1} \alpha_i = 1$, and assume that

$$\mu = \sum_{i=1}^{|\mathcal{G}_{ti}|-1} \alpha_i \mu_i. \tag{8.29}$$

Let $z^{(i)}$ be the solution of the equation (8.7) which corresponds to μ_i. Since this number does not depend on $x \in V$ we see that TISGMs satisfy

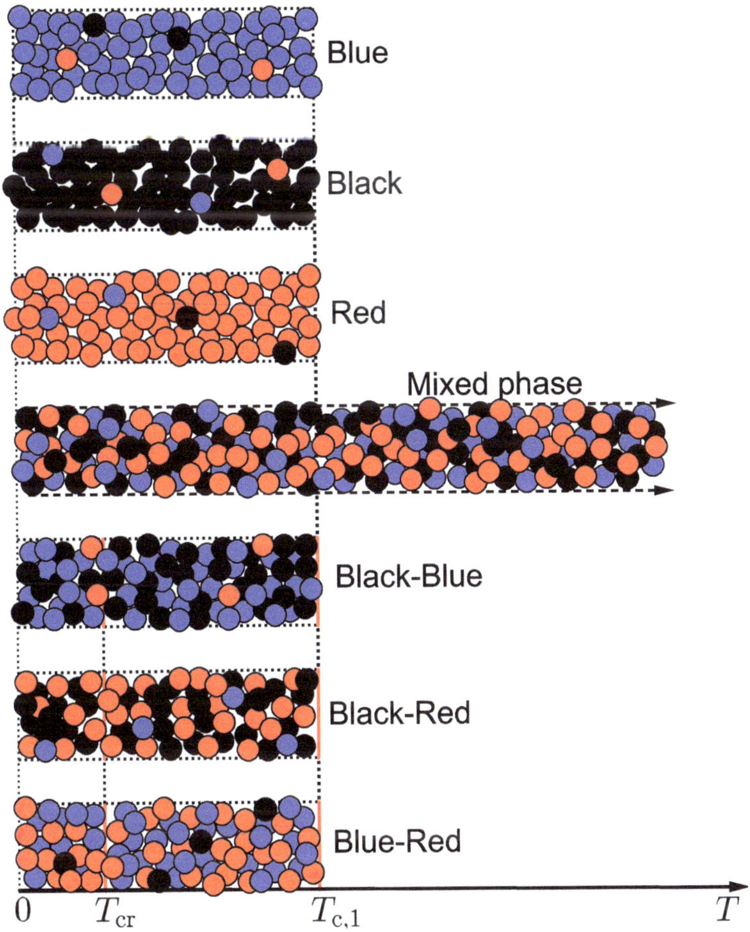

Fig. 8.1 3-state Potts model's typical configurations depending on temperature. Here, states are: red = 1, blue = 2, black = 3. All possible (seven) cases are shown: In the mixed phase all three colors are seen with equal probability (with respect to the unique TISGM, i.e., free Gibbs measure). The mixed phase exists for any $T \geq 0$. The "Red" phase (having many reds and very few other colors) is the typical configuration for the TISGM corresponding to the solution $(z_2, 1, 1)$. Similarly, "Blue" and "Black" phases are the typical configurations for the TISGMs corresponding to the coordinate-permutations of $(z_2, 1, 1)$. The "Black-Blue" phase (having black and blue colors with equal probability, but very few red color) corresponds to $(z_1, 1, 1)$ and the remaining two phases correspond to its coordinate-permutations. Each phase is shown on the values of temperature, where the corresponding TISGM exists. The red vertical segments shown at critical temperatures means that these TISGMs do not exist at the critical temperatures.

conditions of Corollary (12.18) of [98], by this corollary we have that the RHS of the (8.29) can not be SGM (or in language of [98] a Markov chain). But in the LHS we have a SGM μ. This is a contradiction. □

8.3 Boundary conditions for TISGMs

8.3.1 *Setup*

In previous sections it was shown that at sufficiently low temperatures the maximal number of TISGMs is $2^q - 1$. Moreover, it was shown that the number of TISGMs does not depend on the order $k \geq 2$ of the Cayley tree. In the case $k = 2$ the explicit formulae for the critical temperatures and all TISGMs are given. However it is not clear what kind of boundary conditions are needed to get the TISGMs as corresponding limits with the boundary conditions.

In this section following [80] we shall answer this question. It is non-trivial problem since the number of TISGMs (i.e., $2^q - 1$) is larger than the number (i.e., q) of translation-invariant configurations.

Concerning the Ising model, the dependence of TISGMs on boundary conditions has been studied in [111]. Here we first adapt the setup of [111] for the Potts model: two vertices $t, s \in V, (t \neq s)$ are called neighbors if they are connected by an edge. In this case we write $\langle s, t\rangle$. Each vertex of Γ^k has $k+1$ neighbors.

Fix an origin 0 of Γ^k. We write $s \to t$, if $t \neq s$ and the path connecting 0 and t passes through s. If $s \to t$ and s, t are neighbors, we write $s \to_1 t$.

For any finite $A \subset V$, the boundary ∂A of A is

$$\partial A = \{t \in V \setminus A : \exists x \in A, \langle x, t\rangle\}.$$

Recall also that for every $A \subset V$, $\Omega_A = \{1, 2, ..., q\}^A$ denotes the set of all possible spin configurations on A. For brevity denote $\Omega = \Omega_V$.

For every $A \subset V$ define the σ-algebra B_A by[1]

$$B_A = \text{the}\, \sigma - \text{algebra generated by } \{X_t, t \in A\},$$

where $X_t(\sigma) = \sigma(t)$ for all $t \in A, \sigma \in \Omega$. For brevity write B instead of B_V.

[1]Recall definition of a σ-algebra generated by random variables: For a random variable X, define by $\sigma(X) = \{X^{-1}(B);\ B \text{ is a Borel set}\}$ the smallest sigma algebra such that X is a measurable function into $\mathbb{R}$. Similarly, for a set of random variables $\{X_i,\ i \in A\}$, the sigma algebra $\sigma(X_i,\ i \in A)$ generated by these is the smallest sigma algebra such that all X_i are measurable.

Let A be a finite subset of V, $\omega \in \Omega$ and $\sigma \in \Omega_A$. We define an interaction Potts energy on A given the inner configuration σ and the boundary condition ω by

$$E_A^\omega(\sigma) = J \sum_{\substack{t,s\in A:\\ t\to_1 s}} \delta_{\sigma(t)\sigma(s)} - J \sum_{\substack{t\in A,\, s\in\partial A:\\ \langle t,s\rangle}} \delta_{\sigma(t)\omega(s)}, \tag{8.30}$$

where $J \in \mathbb{R}$.

A finite Gibbs measure P_A^ω on Ω_A corresponding to E_A^ω is defined by

$$P_A^\omega(\sigma) = [Z_A^\omega]^{-1} \exp[-E_A^\omega(\sigma)], \sigma \in \Omega_A, \tag{8.31}$$

where $Z_A^\omega = \sum_{\widehat{\sigma}\in\Omega_A} \exp[-E_A^\omega(\widehat{\sigma})]$. As usual P_A^ω can be considered as a probability measure on (Ω, B).

For fixed J, if there is an increasing sequence of finite subsets $\{V_n\}$ such that $V_n \nearrow V$ as $n \to \infty$ and $P^\omega = \mathrm{w} - \lim_{n\to\infty} P_{V_n}^\omega$ (the weak convergence of measures) exists for suitable fixed $\omega \in \Omega$, then P^ω is called a *limiting Gibbs measure* with boundary condition ω for J. On the other hand, a Gibbs measure P for J is defined as a probability measure on (Ω, B) such that for every M in B_A

$$P(M|B_{A^c})(\omega) = P_A^\omega(M). \quad a.s.(P). \tag{8.32}$$

Remark 8.6. 1. As it was mentioned in Chapter 2 (see ([98], [111])) the set $\Im(J)$ of all Gibbs measures for a fixed J is a non-empty, compact convex set. A limiting Gibbs measure is a Gibbs measure for the same J. Conversely, every extremal point of $\Im(J)$ is a limiting Gibbs measure with a suitable boundary condition for the same J.

2. It is known (see Remark 2.7) that any extreme Gibbs measure of a Hamiltonian with nearest-neighbor interactions is a *splitting* Gibbs measure. Consequently, any non-splitting Gibbs measure is not extreme. However, any splitting Gibbs measure (not necessary extreme) is a limiting Gibbs measure, because it corresponds to a (generalized)[2] boundary condition satisfying a compatibility (tree recursion) condition of Kolmogorov's theorem.

3. In [56] it was shown that for non-extremal Gibbs measures on $\mathbb{Z}^d$ a Gibbs measure need not be a limiting Gibbs measure (see [56] and [75] for more details).

---

[2]Recall that added a boundary field at each site of the boundary is called a generalized boundary condition [78] or boundary law [98].

Let $|t|$ denote the distance between 0 and $t \in V$, i.e., $|t| = n$ if there exists a chain $0 \to_1 u_1 \to_1 u_2 \to_1 u_3 \to_1 \dots \to_1 u_{n-1} \to_1 t$. We only consider the sequence of boxes

$$V_n = \{t \in V : |t| \leq n\}, n \geq 1.$$

For every $s \in V$ define

$$\Gamma_s^k = \{s\} \cup \{t \in V : s \to t\}$$

and

$$V_{n,s} = \Gamma_s^k \cap V_n, \quad n \geq 1.$$

Elements of $V_1 \setminus \{0\}$ are numbered $1, 2, \dots, k+1$.

For every $\omega \in \Omega$, $s \in V \setminus \{0\}$ and $n \geq |s|$ define

$$W_{n,s}^{\omega}(l) = \sum_{\sigma \in \Omega_{V_{n,s}} : \sigma(s) = l} \exp[-E_{V_{n,s}}^{\omega}(\sigma) - J\delta_{l\omega(t)}], l = 1, 2, \dots, q, \tag{8.33}$$

$$R_{n,s}^{l}(\omega) = \frac{W_{n,s}^{\omega}(l)}{W_{n,s}^{\omega}(q)}, \quad l = 1, 2, \dots, q, \tag{8.34}$$

here t is the unique vertex such that $t \to_1 s$.

We have

$$\frac{P_{V_n}^{\omega}(\sigma(0) = l)}{P_{V_n}^{\omega}(\sigma(0) = q)} = \prod_{i=1}^{k+1} \frac{(\exp(J) - 1)R_{n,i}^{l}(\omega) + \sum_{p=1}^{q-1} R_{n,i}^{p}(\omega) + 1}{\exp(J) + \sum_{p=1}^{q-1} R_{n,i}^{p}(\omega)}, \tag{8.35}$$

where $l = 1, \dots, q$.

$$W_{n,s}^{\omega}(l) = \prod_{u : s \to_1 u} [(\exp(J) - 1)W_{n,u}^{\omega}(l) + \sum_{p=1}^{q} W_{n,u}^{\omega}(p)], l = 1, 2, \dots, q, \tag{8.36}$$

and for $n > m$, $\eta \in \Omega_{V_m}$, we get

$$P_{V_n}^{\omega}(\{\sigma(s) = \eta(s), s \in V_m\}) = \frac{\exp[-E_{V_{m-1}}^{\eta}(\eta)] \prod_{s \in \partial V_{m-1}} W_{n,s}^{\omega}(\eta(s))}{\sum_{\xi \in \Omega_{V_m}} \exp[-E_{V_{m-1}}^{\xi}(\xi)] \prod_{s \in \partial V_{m-1}} W_{n,s}^{\omega}(\xi(s))}. \tag{8.37}$$

From the above equalities we obtain the following

Lemma 8.4. *Let $\omega \in \Omega$ be given. If there is $N > 0$ such that $R_{n,s}^{l}(\omega)$ converges as $n \to \infty$ for every $s \in V \setminus V_N$ and for every $l = 1, \dots, q$, then $P^{\omega} = w - \lim_{n \to \infty} P_{V_n}^{\omega}$ exists.*

For $n\geq 1$, $p=1,2,...,q$, $i=1,...,k+1$ denote

$$A_n=\{t\in V:|t|=n\},\quad N^{(p)}_{n,i}(\sigma)=|\{x\in A_n\cap V_{n,i}:\sigma(x)=p\}|.$$

Lemma 8.5. *Let ω be a configuration such that*

$$c^l(\omega)=\sum_{s:t\to_1 s}\delta_{l\omega(s)},$$

is independent of $t\in V\setminus\{0\}$. Then $R^l_{n,i}(\omega)=R^l_{n,j}(\omega)$ for any $i,j=1,2,...,k+1$.

Proof. Since $R^l_{n,i}(\omega)=\frac{W^\omega_{n,i}(l)}{W^\omega_{n,i}(q)}$, it suffices to prove that $W^\omega_{n,i}(l)=W^\omega_{n,j}(l)$ for any $i,j=1,2,...,k+1$.

For the Hamiltonian we have

$$E^\omega_{V_{n,i}}(\sigma)=E^\sigma_{V_{n-1,i}}(\sigma)-J\sum_{x\in A_n\cap V_{n,i}}\sum_{x\to_1 y}\delta_{\sigma(x)\omega(y)}.\tag{8.38}$$

By the condition of Lemma 8.5 we obtain

$$\begin{aligned}E^\omega_{V_{n,i}}(\sigma)&=E^\sigma_{V_{n-1,i}}(\sigma)-J\sum_{x\in A_n\cap V_{n,i}}c^{\sigma(x)}(\omega)\\&=E^\sigma_{V_{n-1,i}}(\sigma)-J\sum_{p=1}^{q}N^{(p)}_{n,i}(\sigma)c^p(\omega).\end{aligned}$$

For $W^\omega_{n,i}(l)$ we have

$$W^\omega_{n,i}(l)=\sum_{\sigma\in V_{n,i}:\sigma(i)=l}\exp[E^\sigma_{V_{n-1,i}}(\sigma)-J\sum_{p=1}^{q}N^{(p)}_{n,i}(\sigma)c^p(\omega)-J\delta_{l\omega(t)}].$$

Since $V_{n,i}$ is similar to $V_{n,j}$, for any $i,j\in\{1,\dots,k+1\}$, $n\geq 1$, there is an one-to-one correspondence γ between sets $\Omega_{V_{n,i}}$ and $\Omega_{V_{n,j}}$, which can be obtained by a rotation of the $V_{n,i}$ on the set $V_{n,j}$. We note that the Potts interaction energy (8.30) is translation-invariant and by the condition of the lemma the quantity $c^l(\omega)$ also does not depend on vertices of the tree. Therefore, if $\gamma(\sigma)=\varphi$ then

$$E^\sigma_{V_{n-1,i}}(\sigma)=E^{\gamma(\sigma)}_{V_{n-1,j}}(\gamma(\sigma))=E^\varphi_{V_{n-1,j}}(\varphi),$$

$$N^{(p)}_{n,i}(\sigma)=N^{(p)}_{n,j}(\varphi),\quad\forall i,j=1,\dots,k+1,\ p=1,\dots,q.$$

Using these equalities we get

$$W^{\omega}_{n,i}(l) = \sum_{\sigma\in\Omega_{V_{n,i}}:\sigma(i)=l} \exp\left[E^{\sigma}_{V_{n-1,i}}(\sigma) - J\sum_{p=1}^{q} N^{(p)}_{n,i}(\sigma)c^{p}(\omega) - J\delta_{l\omega(t)}\right]$$

$$= \sum_{\varphi\in\Omega_{V_{n,j}}:\varphi(j)=l} \exp\left[E^{\varphi}_{V_{n-1,j}}(\varphi) - J\sum_{p=1}^{q} N^{(p)}_{n,j}(\varphi)c^{p}(\omega) - J\delta_{l\omega(t)}\right]$$

$$= W^{\omega}_{n,j}(l).$$

Thus $R^{l}_{n,i}(\omega) = R^{l}_{n,j}(\omega)$ for any $i, j = 1, 2, ..., k+1$. □

Remark 8.7. One can construct a configuration ω satisfying condition of Lemma 8.5, for example, by the following steps:

(1) Take arbitrary value $l := \omega(0) \in \{1, \dots, q\}$ and arbitrary configuration $\omega(s)$, for $s : 0 \to_1 s$.
(2) Calculate
$$c := c^{l}(\omega) = \sum_{s:0\to_1 s} \delta_{l\omega(s)} = \text{“the number of } s \text{ such that } \omega(s) = l\text{”}.$$
(3) Extend the configuration of the step 1) to all vertices of V, starting from the set $S(s)$ of the direct successors of each $s : 0 \to_1 s$, putting c times of $\omega(s)$ on $S(s)$ and $k - c$ times any values different from $\omega(s)$. Then iterate this construction for all remaining vertices (see Fig. 8.2 for an example).

From the above proof it follows that $R^{l}_{n,s}(\omega)$ depends only on $n - |s|$, i.e., we have

$$R^{l}_{n,s}(\omega) = R^{l}_{n-|s|+1}(\omega), l = 1, 2, ...q. \tag{8.39}$$

Then from (8.36) we get the following

$$Y^{l}_{n}(\omega) = kF_l(Y^{1}_{n-1}(\omega), Y^{2}_{n-1}(\omega), ..., Y^{q-1}_{n-1}(\omega)), \tag{8.40}$$

where $l = 1, ..., q-1$, $Y^{l}_{n}(\omega) = \ln R^{l}_{n}(\omega)$ and $F = (F_1, ..., F_{q-1})$ with coordinates

$$F_l(x^1, x^2, ..., x^{q-1}) = \ln \frac{(\theta-1)e^{x^l} + \sum_{p=1}^{q-1} e^{x^p} + 1}{\theta + \sum_{p=1}^{q-1} e^{x^p}}, \tag{8.41}$$

where $\theta = \exp(J)$.

It is clear that if $Y^{i}(\omega)$ is a limit point of $Y^{i}_{n}(\omega)$, as $n \to \infty$ then by (8.40) we get

$$Y^{l}(\omega) = kF_l(Y^{1}(\omega), Y^{2}(\omega), ..., Y^{q-1}(\omega)), \quad l = 1, 2, \dots, q-1. \tag{8.42}$$

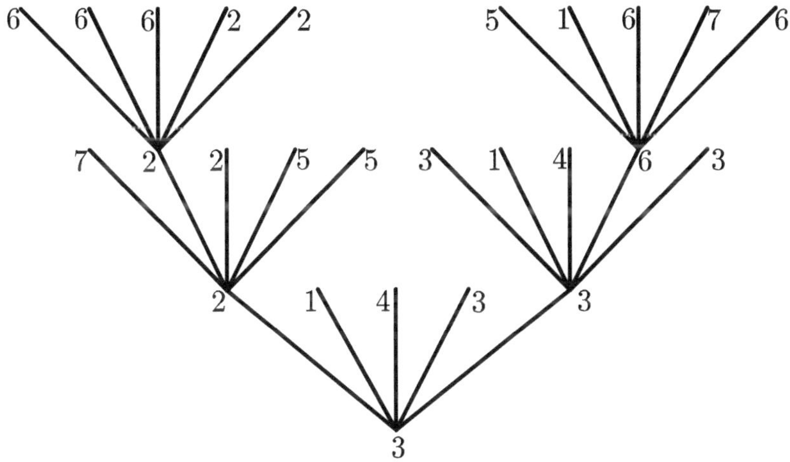

Fig. 8.2 Part of a configuration satisfying condition of Lemma 8.5. Here $k = 5$, $q = 7$, $l = 3$, $c = 2$.

For convenience we denote

$$h_l = Y^l(\omega), \quad l = 1, 2, \ldots, q-1.$$

Then the system (8.42) becomes

$$h_i = k \ln \left(\frac{(\theta - 1)e^{h_i} + \sum_{j=1}^{q-1} e^{h_j} + 1}{\theta + \sum_{j=1}^{q-1} e^{h_j}} \right), \quad i = 1, \ldots, q-1. \tag{8.43}$$

For $k \geq 3$ explicit formulas for the solutions of (8.43) are *not* known. Therefore we consider only the case $k = 2$. In this case we know that for a given $m \leq [\frac{q}{2}]$, a fixed solution generates $\binom{q}{m}$ solution vectors by permuting coordinates of the vector

$$(\underbrace{h_i, h_i, \ldots, h_i}_{m}, \underbrace{0, 0, \ldots, 0}_{q-m}),$$

and giving $\binom{q}{m}$ TISGMs. Thus without loss of generality we can only consider the measure $\mu_i(\theta, m)$ corresponding to vector

$$\mathbf{h}(m, i) = (\underbrace{h_i, h_i, \ldots, h_i}_{m}, \underbrace{0, 0, \ldots, 0}_{q-m-1}),$$

i.e., normalized on qth coordinate.

Denote by $\mu_0 \equiv \mu_0(\theta)$ the TISGM corresponding to solution $h_i \equiv 0$ and by $\mu_i \equiv \mu_i(\theta, m)$ the TISGM corresponding to the solution $h_i(\theta, m)$, $i = 1, 2$, $m = 1, \ldots, [\frac{q}{2}]$ (given in Proposition 8.2). Here we want to obtain all measures μ_i by changing boundary conditions.

8.3.2 *Boundary conditions for TISGMs*

The following lemma can be proved by simple analysis.

Lemma 8.6. *The following assertions hold*

(i) For $k \geq 2$ and $\theta > 1$ the function $f_m(h)$, $h \in \mathbb{R}$ defined in (8.7) has the following properties:

(a) $\{h : f_m(h) = h\} = \{0, h_1, h_2\}$, if $\theta > \theta_m, m \leq [q/2]$;
(b) $a < f_m(h) < A$, where

$$a = k \ln \frac{q-m}{q+\theta-m-1}, \quad A = k \ln \frac{\theta+m-1}{m};$$

(c)

$$\frac{d}{dh} f_m(h) = \frac{k(\theta-1)(\theta+q-1)e^h}{(me^h+\theta+q-m-1)((\theta+m-1)e^h+q-m)} > 0;$$

(ii) If $k = 2$ and $m \leq q/2$ then for solutions h_1 and h_2 mentioned in Proposition 8.2 the following statements hold

$$\begin{aligned}
&0 < h_1 = h_2, \quad \text{if } \theta = \theta_m \\
&0 < h_1 < h_2, \quad \text{if } \theta_m < \theta < \theta_c, \quad \text{with } \theta_c = q+1 \\
&0 = h_1 < h_2, \quad \text{if } \theta = q+1 \\
&h_1 < 0 < h_2, \quad \text{if } q+1 < \theta.
\end{aligned}$$

For each solution $h_i(\theta, m)$ we want to find $\omega = \omega(h_i) \in \Omega$, such that $\mu_i(\theta, m) = P^\omega$, where P^ω is defined in Lemma 8.4.

Consider the dynamical system (8.40) for $k = 2$. Denote $G(h) = 2F(h)$. For a given initial vector $v^{(0)} = (v_1^{(0)}, \dots, v_{q-1}^{(0)})$, we shall study the limit

$$\lim_{n\to\infty} G^{(n)}(v^{(0)}), \tag{8.44}$$

here $G^{(n)}(v) = \underbrace{G(G(...G(v))...)}_{n}$.

Figures 8.3–8.6 show the streamlines of the vector field $G^{(n)}(v)$ for $k = 2$, $q = 3$. These figures also illustrate the limit points of (8.44).

Denote

$$I_m = \{v \in \mathbb{R}^{q-1} : v_1 = \cdots = v_m, \quad v_{m+1} = \cdots = v_{q-1} = 0\}.$$

It is easy to see that the set I_m is invariant with respect to G, i.e., $G(I_m) \subset I_m$.

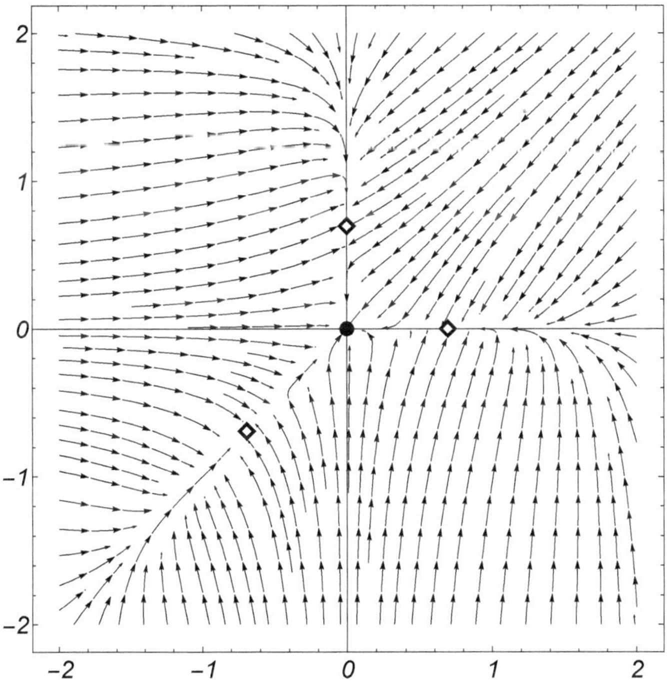

Fig. 8.3 Here and in Figures 8.4-8.6 the streamlines of the vector field $G^{(n)}(v)$ for $k = 2$, $q = 3$ are shown. These figures also illustrate the limit points of (8.44). The plane is formed by a horizontal v_1-axis, and a vertical v_2-axis. This figure applies for the case $\theta = \theta_1 = 1 + 2\sqrt{2}$. Four fixed points. The origin is an attractor. There are 3 saddle fixed points.

The following lemma gives the limits of (8.44) on the invariant I_m (compare with Figs. 8.3–8.6).

Lemma 8.7. *The following hold*

(1) If $\theta = \theta_m$, *for some* $m = 1, \dots, [\frac{q}{2}]$ *then*

$$\lim_{n\to\infty} G^{(n)}(v^{(0)}) = \begin{cases} \mathbf{h}(m,1), & \text{if } v^{(0)} \in I_m \text{ and } v_1^{(0)} \geq h_1 \\ (0,\dots,0), & \text{if } v^{(0)} \in I_m \text{ and } v_1^{(0)} < h_1. \end{cases} \tag{8.45}$$

(2) If $\theta_m < \theta < \theta_c = q + 1$ *then*

$$\lim_{n\to\infty} G^{(n)}(v^{(0)}) = \begin{cases} \mathbf{h}(m,2), & \text{if } v^{(0)} \in I_m \text{ and } v_1^{(0)} > h_1 \\ \mathbf{h}(m,1), & \text{if } v^{(0)} \in I_m \text{ and } v_1^{(0)} = h_1 \\ (0,\dots,0), & \text{if } v^{(0)} \in I_m \text{ and } v_1^{(0)} < h_1. \end{cases} \tag{8.46}$$

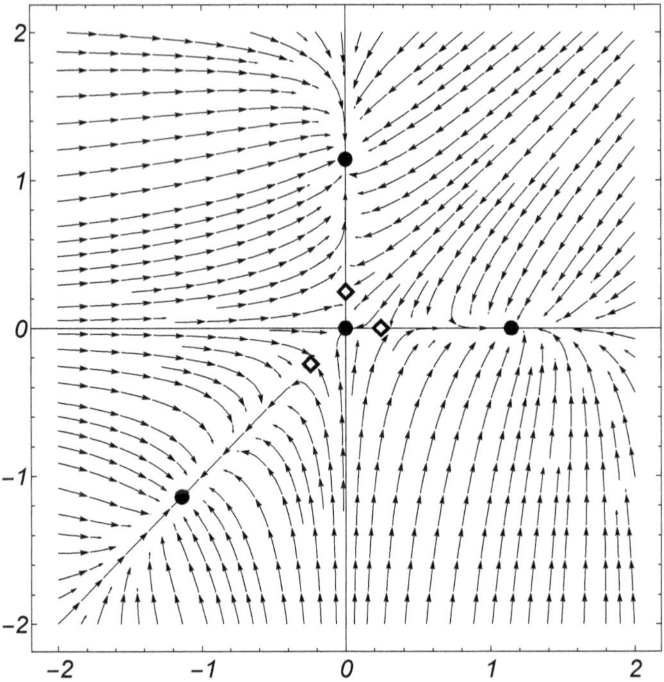

Fig. 8.4 $\theta = 3.9 > \theta_1$. Seven fixed points. Four of them (black dots) are attractors. Three (rectangular dots) points are saddles coming from the saddle points of Fig. 8.3.

(3) If $\theta = \theta_c$ then

$$\lim_{n\to\infty} G^{(n)}(v^{(0)}) = \begin{cases} \mathbf{h}(m,2), & \text{if } v^{(0)} \in I_m \text{ and } v_1^{(0)} > 0 \\ (0,\dots,0), & \text{if } v^{(0)} \in I_m \text{ and } v_1^{(0)} \le 0. \end{cases} \tag{8.47}$$

(4) If $\theta > \theta_c$ then

$$\lim_{n\to\infty} G^{(n)}(v^{(0)}) = \begin{cases} \mathbf{h}(m,2), & \text{if } v^{(0)} \in I_m \text{ and } v_1^{(0)} > 0 \\ \mathbf{h}(m,1), & \text{if } v^{(0)} \in I_m \text{ and } v_1^{(0)} < 0 \\ (0,\dots,0), & \text{if } v^{(0)} \in I_m \text{ and } v_1^{(0)} = 0. \end{cases} \tag{8.48}$$

Proof. By restriction of function $G(h)$ to I_m, one obtains the jth coordinate of $G(h)$ (for any $j = 1,\dots,m$) is equal to $f_m(h)$ which is introduced in (8.7). Moreover, other coordinates of $G(h)$ are equal to 0. By Lemma 8.6 we have that f_m is an increasing function.

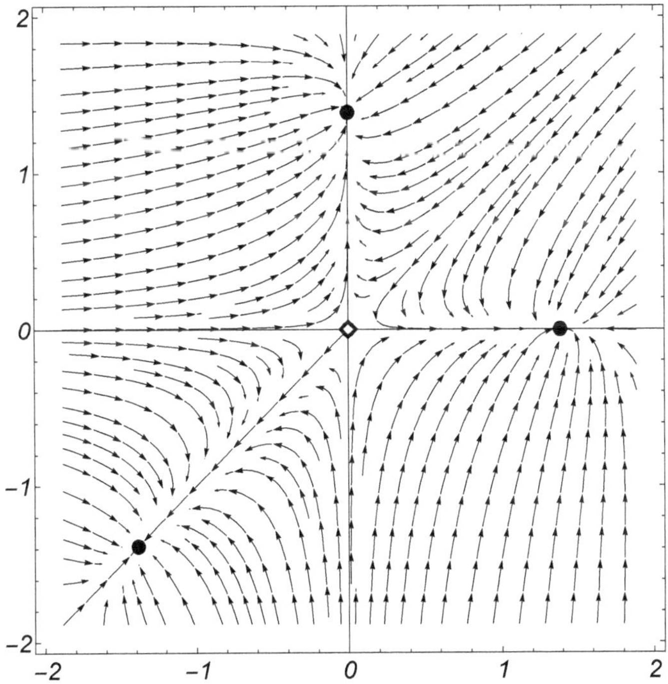

Fig. 8.5 $\theta = 4$. Four fixed points. The origin is a repeller point. Other fixed points are attractors.

In this proof consider the case when the function f_m has three fixed points 0, h_1, h_2. This proof is more simple for cases when f_m has two fixed points. Let us prove the part 2), other parts are similar.

In case 2), by Lemma 8.6 we have that $0 < h_1 < h_2$ and the point h_1 is a repeller, i.e., $f'_m(h_1) > 1$ and the points $0, h_2$ are attractive, i.e., $f'_m(0) < 1$, $f'_m(h_2) < 1$.

Take now an arbitrary $x_0 > 0$ and prove that $x_n = f_m(x_{n-1})$, $n \geq 1$ converges as $n \to \infty$. Consider the following partition

$$(-\infty, +\infty) = (-\infty, 0) \cup \{0\} \cup (0, h_1) \cup \{h_1\} \cup (h_1, h_2) \cup \{h_2\} \cup (h_2, +\infty).$$

For any $x \in (-\infty, 0)$ we have $x < f_m(x) < 0$, since f_m is an increasing function, from the last inequalities we get $x < f_m(x) < f_m^2(x) < f_m(0) = 0$. Iterating this argument we obtain $f_m^{n-1}(x) < f_m^n(x) < 0$, which for any $x_0 \in (-\infty, 0)$ gives $x_{n-1} < x_n < 0$, i.e., x_n converges and its limit is a fixed point of f_m, since f_m has unique fixed point 0 in $(-\infty, 0]$ we conclude that the limit is 0.

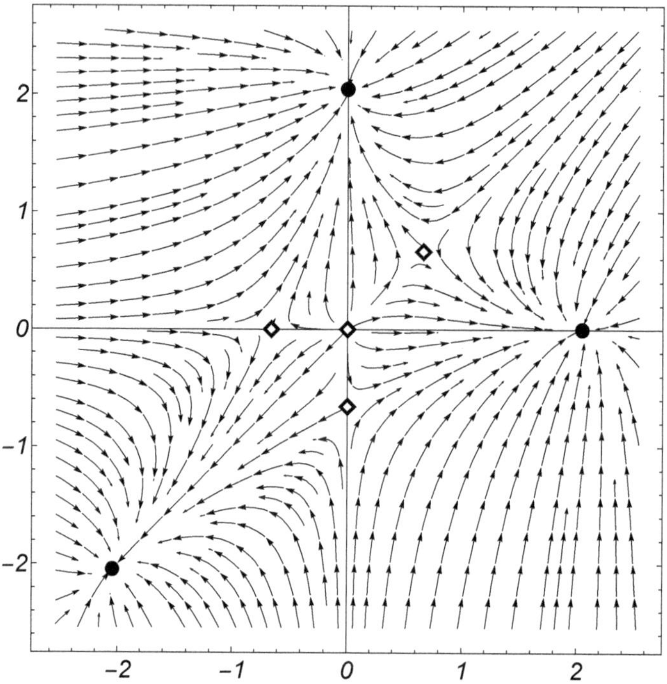

Fig. 8.6 $\theta = 4.5$. Seven fixed points. The origin is a repeller, other rectangular dots are saddles. The black dots are attractors.

For $x \in (0, h_1)$ we have $h_1 > x > f(x) > 0$, consequently $x_n > x_{n+1}$, i.e., x_n converges and its limit is again 0. Similarly, one can show that if $x_0 > h_1$ then $x_n \to h_2$ as $n \to \infty$. □

By (8.40) the asymptotic behavior of the vector

$$Y_n(\omega) = (Y_n^1(\omega), \dots, Y_n^{q-1}(\omega)),$$

depends only on the vector $Y_1(\omega)$, where

$$Y_1^l(\omega) = J\left(c^l(\omega) - c^q(\omega)\right), \quad l = 1, \dots, q-1. \tag{8.49}$$

For a given $m \in \{1, \dots, [\frac{q}{2}]\}$ and $J > 0$ we introduce the following sets of configurations:

$$\mathbb{B}_m = \{\omega \in \Omega : c^1(\omega) = \cdots = c^m(\omega), \quad c^{m+1}(\omega) = \cdots = c^{q-1}(\omega) = c^q(\omega)\},$$

$$\mathbb{B}_{m,0}^+ = \{\omega \in \mathbb{B}_m : c^1(\omega) > c^q(\omega)\},$$

$$\mathbb{B}_{m,0}^0 = \{\omega \in \mathbb{B}_m : c^1(\omega) = c^q(\omega)\},$$

$$\mathbb{B}^{-}_{m,0} = \{\omega \in \mathbb{B}_m : c^1(\omega) < c^q(\omega)\},$$

$$\mathbb{B}^{+}_{m,1} = \{\omega \in \mathbb{B}_m : J\left(c^1(\omega) - c^q(\omega)\right) > h_1\},$$

$$\mathbb{B}^{0}_{m,1} = \{\omega \in \mathbb{B}_m : J\left(c^1(\omega) - c^q(\omega)\right) = h_1\},$$

$$\mathbb{B}^{-}_{m,1} = \{\omega \in \mathbb{B}_m : J\left(c^1(\omega) - c^q(\omega)\right) < h_1\}.$$

Now taking the coordinates of an initial vector as in (8.49) by Lemma 8.4 and Lemma 8.7 we get the following

Theorem 8.6. *The following assertions hold*

(1) If $\theta = \theta_m$, for some $m = 1, \dots, [\frac{q}{2}]$ then

$$P^{\omega} = \begin{cases} \mu_1(\theta, m), \text{ if } \omega \in \mathbb{B}^{+}_{m,1} \cup \mathbb{B}^{0}_{m,1} \\ \mu_0(\theta), \quad \text{ if } \omega \in \mathbb{B}^{-}_{m,1}. \end{cases} \tag{8.50}$$

(2) If $\theta_m < \theta < \theta_c = q + 1$ then

$$P^{\omega} = \begin{cases} \mu_2(\theta, m), \text{ if } \omega \in \mathbb{B}^{+}_{m,1} \\ \mu_1(\theta, m), \text{ if } \omega \in \mathbb{B}^{0}_{m,1} \\ \mu_0(\theta), \quad \text{ if } \omega \in \mathbb{B}^{-}_{m,1} \end{cases} \tag{8.51}$$

(3) If $\theta = \theta_c$ then

$$P^{\omega} = \begin{cases} \mu_2(\theta, m), \text{ if } \omega \in \mathbb{B}^{+}_{m,0} \\ \mu_0(\theta), \quad \text{ if } \omega \in \mathbb{B}^{-}_{m,0} \cup \mathbb{B}^{0}_{m,0} \end{cases} \tag{8.52}$$

(4) If $\theta > \theta_c$ then

$$P^{\omega} = \begin{cases} \mu_2(\theta, m), \text{ if } \omega \in \mathbb{B}^{+}_{m,0} \\ \mu_1(\theta, m), \text{ if } \omega \in \mathbb{B}^{-}_{m,0} \\ \mu_0(\theta), \quad \text{ if } \omega \in \mathbb{B}^{0}_{m,0} \end{cases} . \tag{8.53}$$

8.3.3 *Construction of boundary conditions*

In this subsection using Theorem 8.6 we construct some concrete boundary conditions.

Let $k = 2$, $J > 0$, $q \geq 3$ and $m \in \{1, \dots, [\frac{q}{2}]\}$. Here we give examples of boundary configurations.

For $k = 2$ by Proposition 8.2 there are up to $2^q - 1$ TISGMs.

We shall consider only $\mu_0(\theta)$ corresponding to $h = (0, 0, \dots, 0)$ and $\mu_i(\theta, m)$ corresponding to vector

$$\mathbf{h}(m,i) = (\underbrace{h_i, h_i, \dots, h_i}_{m}, \underbrace{0, 0, \dots, 0}_{q-m-1}), \quad i = 1, 2,$$

with

$$\begin{aligned} h_1 &= 2\ln\frac{\theta - 1 - \sqrt{(\theta-1)^2 - 4m(q-m)}}{2m}, \\ h_2 &= 2\ln\frac{\theta - 1 + \sqrt{(\theta-1)^2 - 4m(q-m)}}{2m}. \end{aligned} \tag{8.54}$$

Using Theorem 8.6 we shall give some boundary conditions for each measure μ_i. Boundary conditions for the remaining measures can be obtained by using the permutation symmetry of the Potts model.

Case μ_0. If $\theta < \theta_1$ then μ_0 is a unique measure, and one can take any boundary configuration ω to have $P^\omega = \mu_0$. But for $\theta \geq \theta_1$ one has to check the conditions of Theorem 8.6 to have the limiting measure equal to μ_0.

For example, if $\theta = \theta_m < q + 1$ for some $m \in \{1, \dots, [\frac{q}{2}]\}$ then we must take $\omega \in \mathbb{B}^-_{m,1}$, i.e.,

$$\begin{aligned} J(c^1(\omega) - c^q(\omega)) &< h_1, \\ c^1(\omega) = \cdots &= c^m(\omega), \\ c^{m+1}(\omega) = \cdots &= c^q(\omega). \end{aligned} \tag{8.55}$$

Note that for a given TISGM to find its boundary condition one has to construct configurations ω which satisfy the system (like (8.55)) derived by corresponding sufficient conditions of Theorem 8.6.

Now we give several examples of such configurations.

It will be clear from our examples that some TISGM may have an infinite set of boundary configurations.

Since $h_1 > 0$ (see Lemma 8.6), the system (8.55) is satisfied, *for example*, if ω satisfies one of the following

- $c^i(\omega) = 0$, $i = 1, 2, \dots, m$, i.e., if $\omega(x) = i$, then on direct successors x_1, x_2 of x one has $\omega(x_1) \neq i$, $\omega(x_2) \neq i$; and $c^j(\omega) = 1$ for each $j = m+1, \dots, q$, i.e., if $\omega(x) = j$ then one has $\omega(x_1) = j$ but $\omega(x_2) \neq j$. See Fig. 8.7 for an example of such a configuration for $q = 5$ and $m = 2$.
- $c^i(\omega) = 1$, $i = 1, 2, \dots, m$, i.e., if $\omega(x) = i$ then $\omega(x_1) = i$ but $\omega(x_2) \neq i$ and it contains $j \in \{m+1, \dots, q\}$ in such a way that if $\omega(x) = j$ then on direct successors x_1, x_2 of x one has $\omega(x_1) = \omega(x_2) = j$. In this case $c^i(\omega) = 2$ for each $i = m+1, \dots, q$. See Fig. 8.8 for an example of such a configuration for $q = 15$ and $m = 3$.

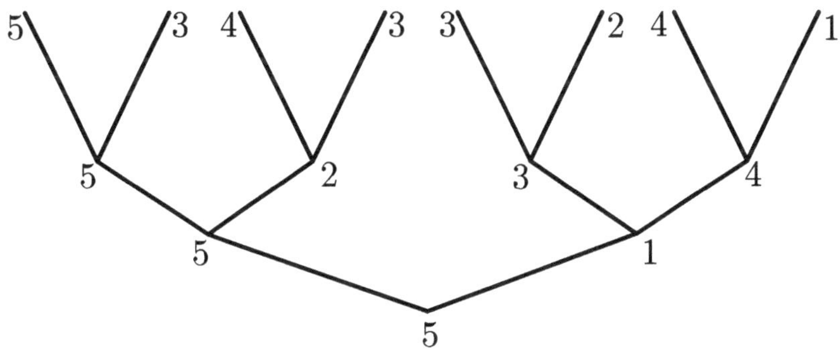

Fig. 8.7 A boundary condition for μ_0, $q = 5$, $m = 2$. Here $c^1(\omega) = c^2(\omega) = 0$, $c^3(\omega) = c^4(\omega) = c^5(\omega) = 1$. The values $1, 2$ occur often satisfying $c^1(\omega) = c^2(\omega) = 0$.

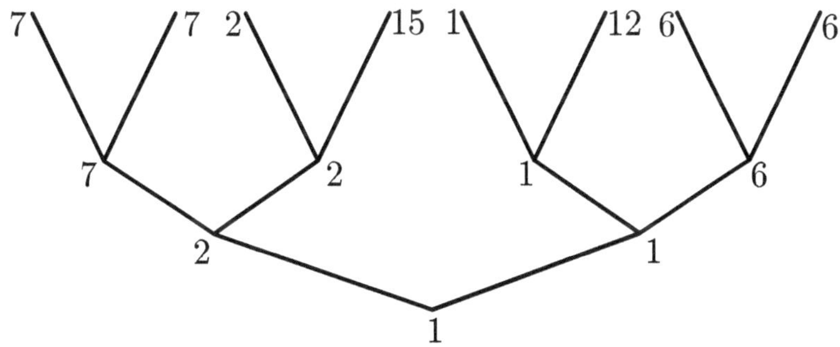

Fig. 8.8 An example of a boundary condition for the TISGM μ_0, for $q = 15$, $m = 3$. Here we have $c^i(\omega) = 1$, $i = 1, 2, 3$ and $c^j(\omega) = 2$, $j = 4, 5, \dots, 15$.

Case μ_1. This measure exists for $\theta \geq \theta_m$.

Subcase $\theta = \theta_m$. By Theorem 8.6 for μ_1 we have the condition $\omega \in \mathbb{B}^+_{m,1} \cup \mathbb{B}^0_{m,1}$, i.e.,

$$\begin{aligned} J(c^1(\omega) - c^q(\omega)) &\geq h_1, \\ c^1(\omega) &= \cdots = c^m(\omega), \\ c^{m+1}(\omega) &= \cdots = c^q(\omega). \end{aligned} \tag{8.56}$$

Note that $J = \ln \theta$. Assume $\ln \theta_m \geq h_1 = 2 \ln \frac{\theta_m - 1}{2m}$ which is equivalent to the following

$$2m\sqrt{m^2+1}(\sqrt{m^2+1} - m) \leq q \leq 2m\sqrt{m^2+1}(\sqrt{m^2+1} + m). \tag{8.57}$$

Under condition (8.57) the system (8.56) is satisfied *for example*, if $m \geq 1$ and ω is such that $c^i(\omega) = 1$, $i = 1, \ldots, m$ and $c^j(\omega) = 0$, $i = m+1, \ldots, q$. See Fig. 8.9 for an example of such a configuration for $q = 5$ and $m = 2$ and Fig. 8.10 for configuration in case $q = 10$, $m = 4$.

Subcase $\theta_m < \theta < q+1$. From Theorem 8.6 for μ_1 we have the condition

$$\begin{aligned} J(c^1(\omega) - c^q(\omega)) &= h_1, \\ c^1(\omega) &= \cdots = c^m(\omega), \\ c^{m+1}(\omega) &= \cdots = c^q(\omega). \end{aligned} \tag{8.58}$$

Assume θ is a solution to the equation $\ln \theta = h_1$. Computer analysis shows that this equation has a solution if for example $q = 17$, $m = 1$ or $q = 55$, $m = 2$. So assuming existence of such a solution $\theta = \theta^*$ one can take a boundary condition configuration as in the previous case (like in Fig. 8.9)

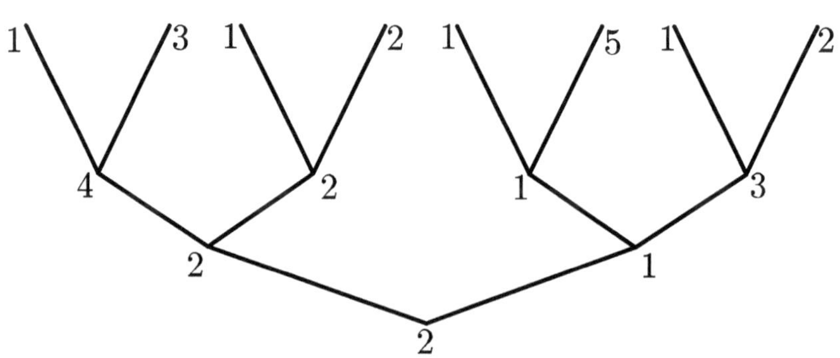

Fig. 8.9 An example of a boundary condition for the TISGM μ_1, $q = 5$, $m = 2$. Here we have $c^1(\omega) = c^2(\omega) = 1$, $c^3(\omega) = c^4(\omega) = c^5(\omega) = 0$.

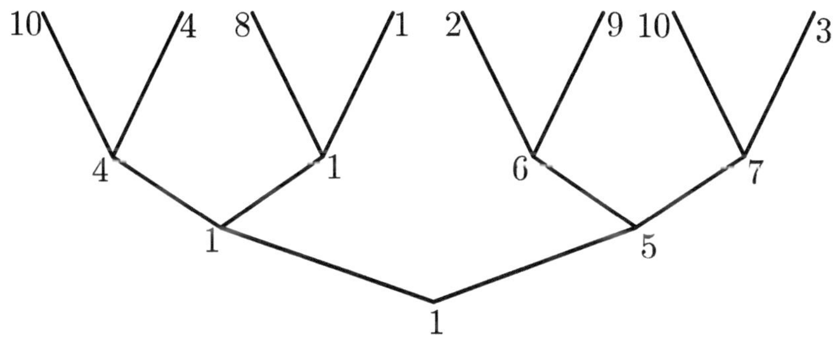

Fig. 8.10 An example of a boundary condition for the TISGM μ_1, $q = 10$, $m = 4$. Here we have $c^i(\omega) = 1$, $i = 1, 2, 3, 4$; $c^j(\omega) = 0$, $j = 5, \dots, 10$.

Subcase $\theta = q+1$. In this case we have $\mu_1 = \mu_0$. Therefore the boundary condition can be taken as in Case μ_0.

Subcase $\theta > q + 1$. For μ_1 we should have

$$c^1(\omega) - c^q(\omega) < 0, \quad c^1(\omega) = \cdots = c^m(\omega), \quad c^{m+1}(\omega) = \cdots = c^q(\omega). \tag{8.59}$$

We can take a configuration ω such that $c^i(\omega) = 0$, $i = 1, \dots, m$ and $c^j(\omega) = 1$, $j = m+1, \dots, q$ (see Fig. 8.7 for such a configuration).

Remark 8.8. From above examples one can see that depending on the temperature (equivalently depending on the parameter θ) a configuration may be the boundary condition for different TISGMs. For example, the configuration given in Fig. 8.7 is a boundary condition for TISGM μ_0 if $\theta = \theta_2 < q + 1$, but the same configuration is the boundary condition for TISGM μ_1 if $\theta > q + 1$.

Case μ_2. Check the conditions of Theorem 8.6:

Subcase $\theta = \theta_m$. In this case we have $\mu_2 = \mu_1$, i.e., the boundary condition is constructed in the previous case.

Subcase $\theta_m < \theta < q+1$. From Theorem 8.6 for μ_2 we have the condition

$$\begin{aligned} J(c^1(\omega) - c^q(\omega)) &> h_1, \\ c^1(\omega) &= \cdots = c^m(\omega), \\ c^{m+1}(\omega) &= \cdots = c^q(\omega). \end{aligned} \tag{8.60}$$

If $\ln\theta > h_1$ then it is easy to see that ω satisfies the condition (8.60) if $m \geq 1$ and ω is such that $c^i(\omega) = 1$, $i \in \{1, \ldots, m\}$ and $c^j(\omega) = 0$, $i \in \{m+1, \ldots, q\}$ (like in Fig. 8.9)

Subcase $\theta \geq q+1$. For μ_2 we should have

$$c^1(\omega) - c^q(\omega) > 0, \quad c^1(\omega) = \cdots = c^m(\omega), \quad c^{m+1}(\omega) = \cdots = c^q(\omega). \tag{8.61}$$

Condition (8.61) is easily checkable. For example, configurations shown in Figs. 8.9 and 8.10 satisfy this condition.

Chapter 9

Extremality of Gibbs measures

In this chapter, for the q-state Potts model on a Cayley tree we give some regions for the temperature parameter ensuring that a given TISGM (described in Chapter 8) is (non-)extreme in the set of all Gibbs measures. In particular it is shown the existence of a temperature interval for which there are at least $2^{q-1}+q$ extremal TISGMs.

9.1 Eigenvalues of transition matrix

In this chapter, we mainly follow [139]. To investigate the (non-)extremality for a given TISGM we need explicit formulas of these solutions, which are known for $k=2$ by Proposition 8.2.

By the symmetry of the Potts model, for each fixed $m \leq [\frac{q}{2}]$, we reduce the study of extremality to at most three TISGMs: μ_0 which corresponds to the solution $h_x=(0,\ldots,0)$ and two TISGMs $\mu_1(\theta,m)$, $\mu_2(\theta,m)$ which correspond to the vectors

$$h_x = h_1 = (\underbrace{\ln z_1, \ln z_1, \ldots, \ln z_1}_{m}, 0, \ldots, 0),$$

$$h_x = h_2 = (\underbrace{\ln z_2, \ln z_2, \ldots, \ln z_2}_{m}, 0, \ldots, 0),$$

where $z_1=z_1(\theta,m)$ and $z_2=z_2(\theta,m)$ are solutions of the equation (8.7).

Thus for each given m all TISGMs belong to three classes:

- The first class contains only μ_0.
- The second class all measures which correspond to vectors obtained by permutations of coordinates of h_1.
- The third class contains all TISGMs which correspond to a vector obtained by permutations of h_2.

Noted that in case $m = q/2$ the solutions h_1 and h_2 define the same TISGMs after a relabeling of indices and a re-normalization of h_1.

In order to study extremality of a TISGM corresponding to a solution $z > 0$, $z \neq 1$ of (8.7), (i.e., to a vector $l = (l_1, \dots, l_q) \in \mathbb{R}^q$ with m coordinates equal to z and $q - m$ coordinates equal to 1) we divide the coordinates of this vector into two classes: take l' which has m coordinates each equal to z and l'' which has $q - m$ coordinates each equal to 1.

The fuzzy Potts model (see [163]) on a graph is obtained from a standard q-state Potts model on the same graph by regrouping the possible spin-values into r classes with $r < q$. The use of the word "fuzzy" suggests that only partial information of the underlying model is kept.

Here we consider special transformations which are adapted to the symmetry of the TISGMs.

Without loss of generality, by relabeling of coordinates, we can take l, l' and l'' as follows:

$$l = (\underbrace{z, z, \dots, z}_{m}, \underbrace{1, 1, \dots, 1}_{q-m}) = ((\underbrace{z, z, \dots, z}_{m}), (\underbrace{1, 1, \dots, 1}_{q-m})) = (l', l'').$$

The fuzzy map $T : \{1, 2, \dots, q\} \to \{1', 2'\}$ is defined as

$$T(i) = \begin{cases} 1', & \text{if } i \leq m \\ 2', & \text{if } i \geq m + 1. \end{cases} \tag{9.1}$$

This map identifies spin-values which have the same value of the boundary law and are treated in an equal fashion by the transition matrix. Extend this map to act on infinite-volume spin configurations and measures in the infinite volume, i.e.,

$$T(\sigma)(x) = T(\sigma(x)), \quad T(\mu)(A) = \mu(T^{-1}(A)).$$

We note that a TISGM corresponding to a vector $l \in \mathbb{R}^q$ is a tree-indexed Markov chain with states $\{1, 2, \dots, q\}$ and transition probabilities matrix $\mathbb{P} = (P_{ij})$ with P_{ij} given by (2.28).

From (2.28) we get

$$P_{ij} = \begin{cases} \theta z/Z_1, & \text{if } i = j, \, i \in \{1, \dots, m\} \\ z/Z_1, & \text{if } i \neq j, \, i, j \in \{1, \dots, m\} \\ 1/Z_1, & \text{if } i \in \{1, \dots, m\}, j \in \{m + 1, \dots, q\} \\ z/Z_2, & \text{if } i \in \{m + 1, \dots, q\}, j \in \{1, \dots, m\} \\ \theta/Z_2, & \text{if } i = j, \, i \in \{m + 1, \dots, q\} \\ 1/Z_2, & \text{if } i \neq j, \, i, j \in \{m + 1, \dots, q\}, \end{cases} \tag{9.2}$$

where

$$Z_1 = (\theta + m - 1)z + q - m, \quad Z_2 = mz + \theta + q - m - 1.$$

Consider the translation-invariant tree-indexed Markov chain μ with transition matrix given by (9.2). Then its image measure $T(\mu)$ under the site-wise application of the fuzzy map T is a tree-indexed Markov chain with local state space $1', 2'$ with the transition matrix given by

$$T(\mathbb{P}) = \hat{\mathbb{P}} = (p_{ij})_{i,j=1',2'}, \tag{9.3}$$

with

$$p_{ij} = \begin{cases} \frac{(\theta+m-1)z}{Z_1}, & \text{if } i = j = 1' \\ \frac{q-m}{Z_1}, & \text{if } i = 1', j = 2' \\ \frac{mz}{Z_2}, & \text{if } i = 2', j = 1' \\ \frac{\theta+q-m-1}{Z_2}, & \text{if } i = j = 2'. \end{cases} \tag{9.4}$$

Note that if a Markov chain μ is extremal in the set of all Gibbs measures for the Potts model then this implies its triviality on the tail-sigma algebra of the q-spin events (see Theorem 2.1). Because, the extremality of Gibbs measures on any graph is equivalently characterized by the triviality of the measure on the tail-sigma algebra. This implies that the mapped chain $T(\mu)$ is trivial on its tail-sigma algebra (since the latter can be identified with the sub-sigma algebra of events in the tail-sigma algebra of the q-spin events which do not distinguish spin-values which have the same values of the fuzzy variable).

Relations between the transition matrices of the Potts model and the mapped model will be very useful below. The explicit formulas of eigenvalues are very important in the checking of (non-)extremality, let us calculate these eigenvalues now.

We note that the matrix $\hat{\mathbb{P}}$ (see (9.3)) has two eigenvalues 1 and

$$\lambda_2(\hat{\mathbb{P}}) = \frac{(\theta + m - 1)z}{Z_1} + \frac{\theta + q - m - 1}{Z_2} - 1. \tag{9.5}$$

Note that the transition matrices have the same type of two-block symmetry as transition matrix $\mathbb{P}$ has. Namely, the matrix $\mathbb{P}$ is a stochastic $q \times q$-matrix, in block-form which has as parameters the block size m and can be written as follows in terms of four real parameters p_1, p_2, q_1, q_2. Below the simple expressions of the parameters in terms of existing quantities z, θ, q, and m are given in the proof of Corollary 9.1. The real parameters are defined as follows

p_1 is the transition rate for going from a state in $1'$ to a different state in $1'$,

q_1 is the transition rate for going from a state in $2'$ to a different state in $2'$,

p_2 is the transition rate for going from a state in $1'$ to a state in $2'$,

q_2 is the transition rate for going from a state in $2'$ to a state in $1'$.

The form is

$$\mathbb{P} = (P_{i,j})_{i,j=1,\ldots,q} = \begin{pmatrix} aE_m & 0 \\ 0 & bE_{q-m} \end{pmatrix} + \begin{pmatrix} p_1 I_{m,m} & p_2 I_{m,q-m} \\ q_2 I_{q-m,m} & q_1 I_{q-m,q-m} \end{pmatrix}, \quad (9.6)$$

where $I_{k,l}$ is the $k \times l$ matrix which has all matrix elements equal to 1 and E_k is the k-dimensional unity matrix. Here necessarily $a = 1 - mp_1 - (q-m)p_2$ and $b = 1 - mq_2 - (q-m)q_1$, so that the matrix is stochastic.

Consider the action of the fuzzy map on probability vectors and introduce the linear map from $\mathbb{R}^q$ to $\mathbb{R}^2$ given by

$$Lv = \begin{pmatrix} \sum_{i=1}^m v_i \\ \sum_{i=m+1}^q v_i \end{pmatrix}.$$

In matrix form:

$$L = \begin{pmatrix} I_{1,m} & 0 \\ 0 & I_{1,q-m} \end{pmatrix}.$$

Lemma 9.1. *We have*

$$L\mathbb{P} = \left(\begin{pmatrix} a & 0 \\ 0 & b \end{pmatrix} + \begin{pmatrix} mp_1 & mp_2 \\ (q-m)q_2 & (q-m)q_1 \end{pmatrix} \right) L =: \mathbb{P}' L,$$

and

$$L\mathbb{P}^t = \left(\begin{pmatrix} a & 0 \\ 0 & b \end{pmatrix} + \begin{pmatrix} mp_1 & mq_2 \\ (q-m)p_2 & (q-m)q_1 \end{pmatrix} \right) L = \hat{\mathbb{P}}^t L,$$

where $\hat{\mathbb{P}}$ (see (9.3), (9.4)) is the transition matrix for the coarse-grained chain with two states $1', 2'$. By $\mathbb{P}^t$ we denote the transpose of the matrix $\mathbb{P}$.

Proof. Straightforward. □

The following proposition describes all eigenvalues of $\mathbb{P}^t$, which we need below.

Proposition 9.1. *The matrix $\mathbb{P}^t$ is diagonalisable and has the eigenvalues $1, \lambda_2(\hat{\mathbb{P}}), a, b$. The dimension of the eigenspace to a is $m-1$. The dimension of the eigenspace to b is $q-m-1$.*

Proof. Suppose v is a right-eigenvector of $\mathbb{P}^t$ for the eigenvalue λ, so $\mathbb{P}^t v = \lambda v$. Then using Lemma 9.1 we get $\hat{\mathbb{P}}^t Lv = L\mathbb{P}^t v = \lambda Lv$.

Two cases are possible:

(1) $Lv \neq 0$ and hence λ is an eigenvalue for $\hat{\mathbb{P}}$, too. The two eigenvalues for the two-by-two matrix can be easily evaluated, one eigenvalue is equal to 1, call the other one $\lambda_2(\hat{\mathbb{P}})$.

(2) $Lv = 0$. Then we must have

$$\begin{aligned} \mathbb{P}^t v \equiv \mathbb{P}^t \begin{pmatrix} u \\ w \end{pmatrix} &= \begin{pmatrix} aE_m & 0 \\ 0 & bE_{q-m} \end{pmatrix} \begin{pmatrix} u \\ w \end{pmatrix} \\ &= \lambda \begin{pmatrix} u \\ w \end{pmatrix} \end{aligned} \tag{9.7}$$

To have an eigenvalue v at least one of the components u or w has to be nonzero. Consequently the possible eigenvalues are a, b.

The eigenvectors corresponding to a are of the form $\begin{pmatrix} u \\ 0 \end{pmatrix}$ with $\sum_{i=1}^m u_i = 0$. We have $m - 1$ linearly independent of them.

The eigenvectors corresponding to b are of the form $\begin{pmatrix} 0 \\ w \end{pmatrix}$ with $\sum_{i=m+1}^q w_i = 0$. Clearly we have $q - m - 1$ of them. □

Below to check the (non-)extremality of a TISGM we need to the explicit formulas for the eigenvalues of the matrix $\mathbb{P}$, which are given by the following corollary of Proposition 9.1.

Corollary 9.1. *The matrix $\mathbb{P}$ defined by (9.2) with $m \leq [q/2]$ has the following eigenvalues*

$$\begin{aligned} &\{1, b, \lambda_2(\hat{\mathbb{P}})\}, \quad \textit{if} \ \ m = 1 \\ &\{1, a, b, \lambda_2(\hat{\mathbb{P}})\}, \quad \textit{if} \ \ m \geq 2, \end{aligned} \tag{9.8}$$

where

$$a = \frac{(\theta - 1)z}{Z_1}, \quad b = \frac{(\theta - 1)\sqrt[k]{z}}{Z_1}, \quad \lambda_2(\hat{\mathbb{P}}) = \frac{[\theta - 1 + (1 - \sqrt[k]{z})m]z}{Z_1}. \tag{9.9}$$

Proof. Since z is a solution to (8.7) we have $Z_1 = \sqrt[k]{z} Z_2$. In case $m = 1$ we have $p_1 = 0$ and the above mentioned condition $Lu = 0$ gives $u = u_1 = 0$, i.e., a can not be an eigenvalue. For $m \geq 2$ using $a + p_1 = \frac{\theta z}{Z_1}$, $p_1 = \frac{z}{Z_1}$, $b + q_1 = \frac{\theta}{Z_2}$, $q_1 = \frac{1}{Z_2}$ and (9.5) we get (9.8). □

For a, b, λ_2 given by (9.8) we denote

$$\widehat{\lambda} = \widehat{\lambda}(k, q, m, \theta) = \max\{a, b, |\lambda_2(\hat{\mathbb{P}})|\}.$$

For $k = 2$, using formula (8.12) of $z \in \{x_1^2, x_2^2\}$, for $m = 1$ we obtain $\widehat{\lambda} = b$ and for $m \geq 2$ we have

$$|\lambda_2(\hat{\mathbb{P}})| < \widehat{\lambda} = \begin{cases} b, & \text{if } z < 1 \\ a, & \text{if } z > 1. \end{cases} \tag{9.10}$$

Define

$$g(z) = \sqrt[k]{f_m(z)} = \frac{(\theta + m - 1)z + q - m}{mz + q - m - 1 + \theta}.$$

The following simple lemma is useful

Lemma 9.2. *We have*

(i) For any $k \geq 2$ and $\theta > 1$ the function $g(z)$, $z > 0$ has the following properties:

(a) $\{z : g^k(z) = f_m(z) = z\} = \{1, z_1, z_2\}$;
(b) $a < g(z) < A$, with $a = \frac{q-m}{q+\theta-m-1}$, $A = \frac{\theta+m-1}{m}$;
(c) $\frac{d}{dz}g(z) = \frac{(\theta-1)(\theta+q-1)}{(mz+\theta+q-m-1)^2} > 0$;
(d) $\frac{d^2}{dz^2}g(z) < 0$, $z > 0$.

(ii) If $k = 2$ and $m \leq q/2$ then, for the solutions z_1 and z_2 mentioned in Proposition 8.2, the following statements hold

$$1 < z_1 = z_2, \quad \text{if } \theta = \theta_m$$

$$1 < z_1 < z_2, \quad \text{if } \theta_m < \theta < \theta_c, \quad \text{with } \theta_c = q + 1$$

$$1 = z_1 < z_2, \quad \text{if } \theta = q + 1$$

$$z_1 < 1 < z_2, \quad \text{if } q + 1 < \theta.$$

In Fig. 9.1 the functions $z_i = x_i^2(m, \theta)$, $i = 1, 2$ are shown for $k = 2$, $q = 8$, $m = 1, 2, 3$.

9.2 Non-extreme TISGMs

Let us first introduce some parameters (as functions of parameters of the Potts model on Cayley trees):

$$\theta_m \equiv \theta_m(k) = 1 + \frac{mx_*^k + q - m}{x_*^{k-1} + x_*^{k-2} + \cdots + x_*}, \tag{9.11}$$

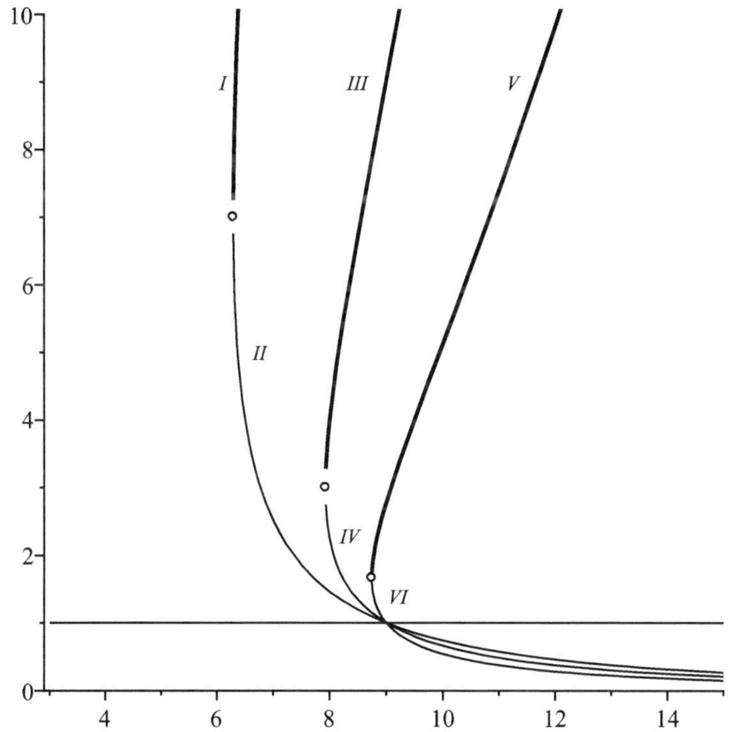

Fig. 9.1 The graphs of the functions $z_i = z_i(m, \theta)$, $i = 1, 2$, for $k = 2$, $q = 8$ and $m = 1, 2, 3$. The horizontal axis presents θ and the vertical axis presents z_i. Circle dots having coordinates $(\theta_m, \frac{q-m}{m})$, $m = 1, 2, 3$ separate graph of z_1 from graph of z_2. The graph of $z_2(1, \theta)$ is the curve I. The graph of $z_1(1, \theta)$ is the curve II. $z_2(2, \theta)$ is III. $z_1(2, \theta)$ is IV, $z_2(3, \theta)$ is V. $z_1(3, \theta)$ is VI. The intersection point of z_1 functions has coordinate $(\theta_c, 1) = (9, 1)$. For $m = 4$ the graphs of the functions $z_i(4, \theta)$ are similar to the above given graphs, but the circle point separating them coincides with the point $(9, 1)$ of the intersections, since $\theta_{[\frac{8}{2}]=4} = \theta_c = 9$.

where x_* is the unique solution (see Step 3 in proof of Theorem 8.3) of the following equation

$$m \sum_{i=1}^{k-1} i x^{2k-i-1} - (q - m) \sum_{i=1}^{k-1} i x^{i-1} = 0. \tag{9.12}$$

In the case of three solutions, we denote the middle solution by z_{mid}, which is unique element of the following set

$$\{1, z_1, z_2\} \setminus \{\min\{1, z_1, z_2\}, \max\{1, z_1, z_2\}\}.$$

By Lemma 9.2, for $2m < q$ we have $z_{\rm mid} = \max\{1, z_1\}$.

Recall that μ_i is the TISGM of the Potts model which corresponds to the boundary law $z_i(m)$, for the branches $i = 1, 2$.

Define the following numbers:

$$\widehat{\theta} = (\sqrt{2}-1)q + 2m + 1, \quad \theta^* = 1 + (\sqrt{2}+1)q - 2m. \tag{9.13}$$

Theorem 9.1. *Let $k = 2$, $2m < q$. Then the following statements hold.*

(i) Assume one of the following conditions is satisfied:

(a) $2 \le m \le q/7$ and $\theta \in [\theta_m, \widehat{\theta})$;
(b) $\theta \in (\theta^*, +\infty)$.

Then $\mu_1(\theta, m)$ is non-extreme.

(ii) Assume one of the following conditions is satisfied:

(c) $2 \le m \le q/7$ and $\theta \ge \theta_m$;
(d) $q < 7m$, $m \ge 2$ and $\theta \in (\widehat{\theta}, +\infty)$.

Then $\mu_2(\theta, m)$ is non-extreme (see Figs. 9.2–9.4).

Proof.

(i) *Case $m \ge 2$.*

Subcase: $\theta_m \le \theta < q+1$. By Lemma 9.2 we have $z_1 > 1$. For $m \ge 2$ using (9.10) and Corollary 9.1 one can see that the second largest, in absolute value, eigenvalue of $\mathbb{P}$ is a so to prove non-extremality one should check the Kesten-Stigum condition $2a^2 > 1$. Since $a > 0$ the last inequality is equivalent to $\sqrt{2}a > 1$. Denote

$$\gamma_1(\theta) = \sqrt{2}a - 1 = \sqrt{2}\frac{(\theta-1)z_1}{(\theta+m-1)z_1 + q - m} - 1,$$

where

$$z_1 = \left(\frac{\theta - 1 - \sqrt{(\theta-1)^2 - 4m(q-m)}}{2m}\right)^2.$$

We have

$$\gamma_1'(\theta) = \frac{\sqrt{2}(\theta-1)z_1}{((\theta+m-1)z_1+q-m)^2}\left(\sqrt{z_1} - \frac{2(q-m)}{\sqrt{(\theta-1)^2-4m(q-m)}}\right)$$
$$= -\frac{\sqrt{2z_1}(\theta-1)^2 z_1}{((\theta+m-1)z_1+q-m)^2\sqrt{(\theta-1)^2-4m(q-m)}}.$$

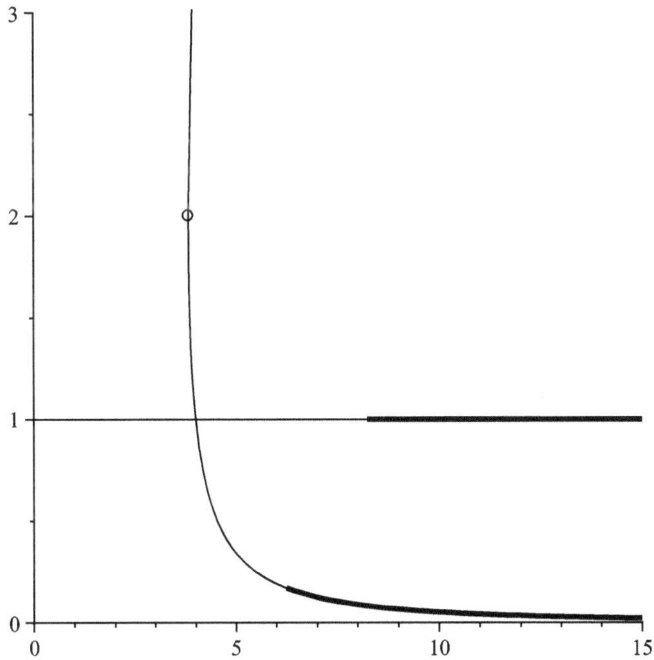

Fig. 9.2 The graphs of the functions $z_i = z_i(m, \theta)$, $i = 1, 2$, for $q = 3$ and $m = 1$. The circle dot having coordinate $(\theta_1, 2)$ separates graph of z_1 from graph of z_2. The bold curves correspond to regions of solutions where the corresponding TISGM is non-extreme. This figure corresponds to Part (i), (b) of Theorem 9.1.

Thus for any $\theta \geq \theta_m$ we have $\gamma_1'(\theta) < 0$. Hence γ_1 is a decreasing function of θ. So to have $\sqrt{2}a > 1$ it is necessary that

$$\gamma_1(\theta_m) = \sqrt{2}\frac{\sqrt{q-m}}{\sqrt{q-m}+\sqrt{m}} - 1 > 0,$$

which is satisfied for $q \geq 7m$. Moreover, we have $\gamma_1(q+1) = -(\frac{\sqrt{2}-1}{\sqrt{2}}) < 0$. Consequently, there exists a unique $\theta = \widehat{\theta}$ such that $\gamma_1(\widehat{\theta}) = 0$. To find $\widehat{\theta}$ we solve $\gamma_1(\widehat{\theta}) = 0$ and by long (but simple) computations, using the explicit formula for $\gamma_1(\theta)$ and the expression for z_1 we get

$$\widehat{\theta} = (\sqrt{2}-1)q + 2m + 1.$$

Thus we proved that $2a^2 > 1$ iff $\theta < \widehat{\theta}$.

Subcase: $\theta \geq q+1$. In this case by Lemma 9.2 we have $z_1 < 1$ so by (9.10) and Corollary 9.1 we must check $2b^2 > 1$. Define

$$\xi_1(\theta) = \sqrt{2}b - 1 = \frac{(\theta-1)\sqrt{2z_1}}{(\theta+m-1)z_1+q-m} - 1,$$

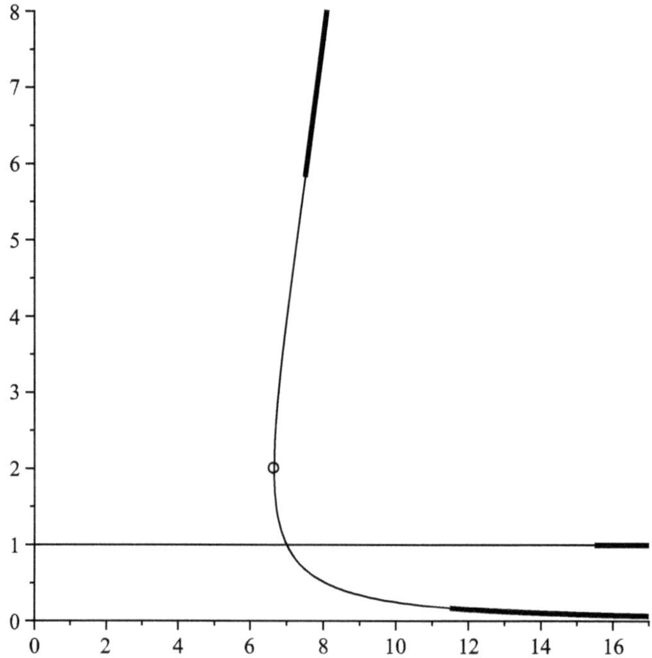

Fig. 9.3 The graphs of the functions $z_i = z_i(m, \theta)$, $i = 1, 2$, for $q = 6$ and $m = 2$. The circle dot having coordinate $(\theta_2, 2)$ separates graph of z_1 from graph of z_2. The bold lines correspond to regions of solutions where corresponding TISGM is non-extreme. The upper bold curve corresponds to part (ii), (d) and the lower bold curve corresponds to part (i), (b) of Theorem 9.1.

We have

$$\xi_1'(\theta) = \frac{\sqrt{2z_1}(\theta-1)^2 z_1}{((\theta+m-1)z_1+q-m)^2\sqrt{(\theta-1)^2-4m(q-m)}} > 0.$$

Hence ξ_1 is an increasing function of θ. Moreover, we have

$$\xi_1(q+1) = \gamma_1(q+1) = -(\frac{\sqrt{2}-1}{\sqrt{2}}) < 0.$$

Since $\xi_1(\theta)$ is increasing it is bounded from above by its limit for $\theta \to +\infty$. So we compute

$$\begin{aligned} \lim_{\theta\to\infty} \xi_1(\theta) &= -1 + \sqrt{2} \lim_{\theta\to\infty} \frac{(\theta-1)\sqrt{z_1}}{(\theta+m-1)z_1+q-m} \\ &= -1 + \sqrt{2} \lim_{\theta\to\infty} \frac{1}{\left(\frac{\theta+m-1}{\theta-1}\right)\sqrt{z_1} + \frac{q-m}{(\theta-1)\sqrt{z_1}}}. \end{aligned} \tag{9.14}$$

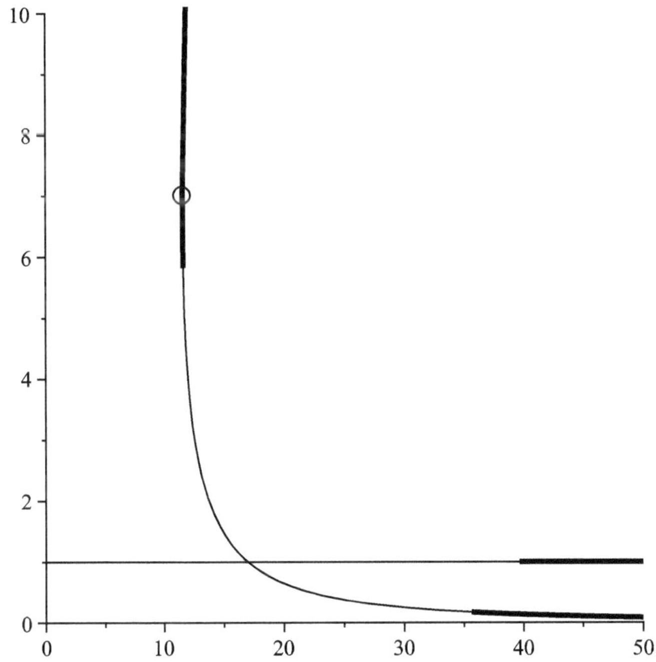

Fig. 9.4 The graphs of the functions $z_i = z_i(m, \theta)$, $i = 1, 2$, for $q = 16$ and $m = 2$. The circle dot having coordinate $(\theta_2, 7)$ separates graph of z_1 from graph of z_2. The bold lines correspond to regions of solutions where corresponding TISGM is non-extreme. The bold z_2 corresponds to Part (ii), (c); Bold parts of z_1 correspond to Part (i), (a) (upper bold curve) and Part (i), (b) (lower bold curve) of Theorem 9.1.

Using formula (8.12) we get

$$\lim_{\theta\to\infty} \sqrt{z_1} = \lim_{\theta\to\infty} \frac{2(q-m)}{\theta - 1 + \sqrt{(\theta-1)^2 - 4m(q-m)}} = 0.$$

$$\lim_{\theta\to\infty} (\theta - 1)\sqrt{z_1} = \lim_{\theta\to\infty} \frac{2(q-m)}{1 + \sqrt{1 - \frac{4m(q-m)}{(\theta-1)^2}}} = q - m.$$

Using these formulas we get from (9.14)

$$\lim_{\theta\to\infty} \xi_1(\theta) = \sqrt{2} - 1 > 0.$$

Consequently, there exists a unique $\theta = \theta^*$ such that $\xi_1(\theta^*) = 0$. To find θ^* we solve $\xi_1(\theta^*) = 0$ and by simple computations using the explicit formula of $\xi_1(\theta)$ we get

$$\theta^* = 1 + (\sqrt{2} + 1)q - 2m.$$

Thus we proved that $2b^2 > 1$ if $\theta > \theta^*$.

Case $m = 1$. In this case, as mentioned before (9.10) the second largest eigenvalue is b, therefore we have to check $\xi_1(\theta) = \sqrt{2}b - 1 > 0$. We note that $\xi_1'(\theta) > 0$ for each $\theta \geq \theta_m$, and

$$\xi_1(\theta_m) = \frac{\sqrt{2m}}{\sqrt{q-m}+\sqrt{m}} - 1 < 0.$$

Hence again we have $\xi_1(\theta) > 0$ iff $\theta \in (\theta^*, +\infty)$.

(ii) *Case* $m \geq 2$. Since $z_2 > 1$ for any $\theta \geq \theta_m$, using (9.10) and Corollary 9.1 we shall check only $2a^2 > 1$. Denote

$$\gamma_2(\theta) = \sqrt{2}a - 1 = \sqrt{2}\frac{(\theta-1)z_2}{(\theta+m-1)z_2+q-m} - 1,$$

where

$$z_2 = \left(\frac{\theta - 1 + \sqrt{(\theta-1)^2 - 4m(q-m)}}{2m}\right)^2.$$

We have

$$\gamma_2'(\theta) = \frac{\sqrt{2}(\theta-1)z_2}{((\theta+m-1)z_2+q-m)^2}\left(\sqrt{z_2} + \frac{2(q-m)}{\sqrt{(\theta-1)^2-4m(q-m)}}\right) > 0.$$

Thus for any $\theta \geq \theta_m$ we have $\gamma_2'(\theta) > 0$. Hence γ_2 is an increasing function of θ. For $q \geq 7m$ we have

$$\gamma_2(\theta_m) = \gamma_1(\theta_m) > 0.$$

Hence $\gamma_2(\theta) > 0$ for all $\theta > \theta_m$. Consequently, μ_2 is non-extreme.

In case $q \leq 6$ or $q \geq 7$ with $[q/7] < m$ we have $\gamma_1(\theta_m) = \gamma_2(\theta_m) < 0$. Since γ_2 is an increasing function $\gamma_2(\theta) = 0$ has a unique solution, which is equal to $\widehat{\theta}$. Thus $\gamma_2(\theta) > 0$ for all $\theta > \widehat{\theta}$.

Case $m = 1$. Define

$$\xi_2(\theta) = \sqrt{2}b - 1 = \frac{(\theta-1)\sqrt{2z_2}}{\theta z_2 + q - 1} - 1.$$

We have

$$\xi_2'(\theta) = -\frac{\sqrt{2z_2}(\theta-1)^2 z_2}{(\theta z_2+q-1)^2\sqrt{(\theta-1)^2-4(q-1)}} < 0.$$

Hence ξ_2 is a decreasing function of θ. Moreover, we have

$$\xi_2(\theta_m) = \xi_1(\theta_m) = -(1 - \frac{\sqrt{2}}{\sqrt{q-1}+1}) < 0.$$

Thus $\xi_2(\theta) < 0$ for all $\theta \geq \theta_m$.

This completes the proof. □

9.3 Extreme TISGMs

In this section following [139], [249] we give sufficient conditions for extremality (or non-reconstructability in information-theoretic language) of the full chains of the Potts model, depending on coupling strength parameterized by θ, the block size m and the branch of the boundary law z.

Recall that by $\mu_i(\theta, m)$ we denote the TISGM which corresponds to the values of the boundary law $z_i(\theta, m)$, $i = 1, 2$, which is a solution to (8.7) and $\theta_1 = 1 + 2\sqrt{q-1}$ (see (8.13)) and $\theta^* = 1 + (\sqrt{2}+1)q - 2m$ (see (9.13)).

Theorem 9.2. *If $k = 2$, $m = 1$ then the following is true.*

(a) If $q = 3, 4, \ldots, 16$ then there exists θ^{**} such that $\theta_c = q + 1 < \theta^{**} < \theta^*$ and the measure $\mu_1(\theta, 1)$ is extreme for any $\theta \in [\theta_1, \theta^{**})$. Moreover θ^{**} is the unique positive solution of the following equation

$$\theta^3 - (q-3)\theta^2 - (2q-7)\theta - (q+5) = 0.$$

If $q \geq 17$ then there are $\bar{\theta}_1, \bar{\theta}_2 \in (\theta_1, \theta_c)$ such that $\bar{\theta}_1 < \bar{\theta}_2$ and the measure $\mu_1(\theta, 1)$ is extreme for any $\theta \in [\theta_1, \bar{\theta}_1) \cup (\bar{\theta}_2, \theta^{**})$. Moreover $\bar{\theta}_1, \bar{\theta}_2$ are positive solutions of the following equation

$$\theta^3 - (q-1)\theta^2 - (2q-3)\theta + (4q^2 - 13q + 11) = 0.$$

(b) The measure $\mu_2(\theta, 1)$ is extreme for any $\theta \geq \theta_1$, $q \geq 2$ (see Fig. 9.5)

Proof. In this section we use the arguments mentioned in Subsection 2.3.4. It is easy to see that the channel $\mathbb{P}$ corresponding to a TISGM of the Potts model is ergodic and permissive. Thus by Theorem 2.3 and Lemma 2.1 the criterion of *extremality* of a TISGM is $k\kappa\gamma < 1$.

The constant γ does not have a clean general formula, but can be estimated in specific models (as Ising, Hard-Core etc.). For example, if $\mathbb{P}$ is the channel of the Potts model then (see Theorem 8.1 in page 160 of [176]):

$$\gamma \leq \frac{\theta - 1}{\theta + 1}. \tag{9.15}$$

Consider the case $z \neq 1$ (where $z = x^2$ and z is a solution to (8.7)), fix a solution of (8.5), which has the form $(\underbrace{z, z, \ldots, z}_{m}, 1, \ldots, 1)$ and the corresponding matrix $\mathbb{P}$ given by (9.2).

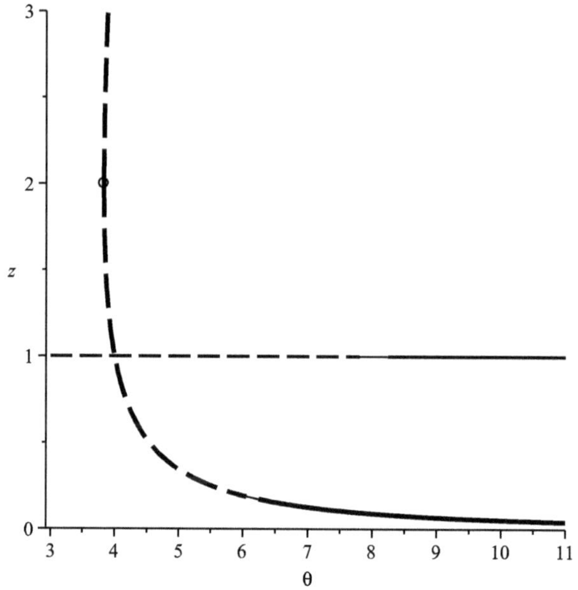

Fig. 9.5 The graphs of the functions $z_i = z_i(m, \theta)$, for $q = 3$, $m = 1$ and the graph of $z(\theta) \equiv 1$. The bold curves correspond to regions of solutions where the corresponding TISGM is non-extreme (corresponds to part (ii), b) of Theorem 9.1). The dashed bold curves correspond to regions of solutions where the corresponding TISGM is extreme (see parts (a) and (b) of Theorem 9.2). The gaps between the two types of curves are given by thin curves.

Let us now compute the constant κ. Recall matrix $\mathbb{P}$ given in (9.2) and Z_1, Z_2 given after formula (9.2). Since z is a solution to (8.7) we have $Z_1 = \sqrt[k]{z} Z_2$. Using this formula and (9.2) we get

$$\frac{1}{2}\sum_{l=1}^{q}|P_{il} - P_{jl}| = \begin{cases} a, & \text{if } \ i, j = 1, \dots, m \\ b, & \text{if } \ i, j = m+1, \dots, q \\ c, & \text{otherwise}, \end{cases}$$

where a and b are defined in (9.9) and

$$c = \frac{1}{2Z_1}\left(z|\theta - \sqrt[k]{z}| + |1 - \theta\sqrt[k]{z}| + (z(m-1) + q - m - 1)|1 - \sqrt[k]{z}|\right). \quad (9.16)$$

Hence we get

$$\kappa = \begin{cases} \max\{b, c\}, & \text{if } \ m = 1 \\ \max\{a, b, c\} & \text{if } \ m \geq 2. \end{cases} \quad (9.17)$$

Write the equation (8.7) for the conditions of theorem: $k = 2$, $m = 1$.

First take the square root on both sides, then obtain a cubic equation for $\sqrt{z}$ which by dividing out the solution $\sqrt{z} = 1$ can be simplified to $z - (\theta - 1)\sqrt{z} + q - 1 = 0$. For the solutions $\sqrt{z_1}$, $\sqrt{z_2}$ of the last equation one has $\sqrt{z_1} + \sqrt{z_2} = \theta - 1$. Using this equality we get

$$\sqrt{z_1} \leq \sqrt{z_1} + \sqrt{z_2} = \theta - 1 < \theta.$$

Similarly, one gets $\theta > \sqrt{z_2}$.

From these inequalities it follows that $|\theta - \sqrt{z_i}| = \theta - \sqrt{z_i}$. Since $\theta > 1$ for $z_i \geq 1$ we get

$$|1 - \theta\sqrt{z_i}| = \theta\sqrt{z_i} - 1, \quad |1 - \sqrt{z_i}| = \sqrt{z_i} - 1.$$

Simplify now the expression for c (see (9.16), for $k = 2$, $m = 1$) using the above mentioned equalities for the absolute values present in c and, using that z_i is a solution to $z - (\theta - 1)\sqrt{z} + q - 1 = 0$ (i.e., replacing z by $(\theta - 1)\sqrt{z} - q + 1$), finally we obtain

$$c = \frac{1}{Z_1}\left((\theta + q - 2)\sqrt{z_i} - (q - 1)\right). \tag{9.18}$$

Recall $b = \frac{(\theta-1)\sqrt{z_i}}{Z_1}$ (see (9.9)). For $z_i \geq 1$, using b and c given by (9.18) it is straightforward to see that $b \leq c$ (with equality if and only if $z_i = 1$). Consequently, using (9.17) under condition $m = 1$ we get $\kappa = c$.

(a) *Case*: $z_1 \geq 1$. For $\theta \in [\theta_1, \theta_c] = [1 + 2\sqrt{q-1}, q + 1]$ we have $z_1 \geq 1$. By (9.15) and $\kappa = c$ we have

$$2\gamma\kappa \leq 2c\frac{\theta - 1}{\theta + 1} = \frac{2(\theta - 1)((\theta + q - 2)\sqrt{z_1} - (q - 1))}{(\theta z_1 + q - 1)(\theta + 1)}.$$

Thus, to check $2\gamma\kappa < 1$ it is sufficient to check

$$\frac{2(\theta - 1)((\theta + q - 2)\sqrt{z_1} - (q - 1))}{(\theta z_1 + q - 1)(\theta + 1)} < 1. \tag{9.19}$$

Since z_1 is a solution to $z - (\theta - 1)\sqrt{z} + q - 1 = 0$ we can replace in (9.19) z_1 by $(\theta - 1)\sqrt{z_1} - q + 1$, simplify and put

$$\sqrt{z_1} = \frac{\theta - 1 - \sqrt{(\theta - 1)^2 - 4(q - 1)}}{2}.$$

Then denoting

$$\psi_1(\theta) = (\theta - 1)(\theta^2 - \theta + 6 - 4q) - (\theta^2 - \theta + 4 - 2q)\sqrt{(\theta - 1)^2 - 4(q - 1)},$$

we obtain that the inequality (9.19) is equivalent to $\psi_1(\theta) > 0$, $\theta \in [\theta_1, \theta_c)$.

By $(\theta-1)^2 > 4(q-1)$, i.e., $\theta > \theta_1$ we get

$$\theta^2 - \theta + 6 - 4q > 0, \quad \theta^2 - \theta + 4 - 2q > 0.$$

Using these inequalities the inequality $\psi_1(\theta) > 0$ can be simplified to

$$u(\theta) = \theta^3 - (q-1)\theta^2 - (2q-3)\theta + (4q^2 - 13q + 11) > 0, \quad \theta \in [\theta_1, \theta_c]. \tag{9.20}$$

The function $u(\theta)$ for $\theta \in (0, \theta_c)$ has a minimum at

$$\bar{\theta} = \bar{\theta}(q) = \frac{1}{3} \cdot (q - 1 + \sqrt{q^2 + 4q - 8}).$$

Now calculate the minimal value $u(\bar{\theta})$. Since $\bar{\theta}$ is a solution to

$$u'(\theta) = 3\theta^2 - 2(q-1)\theta - (2q-3) = 0$$

in $u(\bar{\theta})$ replacing $\bar{\theta}^2$ by $(1/3)(2(q-1)\bar{\theta} + (2q-3))$ (in this way we reduce the powers of $\bar{\theta}$). By replacing $\bar{\theta}^3$ with $(\bar{\theta}/3)(2(q-1)\bar{\theta} + (2q-3))$ and simplifying $u(\bar{\theta})$ one gets

$$u(\bar{\theta}) = (8 - 4q - q^2)\bar{\theta} + (48 - 56q + 17q^2).$$

Now using the formula of $\bar{\theta}$ from the last equality we get

$$u(\bar{\theta}) = -q^3 + 48q^2 - 156q + 136 - (q^2 + 4q - 8)^{3/2}. \tag{9.21}$$

From (9.21) it follows that $u(\bar{\theta}) > 0$ iff $q = 3, 4, \ldots, 16$.

Consequently, $u(\theta) \geq u(\bar{\theta}) > 0$ (i.e., $\psi_1(\theta) > 0$) and $\mu_1(\theta, 1)$ is extreme for *all* $\theta \in [1 + 2\sqrt{q-1}, q+1]$, $q = 3, 4, \ldots, 16$.

For $q \geq 17$, using the explicit formulas for $\theta_1, \bar{\theta}, \theta_c$, one gets $\theta_1 < \bar{\theta} < \theta_c$ and $u(\bar{\theta}) < 0$. Moreover,

$$u(\theta_1) = 4(q + 2\sqrt{q-1}) > 0, \quad u(\theta_c) = 4((q-1)^2 + 3) > 0.$$

Consequently, the equation $u(\theta) = 0$ has two solutions $\bar{\theta}_1$ and $\bar{\theta}_2$ such that

$$\theta_1 < \bar{\theta}_1 < \bar{\theta} < \bar{\theta}_2 < \theta_c.$$

Thus $u(\theta) > 0$ and the measure $\mu_1(\theta, 1)$ is extreme for any

$$\theta \in [\theta_1, \bar{\theta}_1) \cup (\bar{\theta}_2, q+1].$$

Case: $z_1 < 1$. In this case, using the explicit formula for $\sqrt{z_1}$ and the condition $\theta > \theta_1$ we get $\theta\sqrt{z_1} > 1$. Indeed, this inequality is equivalent to $(q-2)\theta^2 + \theta + 1 > 0$, which holds for any $\theta > 1$. Consequently, using $\theta > \sqrt{z_1}$, $\theta\sqrt{z_1} > 1$ and $z_1 < 1$ from formula (9.16) for $m = 1$, for all possible values of $q \geq 2$ we get

$$c = \frac{1}{Z_1}(\theta\sqrt{z_1} - 1).$$

Recall again $b = \frac{(\theta-1)\sqrt{z_1}}{Z_1}$.

Then

$$b \geq c \;\Leftrightarrow\; \frac{(\theta-1)\sqrt{z_1}}{Z_1} \geq \frac{1}{Z_1}(\theta\sqrt{z_1}-1) \;\Leftrightarrow\; (\theta-1)\sqrt{z_1} \geq \theta\sqrt{z_1}-1 \;\Leftrightarrow\; \sqrt{z_1} \leq 1.$$

Thus if $z_1 \leq 1$, i.e., $\theta \in (\theta_c, +\infty)$ then $b \geq c$.

Consequently, $\kappa = b$. Hence

$$2\gamma\kappa \leq 2b \cdot \frac{\theta-1}{\theta+1}.$$

Therefore, it suffices to check

$$2b \cdot \frac{\theta-1}{\theta+1} = \frac{2(\theta-1)^2\sqrt{z_1}}{(\theta+1)(\theta z_1+q-1)} < 1. \tag{9.22}$$

Since z_1 is a solution to $z_1 - (\theta-1)\sqrt{z_1} + q - 1 = 0$, replacing z_1 by $(\theta-1)\sqrt{z_1} - q + 1$ and using the explicit formula for $\sqrt{z_1}$ the inequality (9.22) becomes after some simplifications

$$f(\theta) = \theta^3 - (q+3)\theta^2 - (2q-7)\theta - (q+5) < 0.$$

Using the Descartes rule (compare part (iv) of Theorem 9.1) for $f(\theta)$ we see that the equation $f(\theta) = 0$ has at most one positive solution. Moreover, for any $q \geq 3$ we have $f(q+1) = -4q^2 < 0$.

Now let us show that $f(\theta^*) > 0$, where $\theta^* = (\sqrt{2}+1)q - 1$ (see (9.13) for $m = 1$). Indeed, we have

$$Q(q) \equiv f(\theta^*) = (4+3\sqrt{2})q^3 - 6(3+2\sqrt{2})q^2 + 16(1+\sqrt{2})q - 16,$$

and

$$Q'(q) = 3(4+3\sqrt{2})q^2 - 12(3+2\sqrt{2})q + 16(1+\sqrt{2}).$$

The equation $Q'(q) = 0$ has two positive solutions:

$$q_{\pm} = \frac{2}{3} \cdot \frac{9+6\sqrt{2} \pm \sqrt{33+24\sqrt{2}}}{4+3\sqrt{2}}.$$

We have $q_- < q_+ < 3$. Therefore $Q(q)$ is an increasing function of $q \geq 3$.

Consequently, for any $q \geq 3$ we have

$$Q(q) \geq Q(3) = 21\sqrt{2} - 22 > 0.$$

So there exists a unique θ^{**} with $q+1 < \theta^{**} < \theta^*$ such that $f(\theta^{**}) = 0$. Thus $f(\theta) < 0$ and $\mu_1(\theta, 1)$ is extreme for $\theta \in [q+1, \theta^{**})$. Summarizing the results for the cases $z_1 \geq 1$ and $z_1 < 1$ we get assertion (a).

For $q = 3$ a numerical analysis shows that $\theta^{**} \approx 6.053$. So $\theta^* - \theta^{**} = 0.1896$ (see Fig. 9.6). This is the length of the gap between the Kesten-Stigum threshold beyond which non-extremality certainly holds and our bound below which extremality certainly holds.

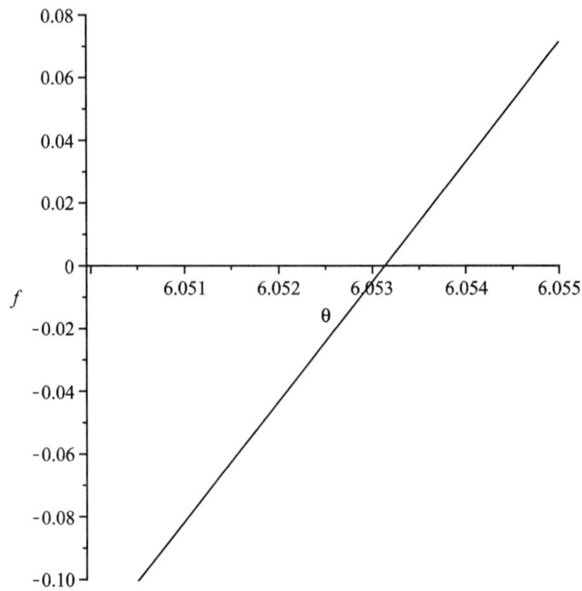

Fig. 9.6 The graph of the function $f(\theta)$, for $q = 3$, given in the interval $[6.05, 6.055]$ indicates that the solution of $f(\theta^{**}) = 0$ is $\theta^{**} \approx 6.053$.

(b) For $\theta \geq \theta_1 = 1 + 2\sqrt{q-1}$ we have $z_2 > 1$, i.e., z_2 is larger than 1 as soon as it exists. Similarly as the case (a) we shall check $2c\frac{\theta-1}{\theta+1} < 1$.

Denoting

$$\psi_2(\theta) = (\theta - 1)(\theta^2 - \theta + 6 - 4q) + (\theta^2 - \theta + 4 - 2q)\sqrt{(\theta-1)^2 - 4(q-1)},$$

and using the fact that z_2 is a solution to $z_2 - (\theta - 1)\sqrt{z_2} + q - 1 = 0$, replacing z_2 by $(\theta - 1)\sqrt{z_2} - q + 1$ and using the explicit formula

$$\sqrt{z_2} = \frac{\theta - 1 + \sqrt{(\theta-1)^2 - 4(q-1)}}{2},$$

after simplifications we get that the last inequality is equivalent to $\psi_2(\theta) > 0$, $\theta \geq \theta_1$.

Note that

$$\theta^2 - \theta + 6 - 4q \leq \theta^2 - \theta + 4 - 2q,$$

so if we get

$$\theta^2 - \theta + 6 - 4q > 0, \tag{9.23}$$

then it follows that $\psi_2(\theta) > 0$. Since $\theta > 1$ the inequality (9.23) is equivalent to $\theta > (1+\sqrt{16q-23})/2$. Moreover, one can easily check that

$$(1+\sqrt{16q-23})/2 < 1+2\sqrt{q-1} = \theta_1.$$

Thus $\psi_2(\theta) > 0$ and $\mu_2(\theta, 1)$ is extreme for *any* $\theta \geq 1+2\sqrt{q-1}$. □

The following theorem corresponds to the case: $m \geq 2$. From condition $2 \leq m \leq [q/2]$ it follows that $q \geq 4$.

Theorem 9.3. *Let $k = 2$.*

(i) If $m = 2$ then the following is true.

(i.1) If $m = 2$ then for each $q = 4, 5, 6, 7, 8$ there exists $\breve{\theta} > \theta_c = q+1$ such that the measure $\mu_1(\theta, 2)$ is extreme for any $\theta \in [\theta_2, \breve{\theta})$.

(i.2) For each $q \geq 9$ there exists $\theta^\dagger \in (\theta_2, q+1)$ such that the measure $\mu_1(\theta, 2)$ is extreme for any $\theta \in [\theta^\dagger, \breve{\theta})$, where $\theta^\dagger = \theta^\dagger(q)$ is the unique solution of

$$\theta^3 - (q+3)\theta^2 + (6q-17)\theta - (9q-19) = 0,$$

and $\breve{\theta} = \breve{\theta}(q)$ is the unique solution of

$$\theta^3 - (q+3)\theta^2 - (2q-15)\theta - (q+13) = 0.$$

(ii) If $m = 2$ then for each $q = 4, 5, 6, 7, 8$ there exists $\grave{\theta} = \grave{\theta}(q)$ such that $\theta_2 < \grave{\theta} \leq q+1$ and $\mu_2(\theta, 2)$ is extreme for $\theta \in [\theta_2, \grave{\theta})$ (see Fig. 9.7).

(iii) If $q < \frac{m+1}{2m}\left[3m+1+\sqrt{m^2+6m+1}\right]$ and $m \geq 2$ then the measure $\mu_1(\theta_m, m) = \mu_2(\theta_m, m)$ is extreme.

Proof. Write the equation (8.7) for $k = 2$, first take the square root on both sides, then obtain a cubic equation for $\sqrt{z}$ which by dividing out the solution $\sqrt{z} = 1$ can be simplified to $mz - (\theta-1)\sqrt{z} + q - m = 0$. Using the fact that z_i is a solution to $mz - (\theta-1)\sqrt{z} + q - m = 0$ we simplify the expression of c given in (9.16) for $m \geq 2$ *and* $z = z_i \geq 1$, $i = 1, 2$. For the solutions $\sqrt{z_1}$, $\sqrt{z_2}$ of the last equation one has $\sqrt{z_1} + \sqrt{z_2} = \frac{\theta-1}{m}$. Using this equality we get

$$\sqrt{z_1} \leq \sqrt{z_1} + \sqrt{z_2} = \frac{\theta-1}{m} < \theta.$$

Similarly, one gets $\theta > \sqrt{z_2}$. From these inequalities it follows that $|\theta - \sqrt{z_i}| = \theta - \sqrt{z_i}$. Since $\theta > 1$ for $z_i \geq 1$ we get

$$|1-\theta\sqrt{z_i}| = \theta\sqrt{z_i} - 1, \quad |1-\sqrt{z_i}| = \sqrt{z_i} - 1.$$

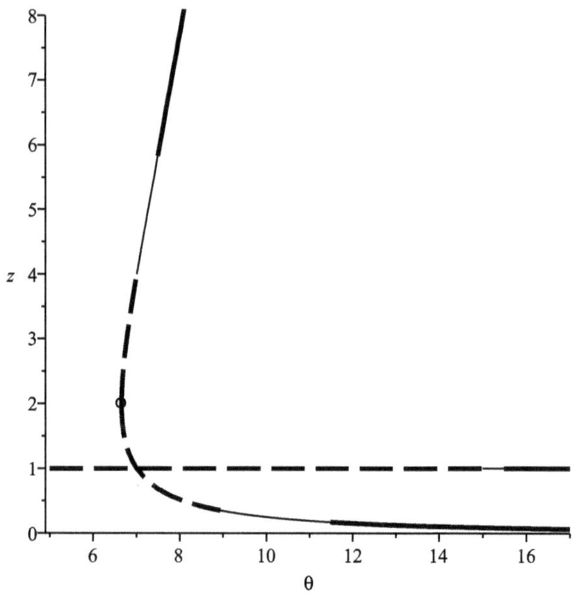

Fig. 9.7 The graphs of the functions $z_i = z_i(m, \theta)$, for $q = 6$, $m = 2$ and the graph of $z(\theta) \equiv 1$. The types of curves corresponding to certain extremality and certain non-extremality are as in Fig. 9.5 (corresponds to Theorem 9.3).

Consequently, repeatedly replacing z_i in c by $(1/m)((\theta-1)\sqrt{z_i} - q + m)$ and simplifying we obtain

$$c = \frac{1}{m^2 Z_1}\left(\{m(\theta^2 - \theta + q - m) - (\theta-1)^2\}\sqrt{z_i} - \{(m-1)\theta + 1\}(q-m)\right).$$

Recall $a = \frac{(\theta-1)z_i}{Z_1}$ (see (9.9)) and by (9.10) for $z_i > 1$ we have $b < a$. Now we shall check that $a > c$: replacing z_i in a by $(1/m)((\theta-1)\sqrt{z_i} - q + m)$ we get the following sequence of equivalent inequalities

$$\begin{aligned} a > c \quad \Leftrightarrow \quad & \frac{(\theta-1)((\theta-1)\sqrt{z_i} - q + m)}{mZ_1} \\ > \; & \frac{\{m(\theta^2 - \theta + q - m) - (\theta-1)^2\}\sqrt{z_i} - \{(m-1)\theta + 1\}(q-m)}{m^2 Z_1} \\ \Leftrightarrow \; & [(m+1)(\theta-1)^2 - m(\theta^2 - \theta + q - m)]\sqrt{z_i} + (m-1)(q-m)(\theta-1) > 0 \\ \Leftrightarrow \; & [\theta^2 - (m+2)\theta - m(q-m-1) + 1]\sqrt{z_i} + (m-1)(q-m)(\theta-1) > 0. \end{aligned}$$

To show that the last inequality is true for any $\theta > \theta_m$ it suffices[1] to prove that

$$\theta^2 - (m+2)\theta - m(q-m-1) + 1 > 0. \qquad (9.24)$$

[1] Note that $(m-1)(q-m)(\theta-1) > 0$, because $m \geq 2$, $q > m$ and $\theta > 1$.

This quadratic inequality (with respect to $\theta > 1$) is true for any

$$\theta > 1 + \frac{1}{2}m + \frac{1}{2}\sqrt{4mq - 3m^2}.$$

Now if we prove that

$$1 + \frac{1}{2}m + \frac{1}{2}\sqrt{4mq - 3m^2} < \theta_m = 1 + 2\sqrt{m(q-m)}, \tag{9.25}$$

then it follows that (9.24) holds for any $\theta > \theta_m$. So let us check (9.25):

$$(9.25) \quad \Leftrightarrow \quad m + \sqrt{4mq - 3m^2} < 4\sqrt{m(q-m)} \quad \Leftrightarrow$$

$$m^2 + 2m\sqrt{4mq - 3m^2} + 4mq - 3m^2 < 16mq - 16m \quad \Leftrightarrow$$

$$\sqrt{4mq - 3m^2} < m + 6q - 8 \quad \Leftrightarrow \quad 3q(3q-8) + (m-2)^2 + 2mq + 12 > 0.$$

The last inequality is true for any $q \geq 2$. Thus we proved that

$$\kappa = \max\{a, b, c\} = a, \quad \text{if} \quad z_i > 1. \tag{9.26}$$

(i) *Case*: $z_1 \geq 1$. If $z = z_1 \geq 1$, i.e., $\theta \in [\theta_m, \theta_c]$ then as shown above we have $\kappa = a$. Hence

$$2\gamma\kappa \leq 2a\frac{\theta - 1}{\theta + 1} = \frac{2(\theta-1)^2 z_1}{((\theta + m - 1)z_1 + (q - m))(\theta + 1)}.$$

Thus to check $2\gamma\kappa \leq 1$ it suffices to show

$$\frac{2(\theta-1)^2 z_1}{((\theta + m - 1)z_1 + (q - m))(\theta + 1)} < 1. \tag{9.27}$$

The last inequality can be written as

$$[-\theta^2 + (m+4)\theta + m - 3]z_1 + (\theta + 1)(q - m) > 0. \tag{9.28}$$

Replacing z_1 in (9.28) by $(1/m)((\theta - 1)\sqrt{z_1} - q + m)$ we get

$$[-\theta^2 + (m+4)\theta + m - 3]((\theta - 1)\sqrt{z_1} - q + m) + m(\theta + 1)(q - m) > 0. \tag{9.29}$$

By formula of $\sqrt{z_1}$ (see (8.12)) we have

$$\sqrt{z_1} = \frac{\theta - 1 - \sqrt{(\theta-1)^2 - 4m(q-m)}}{2m}$$
$$= \frac{\left(\theta - 1 - \sqrt{(\theta-1)^2 - 4m(q-m)}\right)\left(\theta - 1 + \sqrt{(\theta-1)^2 - 4m(q-m)}\right)}{2m\left(\theta - 1 + \sqrt{(\theta-1)^2 - 4m(q-m)}\right)}$$
$$= \frac{2(q-m)}{\theta - 1 + \sqrt{(\theta-1)^2 - 4m(q-m)}}. \tag{9.30}$$

Using formula (9.30) of $\sqrt{z_1}$ we obtain from (9.29)

$$-\theta^3 + (2m+5)\theta^2 - 7\theta - (2m-3)$$
$$+ (\theta^2 - 4\theta + 3)\sqrt{(\theta-1)^2 - 4m(q-m)} > 0. \tag{9.31}$$

We note that

$$-\theta^3 + (2m+5)\theta^2 - 7\theta - (2m-3) = (\theta-1)[-\theta^2 + (2m+4)\theta + 2m - 3],$$

$$\theta^2 - 4\theta + 3 = (\theta-1)(\theta-3).$$

Using these equalities and dividing both side of (9.31) by $-(\theta-1)$ we get

$$\theta^2 - (2m+4)\theta - 2m + 3 - (\theta-3)\sqrt{(\theta-1)^2 - 4m(q-m)} < 0. \tag{9.32}$$

For $m = 2$ from (9.32) we obtain

$$\theta^2 - 8\theta - 1 - (\theta-3)\sqrt{(\theta-1)^2 - 8(q-2)} < 0. \tag{9.33}$$

Subcase $\theta^2 - 8\theta - 1 \leq 0$. Since $\theta \geq \theta_2 = 1 + 2\sqrt{2(q-2)} > 3$, the inequality (9.33) holds if $\theta^2 - 8\theta - 1 \leq 0$. From the last inequality we get $\theta \leq 4 + \sqrt{17}$. Hence we should find $q \geq 4$ such that $1 + 2\sqrt{2(q-2)} \leq 4 + \sqrt{17}$, i.e., $q \leq 8$. Then for $q < 8$ we have $q + 1 < 4 + \sqrt{17}$ and the inequality (9.33) holds for θ satisfying

$$1 + 2\sqrt{2(q-2)} \leq \theta \leq 4 + \sqrt{17}.$$

For $q = 8$ we have

$$1 + 2\sqrt{2(q-2)} \leq \theta \leq 4 + \sqrt{17} < q + 1 = 9.$$

Thus we proved that the measure $\mu_1(\theta, 2)$ is extreme

- If $q < 8$ and $\theta \in [\theta_2, q+1]$.
- If $q = 8$ and $\theta \in [\theta_2, 4 + \sqrt{17}]$.

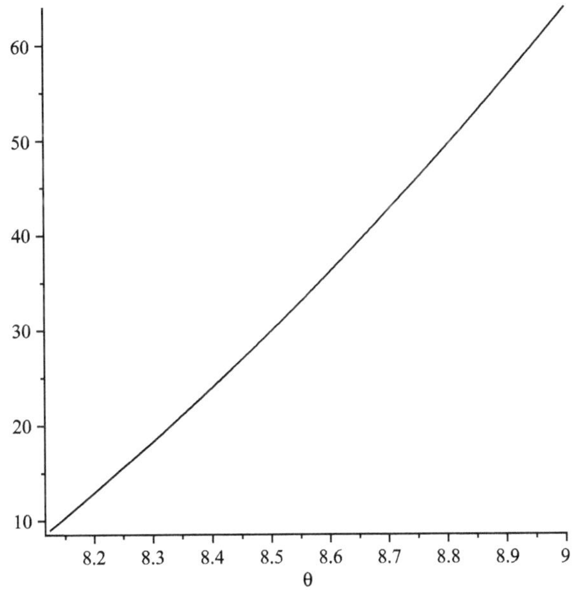

Fig. 9.8 The graph of the function $\varphi(\theta, q)$, for $q = 8$ and $\theta \in [4 + \sqrt{17}, 9]$.

So, for $q = 8$ we should check the inequality (9.33) for $\theta \in (4 + \sqrt{17}, 9]$, since for such θ we have $\theta^2 - 8\theta - 1 > 0$. This verification is included in the following subcase.

Subcase $\theta^2 - 8\theta - 1 > 0$. In this case we have $\theta > 4 + \sqrt{17}$ which, together with $\theta > \theta_2$, gives that we should consider $q \geq 8$. Rewrite (9.33) as

$$\theta^2 - 8\theta - 1 < (\theta - 3)\sqrt{(\theta - 1)^2 - 8(q - 2)}.$$

Take the square on both sides and simplify up to

$$\varphi(\theta, q) = \theta^3 - (q + 3)\theta^2 + (6q - 17)\theta - (9q - 19) > 0.$$

First let us check this inequality for $q = 8$ and $\theta \in [4 + \sqrt{17}, 9]$. The graph of the function $\varphi(\theta, 8)$ is given in Fig. 9.8. Thus we obtain that $\varphi(\theta, 8) > 0$ and $\mu_1(\theta, 2)$ is extreme for $q = 8$ on $[4 + \sqrt{17}, 9]$ too. Summarizing the results of the previous subcase with the present one, we obtain that this measure is extreme on $[\theta_2, q + 1]$ for any $q \leq 8$.

Now we shall consider the case $q \geq 9$. The equation $\varphi'(\theta, q) = 0$ has two solutions:

$$q_\pm = \frac{1}{3} \cdot (q + 3 \pm \sqrt{q^2 - 12q + 60}).$$

Since the coefficient of θ^3 in φ is positive (=1), the function has a maximum at q_- and a minimum at q_+. The maximal value is

$$\varphi(q_-, q) = -\frac{2}{27}[q(q^2 - 18q + 144) - (q^2 - 12q + 60)^{3/2}].$$

Now we prove that $\varphi(q_-, q) < 0$ for any $q \geq 9$:

$$\varphi(q_-, q) < 0 \quad \Leftrightarrow \quad q(q^2 - 18q + 144) > (q^2 - 12q + 60)^{3/2} \quad \Leftrightarrow$$

$$q^2(q^2 - 18q + 144)^2 > (q^2 - 12q + 60)^3 \quad \Leftrightarrow$$

$$\psi(q) = 864q^3 - 15984q^2 + 129600q - 216000 > 0.$$

The last inequality is true for any $q \geq 9$, indeed $\psi'(q) > 0$ and $\psi(9) = 285552$.

Consequently, $\varphi(q_+, q) < \varphi(q_-, q) < 0$.

For any $q \geq 9$ we have

$$\varphi(\theta_2, q) = -4(2q^2 - 3q + 2(9 - 3q)\sqrt{2q - 4}) < 0, \quad \varphi(q+1, q) = 4q(q-6) > 0.$$

It follows from these properties that there exists unique $\theta^\dagger$ such that $\varphi(\theta^\dagger, q) = 0$ and the inequality $\varphi(\theta, q) > 0$ holds for any $\theta \in (\theta^\dagger, q+1]$, $q \geq 9$ (see Fig. 9.9, for $q = 9, 10, 11$). Thus we proved the assertions of (i) corresponding to the case $\theta \leq \theta_c = q + 1$.

Case: $z_1 < 1$. Using the explicit formula (9.30) of $\sqrt{z_1}$ and the condition $\theta > \theta_m$ we get $\theta\sqrt{z_1} > 1$. Indeed,

$$\theta\sqrt{z_1} > 1 \quad \Leftrightarrow \quad \frac{2(q-m)\theta}{\theta - 1 + \sqrt{(\theta-1)^2 - 4m(q-m)}} > 1 \quad \Leftrightarrow$$

$$(2(q-m) - 1)\theta + 1 > \sqrt{(\theta-1)^2 - 4m(q-m)} \quad \Leftrightarrow$$

$$(q - m - 1)\theta^2 + \theta + m > 0.$$

The last inequality is true for any $\theta > 1$. Consequently, using $\theta > \sqrt{z_1}$, $\theta\sqrt{z_1} > 1$ and $z_1 < 1$ from formula (9.16) replacing in it z_1 by $(1/m)((\theta-1)\sqrt{z_1} - q + m)$, we get independently on the values of q and m that

$$c = \frac{1}{Z_1}\left(\theta\sqrt{z_1} - 1\right).$$

For $z_1 < 1$ we have $a < b$ (see (9.10)). Now, to prove that $b > c$ for $z_1 < 1$ (i.e., for $\theta > \theta_c$) we note that

$$b = \frac{(\theta-1)\sqrt{z_1}}{Z_1} > c = \frac{1}{Z_1}\left(\theta\sqrt{z_1} - 1\right) \quad \Leftrightarrow$$

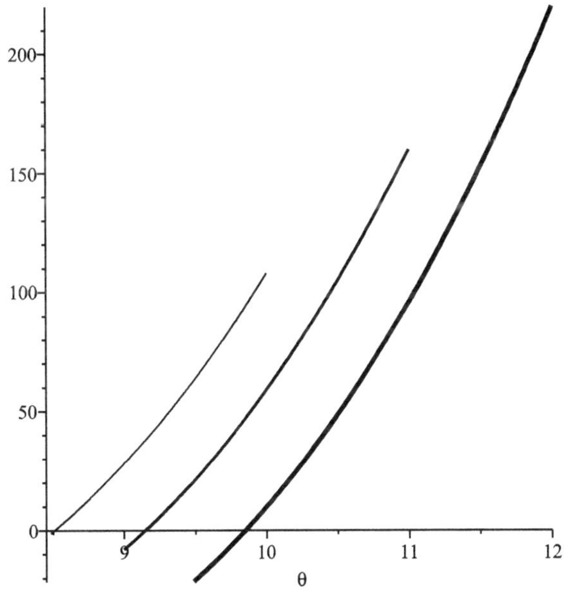

Fig. 9.9 The graphs of the functions $\varphi(\theta, q)$, for $q = 9$ (left curve), $q = 10$ (middle curve) and $q = 11$ (right curve), for $\theta \in [\theta_2(q), q+1]$.

$$(\theta - 1)\sqrt{z_1} > \theta\sqrt{z_1} - 1 \quad \Leftrightarrow \quad 1 > \sqrt{z_1} \quad \Leftrightarrow \quad 1 > z_1.$$

Consequently, $\kappa = \max\{a, b, c\} = b$ and to check $2\gamma\kappa \leq 1$ it suffices that

$$\frac{2(\theta-1)^2\sqrt{z_1}}{(\theta+1)[(\theta+m-1)z_1+q-m]} < 1. \tag{9.34}$$

Again using that z_1 is a solution of $mz_1 - (\theta-1)\sqrt{z_1} + q - m = 0$, for $m = 2$ one can simplify the inequality (9.34) to the following

$$(\theta^2 - 2\theta + 5)\sqrt{z_1} - (q-2)(\theta+1) > 0.$$

Now, from this inequality, using $\sqrt{z_1} = 2(q-2)/[\theta - 1 + \sqrt{(\theta-1)^2 - 8(q-2)}]$ (see (9.30), for $m = 2$) we get

$$g(\theta, q) = \theta^3 - (q+3)\theta^2 - (2q-15)\theta - (q+13) < 0.$$

The equation

$$g'(\theta, q) = 3\theta^2 - 2(q+3)\theta - (2q-15) = 0$$

has two solutions

$$\theta_{\pm} = 1 + \frac{q}{3} \pm \frac{1}{3}\sqrt{q^2 + 12q - 36}.$$

Recall $\theta_c = q+1$. These solutions satisfy

$$\theta_- < \theta_+ < \theta_c = q+1.$$

Thus $g(\theta, q)$ as a function of $\theta \in [\theta_c, +\infty)$ is an increasing function. We have $g(\theta_c, q) = -4q(q-2) < 0$. Now for $q \geq 4$ we show that $g(\theta^*, q) > 0$, where θ^* is defined in (9.13): We compute

$$P(q) \equiv g(\theta^*, q) = (3\sqrt{2}+4)q^3 - 4(5\sqrt{2}+8)q^2 + 4(15\sqrt{2}+14)q - 112,$$

and

$$P'(q) = 3(3\sqrt{2}+4)q^2 - 8(5\sqrt{2}+8)q + 4(15\sqrt{2}+14).$$

The equation $P'(q) = 0$ has two solutions

$$q_\pm = \frac{2}{3} \cdot \frac{10\sqrt{2}+16 \pm \sqrt{18+14\sqrt{2}}}{3\sqrt{2}+4}.$$

These solutions satisfy $q_- < q_+ < 3$. Thus the function $P(q)$ is increasing for $q \geq 4$. Consequently,

$$P(q) \geq P(4) = 16(7\sqrt{2}-9) > 0.$$

This completes the proof of $g(\theta^*, q) > 0$, for any $q \geq 4$.

Consequently $g(\theta, q) = 0$ has the unique solution $\breve{\theta} = \breve{\theta}(q)$ for each $q \geq 4$, such that $\theta_c < \breve{\theta} < \theta^*$. Thus $g(\theta, q) < 0$ and $\mu_1(\theta, 2)$ is extreme for $\theta \in [\theta_c, \breve{\theta})$.

Summarizing the results for the case $z_1 \geq 1$ and the case $z_1 < 1$ we get the desired assertions of (i).

Note that for $q = 6$ we have $\theta^* = 3(1+2\sqrt{2}) \approx 11.485$ and a numerical analysis shows that $\breve{\theta} \approx 8.903$ (see Fig. 9.10). So $\theta^* - \breve{\theta} \approx 2.582$.

(ii) Note that $z = z_2 > 1$, for each $\theta \in [\theta_m, +\infty)$, where $\theta_m = 1 + 2\sqrt{m(q-m)}$. Recall that for $z_2 > 1$ we have $\kappa = a$ (see (9.26)). Therefore, similarly as in (9.32) we have to check

$$\theta^2 - (2m+4)\theta - 2m + 3 + (\theta - 3)\sqrt{(\theta-1)^2 - 4m(q-m)} < 0. \quad (9.35)$$

For $m = 2$ and a given value of $q < 9$, from (9.35) we obtain the assertions of (ii).

In the case $q = 6$, $m = 2$ it is easy to see that the LHS of (9.35) is zero for $\theta = \grave{\theta} = 7$ and the inequality is true for any $\theta \in [\theta_2, 7)$.

In Theorem 9.1 we proved that $\mu_2(\theta, m)$ is non-extreme if $\theta > \widehat{\theta}$ where $\widehat{\theta}$ is defined in (9.13) which now is $\widehat{\theta} = 6\sqrt{2} - 1 \approx 7.4852$. So $\widehat{\theta} - 7 \approx 0.4852$. This is the length of the gap between the Kesten-Stigum bound of non-extremality and our bound of extremality.

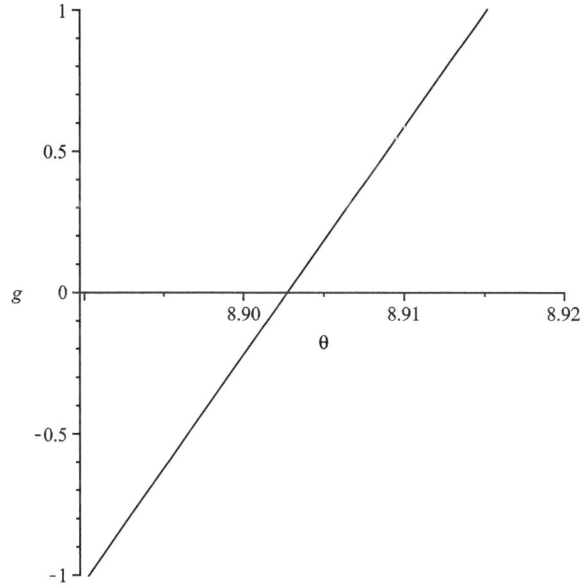

Fig. 9.10 The graph of the function $g(\theta, q)$, for $q = 6$, given in the small interval $[8.88, 8.92]$, shows that the solution of $g(\breve{\theta}, 6) = 0$ is $\breve{\theta} \approx 8.903$.

(iii) For $\theta = \theta_m$ we have $z_1 = z_2 > 1$ and from (9.32) we get

$$\theta_m^2 - (2m+4)\theta_m - 2m + 3 < 0. \tag{9.36}$$

Using the formula $\theta_m = 1 + 2\sqrt{m(q-m)}$ from (9.36) we obtain

$$q < \frac{m+1}{2m}\left[3m + 1 + \sqrt{m^2 + 6m + 1}\right], \quad m \geq 2. \tag{9.37}$$

This completes the proof. □

Now following [249] we give a theorem which is more general than Theorem 9.3.

Theorem 9.4. *Let $k = 2$, $m \geq 2$. Then*

1. *If $2m \leq q < \frac{m+1}{2m}[3m + 1 + \sqrt{m^2 + 6m + 1}]$, then there is $\ddot{\theta} > q + 1$ such that the measure $\mu_1(\theta, m)$ is extreme for any $\theta \in [\theta_m, \ddot{\theta})$;*
2. *If $q > \frac{m+1}{2m}[3m + 1 + \sqrt{m^2 + 6m + 1}]$, then there exists $\overline{\theta} \in (\theta_m, \theta_c)$ such that the measure $\mu_1(\theta, m)$ is extreme for any $\theta \in (\overline{\theta}, \ddot{\theta})$;*
3. *If $2m \leq q < m + \frac{1}{4m}(m + 1 + \sqrt{m^2 + 2m + 7})^2$ then there is $\bar{\bar{\theta}} \in (\theta_m, +\infty)$ such that the measure $\mu_2(\theta, m)$ is extreme for any $\theta \in [\theta_m, \bar{\bar{\theta}})$.*

Proof. This proof is based on the argument of the proof of the previous theorem.

1. *Case* $z_1 \geq 1$. In this case $\theta \in [\theta_m, \theta_c]$ and $\kappa = a$. We shall check extremality of $\mu_1(\theta, m)$. We should verify the inequality

$$2\kappa\gamma \leq 2a\frac{\theta - 1}{\theta + 1} < 1.$$

Since

$$\sqrt{z_1} = \frac{2(q-m)}{\theta - 1 + \sqrt{(\theta-1)^2 - 4m(q-m)}},$$

we get

$$\theta^2 - (2m+4)\theta - 2m + 3 - (\theta - 3)\sqrt{(\theta-1)^2 - 4m(q-m)} < 0,$$

which is equivalent to

$$u(\theta) = \theta^3 - (q+3)\theta^2 + (8q - 10m - 1)\theta - (9q - 8m - 12) > 0. \quad (9.38)$$

Denote

$$\alpha(m) = \frac{m+1}{2m}[3m + 1 + \sqrt{m^2 + 6m + 1}].$$

It is clear that $u(\theta_m) > 0$ for $q < \alpha(m)$. Moreover, the function $u(\theta)$ increases on the set $[\theta_m^{(1)}, +\infty]$, here

$$\theta_m^{(1)} = \frac{q + 3 + \sqrt{(q+3)^2 - 3(8q - 10m - 1)}}{3}.$$

From $\left[\frac{q}{2}\right] \geq m$ it follows that $\theta_m^{(1)} \leq \theta_m$. Indeed, consider the inequality

$$\frac{q + 3 + \sqrt{(q+3)^2 - 3(8q - 10m - 1)}}{3} \leq 1 + 2\sqrt{m(q-m)},$$

that is

$$\xi(q) = 2q\sqrt{m(q-m)} + 5m + 2 - 3q - 6m(q-m) \leq 0. \quad (9.39)$$

The derivative of this function is

$$\xi'(q) = 2\sqrt{m(q-m)} + \frac{mq}{\sqrt{m(q-m)}} - 3 - 6m.$$

Thus the function $\xi(q)$ increases for $m \leq \left[\frac{q}{2}\right]$ and

$$q < \frac{m+1}{2m}[3m + 1 + \sqrt{m^2 + 6m + 1}].$$

The last inequality gives

$$q < \frac{2(m+1)^2}{m} \leq 2(m+3).$$

Consequently, $\xi(q) < \xi(2(m+3))$. One can also see that $\xi(2(m+3)) < 0$ if $m \geq 2$ and $\theta_m^{(1)} \leq \theta_m$. Hence, the function $u(\theta)$ is increasing on the set $[\theta_m, \theta_c]$, i.e., $u(\theta) > 0$ for any $\theta \in [\theta_m, \theta_c]$. Consequently, the measure $\mu_1(\theta, m)$ is extreme for $z_1 > 1$ and $\theta \in [\theta_m, \theta_c]$.

Case: $z_1 < 1$. Now independently on values of q and m, we have

$$c = \frac{1}{Z_1}\left(\theta\sqrt{z_1} - 1\right).$$

One can see that $b \geq c$ if $z_1 < 1$, i.e., if $\theta > \theta_c$. Consequently, $\kappa = b$ and we should verify the following

$$2\kappa\gamma \leq \frac{2(\theta-1)\sqrt{z_1}}{\theta z_1 + q - 1} \cdot \frac{\theta - 1}{\theta + 1} < 1.$$

Denote

$$g(\theta, q, m) = \frac{2(\theta-1)\sqrt{z_1}}{\theta z_1 + q - 1} \cdot \frac{\theta - 1}{\theta + 1} - 1.$$

Since

$$\sqrt{z_1} = \frac{2(q-m)}{\theta - 1 + \sqrt{(\theta-1)^2 - 4m(q-m)}}$$

we get

$$g(\theta_c, q, m) = \frac{\theta - 1}{\theta + 1} - 1 < 0, \quad \lim_{\theta \to +\infty} g(\theta, q, m) = \frac{q - 2m + 1}{q - 1} > 0.$$

Thus there exists $\theta'' \in (\theta_c, +\infty)$ such that

$$g(\theta'', q, m) = 0, \quad \ddot{\theta} = \min\{\theta'' : g(\theta'', q, m) = 0\}.$$

Then $g(\theta, q, m) < 0$ for any $\theta \in [\theta_c, \ddot{\theta})$. Hence, under the condition, the measure $\mu_1(\theta, m)$ is extreme.

2. *Case* $z_1 \geq 1$. For $q > \frac{m+1}{2m}[3m+1+\sqrt{m^2+6m+1}]]$, we have $u(\theta_m) < 0$. Note that $u(\theta_c) > 0$. Moreover, the function $u(\theta)$ is increasing on $[\theta_m, \theta_c]$. Consequently, there exists a unique $\overline{\theta} \in (\theta_m, \theta_c]$ such that $u(\overline{\theta}) = 0$. Hence, from the condition (9.38) it follows that $\mu_1(\theta, m)$ is extreme for any $(\overline{\theta}, \theta_c)$.

Case $z_1 < 1$. Is similar to the proof of part 1 of this theorem and one gets that the measure $\mu_1(\theta, m)$ is extreme for any $\theta \in (\overline{\theta}, \ddot{\theta})$.

3. In case $\theta \geq \theta_m$ we have $z_2 \geq 1$. Verify extremality condition for $\mu_2(\theta, m)$:

$$2\kappa\gamma \leq 2c\frac{\theta - 1}{\theta + 1} < 1,$$

which can be rewritten as

$$v(\theta) = \theta^2 - (2m+4)\theta - 2m + 3 + (\theta - 3)\sqrt{(\theta-1)^2 - 4m(q-m)} < 0. \quad (9.40)$$

The inequality (9.40) has a solution if the following one has a solution:

$$\theta^2 - (2m+4)\theta - (2m-3) < 0.$$

This inequality has solutions:

$$\theta < m + 2 + \sqrt{m^2 + 2m + 7}.$$

Since $\theta \geq \theta_m = 1 + 2\sqrt{m(q-m)}$, we get

$$1 + 2\sqrt{m(q-m)} < m + 2 + \sqrt{m^2 + 2m + 7}.$$

Consequently,

$$2m \leq q < m + \frac{1}{4m}(m + 1 + \sqrt{m^2 + 2m + 7})^2.$$

and

$$\lim_{\theta \to +\infty} v(\theta) = +\infty.$$

Hence, there is $\bar{\bar{\theta}} \in (\theta_m, +\infty)$ such that $v(\bar{\bar{\theta}}) = 0$. Thus, the measure $\mu_2(\theta, m)$ is extreme for any $\theta \in (\theta_m, \bar{\bar{\theta}})$. The theorem is proved. □

Remark 9.1. Note that

- If $q > m + \frac{1}{4m}(m + 1 + \sqrt{m^2 + 2m + 7})^2$, then condition (9.40) does not hold for any θ, i.e., it is unknown whether the measure $\mu_2(\theta, m)$ is extreme in this case.
- From the graphs of the functions one can see that $f'(\theta, q) > 0$, $g'(\theta, q, m) > 0$, for finite q and m, i.e., the functions $f(\theta, q)$ and $g(\theta, q, m)$ are increasing (it is difficult to prove this analytically). Therefore the values θ', θ'' are uniquely defined $\tilde{\theta} = \theta'$, $\ddot{\theta} = \theta''$.
- In [249] for the 3-state Potts model on the Cayley tree of order $k = 3$, the extremality questions are studied for certain TISGMs. The case $k \geq 4$ is not considered yet. Because in this case there is not explicit formula of nontrivial solutions to the equation (8.7).

Now we give results which are valid for θ close to the critical value θ_c at which the lower branches of the boundary law z degenerate into the free value $z = 1$ and the corresponding Markov chains become close to the free chain.

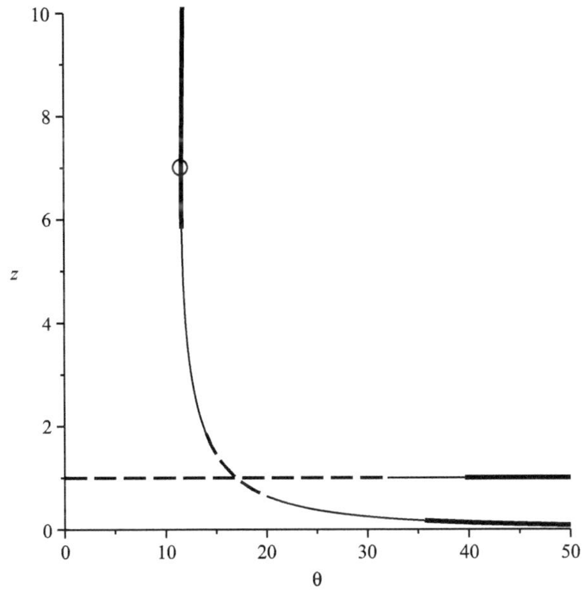

Fig. 9.11 The graphs of the functions $z_i = z_i(m, \theta)$, for $q = 16$, $m = 2$ and the graph of $z(\theta) \equiv 1$. The types of curves are again as in Fig. 9.5 (corresponds to part a) of Theorem 9.5).

Theorem 9.5. *Let $k = 2$.*

(a) For each $m \leq [q/2]$ there exists a neighborhood $U_m(\theta_c)$ of θ_c such that the measure $\mu_1(\theta, m)$ is extreme if $\theta \in U_m(\theta_c)$ (see Fig. 9.11).

(b) If $m = 1$ or condition (iii) of Theorem 9.3 is satisfied then there exists a neighborhood $V_m(\theta_m)$ of θ_m such that measures $\mu_i(\theta, m)$, $i = 1, 2$ are extreme if $\theta \in V_m(\theta_m)$.

Proof. (a) As it was shown above γ and κ are bounded by continuous functions of θ. For $\theta = \theta_c = q + 1$ we have $z_1 = 1$, $a = b = c = (\theta_c - 1/\theta_c - 1 + q) = 1/2$, $\gamma \leq (\theta_c - 1/\theta_c + 1) = (q/q + 2)$ and $\kappa = a = b = c = 1/2$. The measure μ_0 (corresponding to the solution $z = (1, 1, \ldots, 1)$) is extreme at $\theta = \theta_c$ if $2\gamma\kappa \leq 1$. We have

$$2\gamma\kappa \leq 2a\frac{\theta_c - 1}{\theta_c + 1} = \frac{q}{q + 2} < 1, \quad \text{for all } q \geq 2.$$

Hence μ_0 is extreme at $\theta = \theta_c$. Since $\mu_1(\theta_c, m) = \mu_0$, by the above-mentioned continuity in a sufficiently small neighborhood of θ_c the measure $\mu_1(\theta, m)$ is extreme.

(b) This follows from Theorem 9.3 using the continuity similarly as in case a). □

Recall that the model is called ferromagnetic if $\theta > 1$, and as usual a punctured neighborhood of θ_c is an open neighborhood from which the value θ_c was removed.

Theorem 9.6. *For the ferromagnetic q-state Potts model (with $q \geq 3$) on the Cayley tree of order two, there exists a punctured neighborhood $U(\theta_c)$ of θ_c such that there are at least $2^{q-1}+q$ extreme TISGMs for each $\theta \in U(\theta_c)$.*

Proof. By Proposition 8.2 we know that for $\theta_{[q/2]} < \theta \neq \theta_c$ there are $2^q - 1$ TISGMs. One of them corresponds to $z = 1$. Half of the remaining $2^q - 2$, i.e., $2^{q-1} - 1$, TISGMs are generated by z_1 and the other $2^{q-1} - 1$ TISGMs are generated by z_2. By the above-mentioned results we know that $\mu_2(\theta, 1)$ is extreme. The number of such measures is q. By Theorem 9.5, in a sufficiently small punctured neighborhood of θ_c all measures $\mu_1(\theta, m)$ are extreme. Thus the total number of extremal TISGM is at least $2^{q-1} + q$. □

Chapter 10

The model with external field

In this chapter, we consider the q-state Potts model on a Cayley tree of degree $k \geq 2$ in an external (non-zero) field. For a special choice of the external field, we give (non-)uniqueness conditions of translation-invariant splitting Gibbs measures (TISGMs) by identifying the phase diagrams on the "temperature–field" plane for different values of the parameters q and k. In some particular cases, the maximal number of TISGMs in this model is shown to be $2^q - 1$. In case of zero-external field by Chapter 9 we know that the maximal value is reached. In case of non-zero external field a computer calculations supports the conjecture that this bound is valid for all $q \geq 2$ and $k \geq 2$.

10.1 One non-zero coordinate

This section is based on the recent paper [28].

10.1.1 *Criterion of uniqueness of TISGM*

In this chapter we consider the Hamiltonian (2.6) in case of a special external field and give results concerning TISGMs $\mu^h = \mu^h_{\theta,\xi}$, (see Corollary 2.1) that is, with the external field $\{\boldsymbol{\xi}(x)\}_{x\in V}$ and $\{\boldsymbol{h}(x)\}_{x\in V}$ satisfying the homogeneity conditions of Corollary 2.1.

Assume that all coordinates of the (reduced) vector $\check{\boldsymbol{\xi}}^0$ (see Subsection 2.3.1) are zero except one. Then without loss of generality, due to permutational symmetry, we take $\check{\xi}^0_1 = \alpha \in \mathbb{R}$ and $\check{\xi}^0_2 = \cdots = \check{\xi}^0_{q-1} = 0$,

$$\check{\boldsymbol{\xi}}^0 = (\alpha, 0, \ldots, 0) \in \mathbb{R}^{q-1}. \tag{10.1}$$

Denote

$$\check{\boldsymbol{h}}^0 = (\check{h}_1^0, \ldots, \check{h}_{q-1}^0) \in \mathbb{R}^{q-1}.$$

Then, introducing $z_i := \theta^{\check{h}_i^0/k}$ $(i = 1, \ldots, q-1)$, the compatibility equations (2.18) are equivalently rewritten in the form

$$\begin{cases} z_1 = 1 + \dfrac{(\theta-1)(\theta^\alpha z_1^k - 1)}{\theta + \theta^\alpha z_1^k + \sum_{j=2}^{q-1} z_j^k}, \\ z_i = 1 + \dfrac{(\theta-1)(z_i^k - 1)}{\theta + \theta^\alpha z_1^k + \sum_{j=2}^{q-1} z_j^k}, \qquad i = 2, \ldots, q-1. \end{cases} \tag{10.2}$$

To solve the system (10.2), we want to characterize the uniqueness of its solution, which in turn gives a criterion of the uniqueness of completely homogeneous SGM in the Potts model with non-zero external field.

The case $\theta = 1$ is trivial, as the system (10.2) will then have the unique solution $z_1 = \cdots = z_{q-1} = 1$. The case $\theta > 1$ and $\alpha = 0$ has been studied in Chapter 9. Here we consider the case $\alpha \neq 0$.

Denote

$$p_k(z) := z^{k-1} + \cdots + z, \tag{10.3}$$

so that

$$z^k - 1 = (z-1)\big(p_k(z) + 1\big). \tag{10.4}$$

Lemma 10.1. *Let $\theta > 1$, $\alpha \neq 0$ and $(z_1, \ldots, z_{q-1})$, (with $z_i > 0$, $\forall i$) be a solution to* (10.2). *Then*

(i) either $z_1 = u$ and $z_2 = \cdots = z_{q-1} = 1$, where $u = u(\theta, \alpha)$ satisfies the equation

$$u = 1 + \frac{(\theta-1)(\theta^\alpha u^k - 1)}{\theta + \theta^\alpha u^k + q - 2}, \tag{10.5}$$

and, in particular, $u \neq 1$;

(ii) or, provided that $q \geq 3$, there is a non-empty subset $\mathcal{I}_1 \subseteq \{2, \ldots, q-1\}$, with $m := |\mathcal{I}_1|$ ranging from 1 to $q-2$, such that

$$z_i = \begin{cases} u & \text{if } i = 1, \\ v & \text{if } i \in \mathcal{I}_1, \\ 1 & \text{otherwise}, \end{cases}$$

where $u = u(\theta, \alpha, m)$ and $v = v(\theta, \alpha, m) \neq 1$ satisfy the set of equations

$$\begin{cases} u = 1 + \dfrac{(\theta - 1)(\theta^\alpha u^k - 1)}{\theta + \theta^\alpha u^k + m v^k + q - 2 - m}, \\ 1 = \dfrac{(\theta - 1)(p_k(v) + 1)}{\theta + \theta^\alpha u^k + m v^k + q - 2 - m}, \end{cases} \tag{10.6}$$

and, in particular, $u \neq 1$ and $u \neq v$.

Proof. As a general remark, observe that $z_i = 1$ solves the i-th equation of the system (10.2) regardless of all other z_j with $j \neq i$.

Note that $u := z_1 \neq 1$, for otherwise the first equation in (10.2) is not satisfied unless $\theta = 1$ or $\alpha = 0$, either of which is ruled out. Next, if $z_2 = \cdots = z_{q-1} = 1$ then the first equation for $z_1 = u$ in (10.2) specializes to (10.5), as stated.

Suppose now that $\mathcal{I}_1 := \{i \geq 2 \colon z_i \neq 1\} \neq \emptyset$, then similarly as above we show that $z_i = \text{const}$ $(i \in \mathcal{I}_1)$, and the system (10.2) specializes to equations (10.6) with $z_1 = u$ and $z_i = v$ $(i \in \mathcal{I}_1)$.

Finally, assuming to the contrary that $u = v$ and comparing the equations in (10.6), we would conclude that $\theta^\alpha u^k = u^k$, that is, $\alpha = 0$, which is ruled out. Hence, $u \neq v$ as claimed. □

By Lemma 10.1, the system (10.2) with $\theta > 1$ is reduced either to a single equation

$$u = 1 + \frac{(\theta - 1)(\theta^\alpha u^k - 1)}{\theta + \theta^\alpha u^k + q - 2}, \tag{10.7}$$

or to the system of equations (indexed by $m = 1, \ldots, q - 2$)

$$\begin{cases} u = 1 + \dfrac{(\theta - 1)(\theta^\alpha u^k - 1)}{\theta + \theta^\alpha u^k + m v^k + q - 2 - m}, \\ 1 = \dfrac{(\theta - 1)(1 + v + \cdots + v^{k-1})}{\theta + \theta^\alpha u^k + m v^k + q - 2 - m}, \end{cases} \tag{10.8}$$

subject to the condition

$$v \neq 1. \tag{10.9}$$

Consider first the equation (10.7). Denote

$$\theta_{\mathrm{c}} = \theta_{\mathrm{c}}(k, q) := \frac{1}{2}\left(\sqrt{(q-2)^2 + 4(q-1)\left(\frac{k+1}{k-1}\right)^2} - (q-2)\right). \tag{10.10}$$

and

$$b = b(\theta) := \frac{\theta(\theta + q - 2)}{q - 1}. \tag{10.11}$$

We have $b(1) = 1$ and $b(\theta) > 1$ for $\theta > 1$. Comparing (10.10) and (10.11) we get $b(\theta_c) = \left(\frac{k+1}{k-1}\right)^2$ and $b(\theta) > \left(\frac{k+1}{k-1}\right)^2$ for $\theta > \theta_c$.

For $\theta \geq \theta_c$, denote by $x_\pm = x_\pm(\theta)$ the roots of the quadratic equation

$$(b + x)(1 + x) = k(b - 1)x, \tag{10.12}$$

with discriminant

$$\begin{aligned} D = D(\theta) &:= \big(k(b-1) - (b+1)\big)^2 - 4b \\ &= (b-1)(k-1)^2 \left(b - \left(\frac{k+1}{k-1}\right)^2\right), \end{aligned} \tag{10.13}$$

that is,

$$x_\pm = x_\pm(\theta) := \frac{(b-1)(k-1) - 2 \pm \sqrt{D}}{2}. \tag{10.14}$$

Introduce the notation

$$a_\pm = a_\pm(\theta) := \frac{1}{x_\pm}\left(\frac{1 + x_\pm}{b + x_\pm}\right)^k, \qquad \theta \geq \theta_c. \tag{10.15}$$

One can show that $a_-(\theta) < a_+(\theta)$ for all $\theta > \theta_c$ (see the proof of Theorem 10.1 below). Finally, denote

$$\alpha_\pm = \alpha_\pm(\theta) := -(k+1) + \frac{1}{\ln\theta} \ln\frac{q-1}{a_\mp}, \qquad \theta \geq \theta_c, \tag{10.16}$$

so that $\alpha_-(\theta_c) = \alpha_+(\theta_c)$ and $\alpha_-(\theta) < \alpha_+(\theta)$ for $\theta > \theta_c$.

Let $\nu_0(\theta, \alpha)$ denote the number of solutions $u > 0$ of the equation (10.7).

Theorem 10.1. *The following equality holds*

$$\nu_0(\theta, \alpha) = \begin{cases} 1 & \text{if } \theta \leq \theta_c \text{ or } \theta > \theta_c \text{ and } \alpha \notin [\alpha_-, \alpha_+], \\ 2 & \text{if } \theta > \theta_c \text{ and } \alpha \in \{\alpha_-, \alpha_+\}, \\ 3 & \text{if } \theta > \theta_c \text{ and } \alpha \in (\alpha_-, \alpha_+). \end{cases}$$

Proof. By the substitution

$$u^k = \frac{(q-1)x}{\theta^{\alpha+1}}, \qquad x > 0, \tag{10.17}$$

equation (10.7) can be represented in the form (3.20) with the coefficient $b > 1$ defined in (10.11) and

$$a = a(\theta) := \frac{q-1}{\theta^{k+1+\alpha}} > 0. \tag{10.18}$$

By Proposition 3.1 it is known that if $b \le \left(\frac{k+1}{k-1}\right)^2$ then the equation (3.20) has a unique solution, whereas if $b > \left(\frac{k+1}{k-1}\right)^2$ then there are one, two or three solutions according as $a \notin [a_-, a_+]$, $a \in \{a_-, a_+\}$ or $a \in (a_-, a_+)$, respectively. Adapting these results to equation (10.7), in view of the second formula in (10.18) the condition $b(\theta) > \left(\frac{k+1}{k-1}\right)^2$ is equivalent to $\theta > \theta_c$, with $\theta_c = \theta_c(k, q)$ defined in (10.10).

The corresponding critical values $\alpha_\pm$ of the field parameter α are determined by the first formula in (10.18), that is,

$$\theta^{k+1+\alpha_\pm} = \frac{q-1}{a_\mp}, \tag{10.19}$$

leading to formula (10.16). This completes the proof. □

Now consider the system of equations (10.8). For each $m \in \{1, \dots, q-2\}$, consider the functions

$$L_m(v;\theta) := (\theta - 1)\left(v^{k-1} + \cdots + v\right) - mv^k - (q - 1 - m), \tag{10.20}$$

$$K_m(v;\theta) := \frac{\left(v^{k-1} + \cdots + v + 1\right)^k L_m(v;\theta)}{\left(v^{k-1} + \cdots + v + L_m(v;\theta)\right)^k}. \tag{10.21}$$

For $m \in \{1, \dots, q-2\}$, denote by $m' := q-1-m$ the "conjugate" index, $m' \in \{1, \dots, q-2\}$.

Lemma 10.2.

(a) *For every $\theta > 1$, there is $v_m = v_m(\theta) > 0$ such that the function $v \mapsto L_m(v;\theta)$ is increasing for $0 < v < v_m$ and decreasing for $v > v_m$, thus attaining its unique maximum value at $v = v_m$, denoted by*

$$L_m^*(\theta) := L_m(v_m(\theta);\theta) = \max_{v>0} L_m(v;\theta), \qquad \theta > 1. \tag{10.22}$$

(b) *For each $m \ge 1$, the function $\theta \mapsto L_m^*(\theta)$ is continuous and monotone increasing, with $\lim_{\theta\to\infty} L_m^*(\theta) = \infty$. Moreover, $L_m^*(\theta)$ has a unique zero $\theta_m > 1$, i.e.,*

$$L_m^*(\theta_m) = L_m(v_m(\theta_m);\theta_m) = 0. \tag{10.23}$$

(c) *The value* $v_m^* := v_m(\theta_m)$ *is the unique positive root of the equation*

$$m\sum_{i=1}^{k-1} i v^{k-i} - m' \sum_{i=1}^{k-1} i v^{i-k} = 0. \tag{10.24}$$

In particular, $v_m^* = 1$ *if* $m = \frac{1}{2}(q-1)$ *and* $v_m^* > 1$ *if* $m < \frac{1}{2}(q-1)$.

Proof. (a) We have

$$\begin{aligned}\frac{\partial L_m(v;\theta)}{\partial v} &= (\theta-1)p_k'(v) - kmv^{k-1} \\ &= v^{k-1}\left((\theta-1)\sum_{i=1}^{k-1}\frac{k-i}{v^i} - km\right).\end{aligned} \tag{10.25}$$

The function in the parentheses in (10.25) is continuous and monotone decreasing in $v > 0$, with the limiting values $+\infty$ as $v \downarrow 0$ and $-km < 0$ as $v \to \infty$. Hence, there is a unique root $v_m = v_m(\theta)$ of the equation $\partial L_m(v;\theta)/\partial v = 0$, that is,

$$(\theta-1)p_k'(v_m) - kmv_m^{k-1} = 0, \tag{10.26}$$

and, moreover, $\partial L_m/\partial v > 0$ for $0 < v < v_m$ and $\partial L_m/\partial v < 0$ for $v > v_m$. Thus, claim (a) is proved.

(b) Note that the derivative $v_m'(\theta)$ exists by the inverse function theorem applied to equation (10.25). Differentiating (10.22) and using (10.26), we get

$$\begin{aligned}\frac{\mathrm{d}L_m^*(\theta)}{\mathrm{d}\theta} &= \left.\frac{\partial L_m(v;\theta)}{\partial v}\right|_{v=v_m(\theta)} \times \frac{\mathrm{d}v_m(\theta)}{\mathrm{d}\theta} + \left.\frac{\partial L_m(v;\theta)}{\partial \theta}\right|_{v=v_m(\theta)}. \\ &= p_k(v_m(\theta)) > 0.\end{aligned}$$

Thus, $L_m^*(\theta)$ is continuously differentiable and (strictly) increasing, because the function $\theta \mapsto L_m(v;\theta)$ is monotone increasing for each $v > 0$, since $\partial L_m/\partial\theta = p_k(v) > 0$. Note from (10.22) that

$$\begin{aligned}L_m^*(\theta) &\ge L_m(v;\theta)|_{v=1} \\ &= (\theta-1)p_k(1) - m - m' \\ &= (\theta-1)(k-1) - (q-1) \to +\infty, \qquad \theta\to\infty.\end{aligned} \tag{10.27}$$

On the other hand, from (10.25) we see that if $1 < \theta < \frac{k}{k-1}$ then, for all $m \ge 1$,

$$\begin{aligned}\left.\frac{\partial L_m(v;\theta)}{\partial v}\right|_{v=1} &= (\theta-1)\sum_{i=1}^{k-1}(k-i) - km \\ &= (\theta-1)\frac{(k-1)k}{2} - km \\ &< k\left(\frac{1}{2} - m\right) < 0.\end{aligned}$$

Therefore, by part (a), for such θ we have $0 < v_m(\theta) < 1$, hence, for all $m \le q-2$,

$$\begin{aligned} L_m^*(\theta) &= (\theta - 1)\, p_k(v_m) - m v_m^k - m' \\ &< \left(\frac{k}{k-1} - 1\right) p_k(1) - m' \\ &= 1 - m' \le 0. \end{aligned} \tag{10.28}$$

Thus, combining (10.27) and (10.28), it follows that there is a unique root $\theta = \theta_m$ of the equation $L_m^*(\theta) = 0$, which proves claim (b).

(c) Elimination of $\theta = \theta_m$ from the system of equations (10.23) and (10.26) gives for $v_m^* = v_m(\theta_m)$ a closed equation,

$$m k v^{k-1} p_k(v) - \left(m v^k + m'\right) p_k'(v) = 0, \tag{10.29}$$

which can be rearranged to a more symmetric form (10.24). The uniqueness of the root v_m^* is obvious, because the left-hand side of (10.24) is a continuous, increasing function in $v > 0$, with the range from $-\infty$ to $+\infty$. Finally, observe that for $v = 1$ the left-hand side of (10.24) is reduced to $(m - m') \cdot k(k-1)/2$, which vanishes if $m = m'$ and is negative if $m < m'$, so that, respectively, $v_m^* = 1$ or $v_m^* > 1$, as claimed. □

By Lemma 10.2, for any $\theta > 1$ there is a unique value $v_m = v_m(\theta) > 0$ such that

$$L_m^*(\theta) := L_m(v_m; \theta) = \max_{v>0} L_m(v; \theta),$$

and moreover, the function $\theta \mapsto L_m^*(\theta)$ is strictly increasing. Denote by θ_m the (unique) value of $\theta > 1$ such that

$$L_m^*(\theta_m) = 0. \tag{10.30}$$

Thus, for any $\theta > \theta_m$ the range of the functions $v \mapsto L_m(v;\theta)$ and $v \mapsto K_m(v;\theta)$ includes positive values,

$$\mathcal{V}_m^+(\theta) := \{v > 0 \colon L_m(v;\theta) > 0\} \equiv \{v > 0 \colon K_m(v;\theta) > 0\} \ne \emptyset, \tag{10.31}$$

and, therefore, the function

$$\alpha_m(\theta) := \frac{1}{\ln \theta} \max_{v \in \mathcal{V}_m^+(\theta)} \ln K_m(v;\theta) = \frac{\ln K_m^*(\theta)}{\ln \theta}, \qquad \theta > \theta_m, \tag{10.32}$$

is well defined, where

$$K_m^*(\theta) := \max_{v \in \mathcal{V}_m^+(\theta)} K_m(v;\theta).$$

Remark 10.1. For $k = 2$, from (10.20) and (10.30) we obtain explicit formula

$$\theta_m = 1 + 2\sqrt{m(q-m-1)}, \qquad m = 1, \dots, q-2. \tag{10.33}$$

Comparing (10.10) and (10.33), one can see that $\theta_{\mathrm{c}} \le \theta_1$ if and only if $q \ge 6$. For example, for $q = 5$ we have $\theta_1 = 1 + 2\sqrt{3} \approx 4.4641$, $\theta_2 = 5$, $\theta_{\mathrm{c}} = \frac{1}{2}(\sqrt{153} - 3) \approx 4.6847$, that is, $\theta_1 < \theta_{\mathrm{c}} < \theta_2$, whereas for $q = 6$ and $q = 7$ we compute $\theta_1 = \theta_{\mathrm{c}} = 5$ and $\theta_1 = 1 + 2\sqrt{5} \approx 5.4721 > \theta_{\mathrm{c}} = \frac{1}{2}(\sqrt{241} - 5) \approx 5.2621$, respectively. In case $q = 3$ (with $m = 1$ and any $k \ge 2$) it is easy to see that the condition (10.30) is satisfied with $v_1^* = v_1(\theta_1) = 1$ (cf. Lemma 10.2(c)), whence find $\theta_1 = 1 + \frac{2}{k-1}$.

For each $m \in \{1, \dots, q-2\}$, let $\nu_m(\theta, \alpha)$ denote the number of positive solutions (u, v) of the system (10.8).

Now we are ready to formulate the following result.

Theorem 10.2. *The inequality $\nu_m(\theta, \alpha) \ge 1$ holds if and only if $\theta > \theta_m$ and $\alpha \le \alpha_m(\theta)$.*

Proof. The second equation in the system (10.6) is reduced to

$$\theta + \theta^\alpha u^k + m v^k + m' - 1 = (\theta - 1)\big(p_k(v) + 1\big), \tag{10.34}$$

which can be rewritten, in the form

$$\theta^\alpha u^k = L_m(v; \theta). \tag{10.35}$$

Substituting (10.34) and (10.35) into the denominator and numerator, respectively, of the ratio in the first equation of (10.6), we get

$$u = \frac{p_k(v) + L_m(v; \theta)}{p_k(v) + 1}. \tag{10.36}$$

Consequently, substituting (10.36) back into (10.35), one obtains the equation

$$\theta^\alpha = K_m(v; \theta), \tag{10.37}$$

where (cf. (10.21))

$$K_m(v; \theta) := \frac{L_m(v; \theta)(p_k(v) + 1)^k}{(p_k(v) + L_m(v; \theta))^k}. \tag{10.38}$$

Since all steps above are reversible, the equations (10.36) and (10.37) imply the system (10.6).

Note from (10.37) that $v > 0$ must satisfy the condition $K_m(v; \theta) > 0$, that is, $v \in \mathcal{V}_m^+(\theta)$ (see (10.31)); by Lemma 10.2(b), this is possible if and only if $\theta > \theta_m$. Moreover, the equation (10.37) has a solution $v > 0$ if and only if $\alpha \le \alpha_m(\theta)$, with the critical threshold $\alpha_m(\theta)$ defined in (10.32). This completes the proof. □

The function $\alpha_+(\theta)$ has a unique zero given by $\theta_0^+ = 1 + \frac{q}{k-1}$ (see below Proposition 10.8(b)), whereas $\alpha_-(\theta)$ has a unique zero θ_0^-, which coincides with the zero θ_1^0 of $\alpha_1(\theta)$. It also holds that $\alpha_1(\theta)$ is a majorant of the family of functions $\{\alpha_m(\theta)\}$ (see Propositions 10.4 and 10.5).

10.1.2 *Number of TISGMs*

The following lemma says that in the case $q = 3$, there appears an additional critical value denoted by

$$\tilde{\theta}_1 = \tilde{\theta}_1(k) := \frac{5 - k + \sqrt{49k^2 + 62k + 49}}{6(k-1)}. \tag{10.39}$$

Lemma 10.3. *Let $q = 3$ and $k \geq 2$.*

(a) For all $\theta > \tilde{\theta}_1$, we have $K_1(v;\theta)|_{v=1} < \max_{v\in\mathcal{V}_1^+(\theta)} K_1(v;\theta)$.

(b) Let $k \in \{2,3,4\}$. If $1 < \theta \leq \tilde{\theta}_1$ then the function $v \mapsto K_1(v;\theta)$ has the unique maximum at $v = 1$, that is, $K_1(v;\theta) < K_1(1;\theta)$ for any $v \neq 1$.

Proof. (a) For $q = 3$ the partial derivative $\partial K_1/\partial v$ can be represented as

$$\begin{aligned}\frac{\partial K_1}{\partial v} = \frac{(p_k+1)^{k-1}}{(p_k+L_m)^{k+1}} \Big\{ &\big(p_k + 1 + (1-L_1)(k-1)\big) L_1' \big(p_k - (k-1)\big) \\ &- (1-L_1)kL_1\left(p_k' - \frac{k(k-1)}{2}\right) \\ &+ \frac{k(k-1)}{2}(1-L_1)\big(2L_1' - kL_1\big)\Big\},\end{aligned} \tag{10.40}$$

where $t := \theta - 1$, $L_1 = tp_k - v^k - 1$ and $L_1' = \partial L_1/\partial v = tp_k' - kv^{k-1}$.

We have $\partial K_1/\partial v|_{v=1} = 0$; this follows without calculations from the scaling property $K_1(v;t) = K_1(v^{-1};t)$ (see (10.59) below). Namely, using the formulas

$$p_k|_{v=1} = k - 1, \qquad p_k'|_{v=1} = \frac{k(k-1)}{2}, \tag{10.41}$$

it is easy to see that the terms in (10.40) vanish at $v = 1$.

Differentiating (at $v = 1$) the identity

$$v^k - 1 \equiv (v-1)(p_k + 1),$$

we obtain the formula

$$p_k''|_{v=1} = \frac{k(k-1)(k-2)}{3}. \tag{10.42}$$

The second-order derivative $\partial^2 K_1/\partial v^2$ at $v = 1$, is obtained by differentiating in (10.40) only the factors that vanish at $v = 1$, setting $v = 1$ elsewhere:

$$\left.\frac{\partial^2 K_1}{\partial v^2}\right|_{v=1} = \frac{(p_k+1)^{k-1}}{(p_k+L_1)^{k+1}}\left\{(p_k+1)L_1'p_k' + (1-L_1)\left[(k-1)L_1'p_k' \right.\right.$$
$$\left.\left.\left. -kL_1p_k'' + \frac{k(k-1)}{2}(2L_1''-kL_1')\right]\right\}\right|_{v=1}. \qquad (10.43)$$

Using (10.41) and (10.42), we find

$$L_1|_{v=1} = tp_k(1) - 2 = t(k-1) - 2, \qquad (10.44)$$

$$L_1'|_{v=1} = tp_k'(1) - k = t\frac{k(k-1)}{2} - k, \qquad (10.45)$$

$$L_1''|_{v=1} = tp_k''(1) - k(k-1) = t\frac{k(k-1)(k-2)}{3} - k(k-1). \qquad (10.46)$$

Finally, substituting formulas (10.41), (10.42) and (10.44)–(10.46) into (10.43), after simplifications we get

$$\left.\frac{\partial^2 K_1}{\partial v^2}\right|_{v=1} = \left(\frac{k}{(k-1)(t+1)-2}\right)^{k+1}\frac{k-1}{2}$$
$$\times\left(\frac{(k-1)^2}{2}t^2 + \frac{(k-1)(7k-11)}{6}t - 3k + 1\right). \qquad (10.47)$$

The quadratic polynomial in (10.47) has one positive zero $t = t^*$,

$$t^* = \frac{11 - 7k + \sqrt{49k^2+62k+49}}{6(k-1)},$$

which corresponds to $\tilde{\theta}_1 = t^* + 1$ as defined in (10.39). Hence, $\partial^2 K_1/\partial v^2|_{v=1} > 0$ for $\theta > \tilde{\theta}_1$, which implies that $v = 1$ is a local minimum of the function $v \mapsto K_1(v;\theta)$. This completes the proof of Lemma 10.3(a).

(b) For $1 \le \theta \le \tilde{\theta}_1$ it suffices to show that $v = 1$ is the sole root of the equation $\partial K_1/\partial v = 0$. The proof of the latter statement is given by the following steps:

(1) First, the condition $\partial K_1/\partial v = 0$ is reduced to $P(v;t) = 0$, where $P(v;t)$ is a polynomial in v (and also a quadratic polynomial in t).
(2) Since $P(1;t) = 0$, the quotient $R(v;t) = P(v;t)/(v-1)$ is a polynomial in v, and one should prove that $R(v;t) < 0$ for any $t \le t^*$ and all $v \ne 1$.
(3) According to the proof of part (a), the condition $\partial^2 K_1/\partial v^2|_{v=1} \le 0$ is satisfied whenever $R(1;t) \le 0$; moreover, the critical value t^* is determined by the quadratic equation $R(1;t^*) = 0$, which implies that $R(1;t) \le 0$ for $t \in [0, t^*]$.

(4) Bearing in mind the invariance of $K_1(v;t)$ under the map $v \mapsto v^{-1}$, it should be possible to represent the polynomial $R(v;t)$ in the form

$$R(v;t) = \chi(v) \cdot \widetilde{R}(y;t), \qquad y = v + v^{-1} \geq 2,$$

where $\chi(v) > 0$ and $\widetilde{R}(y;t)$ is a polynomial in y.

(5) The crucial step is to show that, for each $t \in [0, t^*]$, the function $y \mapsto \widetilde{R}(y;t)$ is *monotone decreasing.*

(6) Finally, using steps (3) to (5), for any $y > 2$ (i.e., $v \neq 1$) we have

$$\widetilde{R}(y;t) < \widetilde{R}(2;t) = \frac{R(1;t)}{\chi(1)} \leq 0, \qquad 0 \leq t \leq t^*.$$

Hence, for all $t \in [0, t^*]$ and $v \neq 1$, we get

$$R(v;t) = \chi(v) \cdot \widetilde{R}(y;t) < 0,$$

as required. □

For $q \geq 2$, consider the following subsets of the half-plane

$$\{\theta \geq 1\} = \{(\theta, \alpha) \colon \theta \geq 1, \, \alpha \in \mathbb{R}\},$$

$$A_q := \{\theta > \theta_{\rm c}, \, \alpha_-(\theta) \leq \alpha \leq \alpha_+(\theta)\},$$

$$B_q := \begin{cases} \emptyset & \text{if } q = 2, \\ \{\theta > \theta_1, \, \alpha < \alpha_1(\theta)\} \cup \{\theta > \tilde{\theta}_1, \, \alpha = \alpha_1(\theta)\} & \text{if } q = 3, \\ \{\theta > \theta_1, \, \alpha \leq \alpha_1(\theta)\} & \text{if } q \geq 4. \end{cases} \tag{10.48}$$

Denote the total number of positive solutions $\boldsymbol{z} = (z_1, \ldots, z_{q-1})$ of the system of equations (10.2) by $\nu(\theta, \alpha)$ ($\theta \geq 1$, $\alpha \in \mathbb{R}$); note also that this number depends on k and q.

Theorems 10.1 and 10.2 can now be summarized to the following theorem of non-uniqueness:

Theorem 10.3.

(a) *If* $q = 2$ *then*[1] $\nu(\theta, \alpha) \geq 2$ *if and only if* $(\theta, \alpha) \in A_2 \cup B_2 = A_2$.

(b) *If* $q = 3$ *then* $\nu(\theta, \alpha) \geq 2$ *if* $(\theta, \alpha) \in A_3 \cup B_3$. *The "only if" statement holds true at least for* $k = 2, 3, 4$.

(c) *If* $q \geq 4$ *then* $\nu(\theta, \alpha) \geq 2$ *if and only if* $(\theta, \alpha) \in A_q \cup B_q$.

[1] This is the case of Ising model with an external field.

Proof. If $q = 2$ then the only solutions of the system (10.2) are provided by equation (10.7); therefore, Theorem 10.3(a) readily follows from Theorem 10.1.

For $q \geq 3$, in order that $\nu(\theta, \alpha) \geq 2$, either there must be at least two solutions of equation (10.7), that is, $(\theta, \alpha) \in A_q$ (see Theorem 10.1), or, since we always have $\nu_0(\theta, \alpha) \geq 1$, there should exist at least one solution (u, v) of the system (10.8). By Theorem 10.2, such solutions exist if $\alpha \leq \alpha_m(\theta)$ for some m; since $\alpha_1(\theta)$ is a majorant of the family $\{\alpha_m(\theta)\}$ (see Proposition 10.5 below), the latter condition is reduced to $\alpha \leq \alpha_1(\theta)$, which leads to the inclusion $(\theta, \alpha) \in B_q$. By Lemma 10.10 (see below), this solution also satisfies the constraint $v \neq 1$ if $m = 1 < \frac{1}{2}(q-1)$, i.e., $q > 3$, which proves Theorem 10.3(c).

Part (b) of theorem follows by the Lemma 10.3 about the maximum of the function $v \mapsto K_1(v; \theta)$ over the domain $v \in \mathcal{V}_1^+(\theta)$ (see (10.31)). □

Theorem 10.3 provides a sufficient and (almost) necessary condition for the uniqueness of solution of (10.2), illustrated in Fig. 10.1 for $q = 5$ and $q = 3$, both with $k = 2$.

The maximum of the function $v \mapsto K_m(v; \theta)$, as well as the number of solutions $v > 0$ of the equation (10.37) for various values of parameters are illustrated in Fig. 10.2 (for the regular case $q \geq 4$) and in Fig. 10.3 (for the special case $q = 3$).

As mentioned above, the case $q = 3$ is truly critical with regard to the uniqueness. Recall that $\nu(\theta, \alpha)$ denotes the number of positive solutions of the system (10.2); the function $\alpha_1(\theta)$ is defined in (10.32) and θ_1^0 is its zero.

Proposition 10.1. *Let $q = 3$ and $k \geq 2$. There exists $\varepsilon > 0$ small enough such that*

(a) $\nu(\theta, \alpha_1(\theta)) = 1$ *if* $1 \leq \theta < \theta_1 + \varepsilon$;
(b) $\nu(\theta, \alpha_1(\theta)) \geq 2$ *for all* $\theta > \theta_1^0 - \varepsilon$.

Proof. The proof relies on some lemmas that will be proved later, in Sections 10.1.3 and 10.1.3.

(a) By the proof of Lemma 10.3, we have $\partial^2 K_1/\partial v^2|_{v=1} < 0$ for all $\theta \in [\theta_1, \tilde{\theta}_1)$; in particular, $\partial^2 K_1(\theta_1, v)/\partial v^2|_{v=1} < 0$. By Lemma 10.2(c), the set $\mathcal{V}_1^+(\theta)$ (see (10.31)) is reduced for $\theta = \theta_1$ to the single point $v = 1$, and $K_1^*(\theta_1) = \max_{v \in \mathcal{V}_1^+(\theta_1)} K_1(v; \theta_1) = K_1(1; \theta_1) = 0$. By continuity, it follows that $\partial^2 K_1(\theta; v)/\partial v^2 < 0$ for each $\theta \in [\theta_1, \theta_1 + \varepsilon)$ and all $v \in \mathcal{V}_1^+(\theta)$; that is, the function $v \mapsto K_1(v; \theta)$ is concave on $\mathcal{V}_1^+(\theta)$ and therefore has a unique

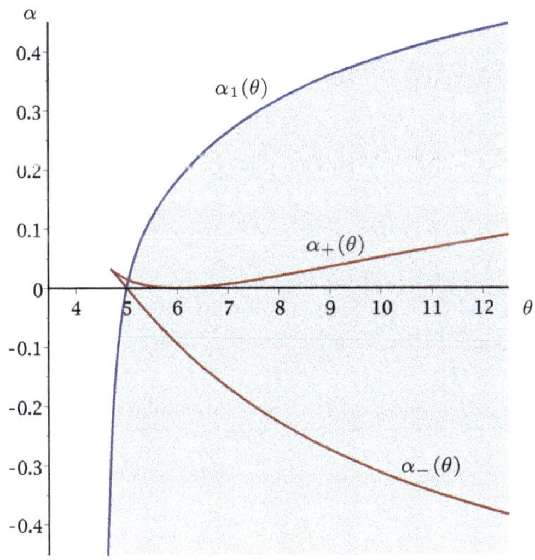

(a) $q = 5$ $(k = 2)$

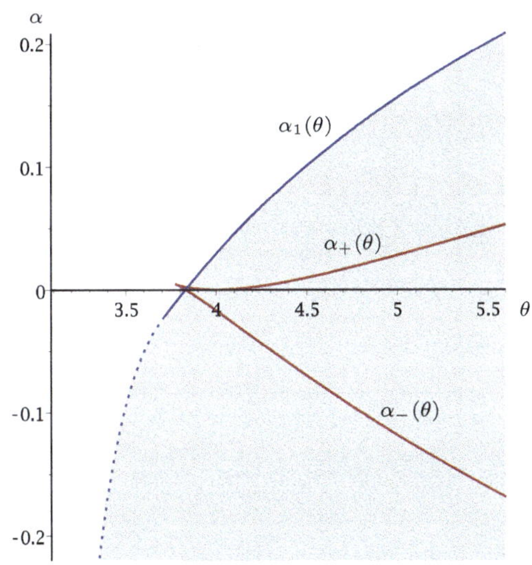

(b) $q = 3$ $(k = 2)$

Fig. 10.1 The phase diagram for the Potts model (10.1) showing the non-uniqueness region (shaded in grey) according to Theorem 10.3: (a) regular case $q \geq 4$ (shown here for $q = 5$); (b) special case $q = 3$, both with $k = 2$. The critical boundaries are determined by (parts of) the graphs of the functions $\alpha_{\pm}(\theta)$ and $\alpha_1(\theta)$ defined in (10.16) and (10.32), respectively. The dotted part of the boundary on panel (b), given by $\alpha = \alpha_1(\theta)$, $\theta \in (\theta_1, \tilde{\theta}_1]$ (see formula (10.48) with $q = 3$), is excluded from the shaded region (see Theorem 10.3(b) and Conjecture 10.1, proved for $2 \leq k \leq 4$); here, $\theta_1 = 3$ and $\tilde{\theta}_1 = \frac{1}{2}(1 + \sqrt{41}) \approx 3.7016$.

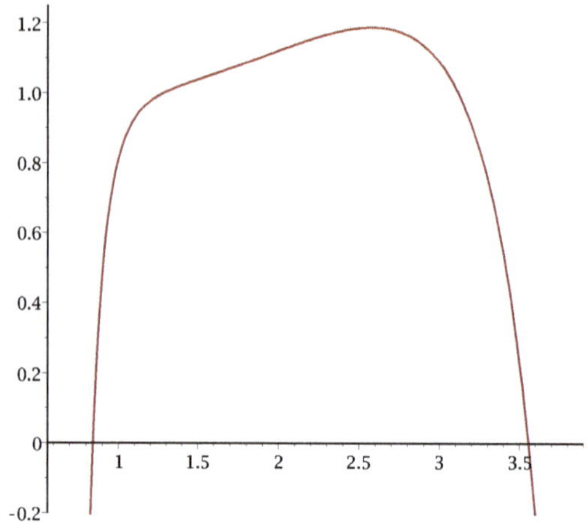

(a) $\theta = 5.4$

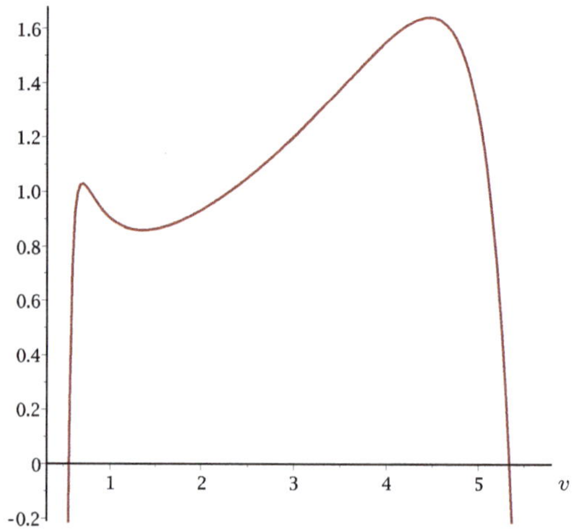

(b) $\theta = 6.9$

Fig. 10.2 The graph of the function $v \mapsto K_1(v;\theta)$ (i.e., with $m = 1$) for $k = 2$, $q = 5$ and various values of $\theta > \theta_1 = 1 + 2\sqrt{3} \approx 4.4641$, illustrating different possible numbers of solutions $\nu_1 = \nu_1(\theta, \alpha)$ of the equation (10.37): (a) $\theta = 5.4$, $0 \le \nu_1 \le 2$; (b) $\theta = 6.9$, $0 \le \nu_1 \le 4$.

maximum located at $v = 1$, recall that $\partial K_1/\partial v|_{v=1} = 0$. But the solution (u, v) with $v = 1$ is not admissible (see (10.9)), hence $\nu(\theta, \alpha_1(\theta)) = 1$, with the only solution of (10.2) coming from equation (10.7).

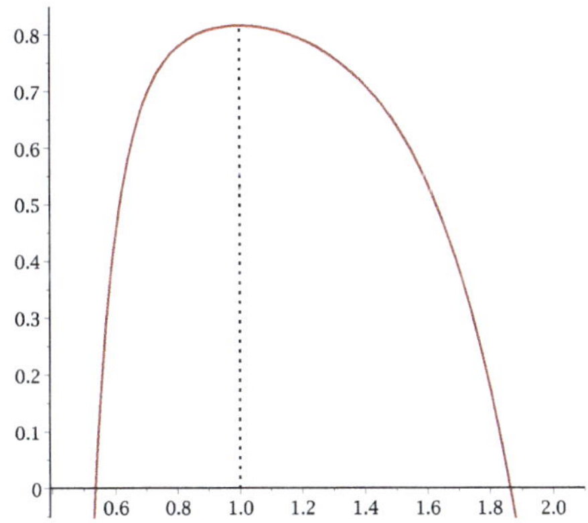

(a) $\theta = 3.4$

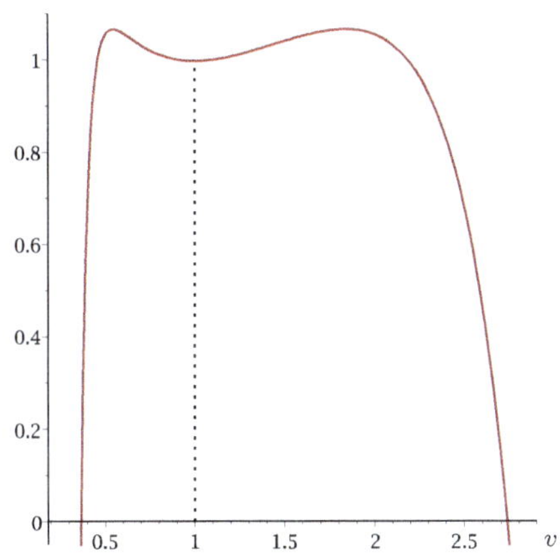

(b) $\theta = 4.1$

Fig. 10.3 The graph of the function $v \mapsto K_1(v;\theta)$ for $q = 3$ and $k = 2$, illustrating the location of its maximum point depending on whether $\theta_c < \theta < \tilde{\theta}_1$ or $\theta > \tilde{\theta}_1$; here, $\theta_c = 3$ and $\tilde{\theta}_1 = \frac{1}{2}(1 + \sqrt{41}) \approx 3.7016$.

(b) Since $\alpha_1(\theta_1^0) = 0$, we have $K_1^*(\theta_1^0) = 1$ (see (10.32)) and, by Lemma 10.11(b), it follows that $L_1^*(\theta_1^0) = 1$. If $v_1 = v_1(\theta_1^0)$ is the point where the latter maximum is attained, that is, $L_1(v_1;\theta_1^0) = L_1^*(\theta_1^0) = 1$, then it holds that $v_1 > 1$ (see Lemma 10.8 below). Furthermore, according to Lemma 10.2(a), v_1 is the unique maximum of the function $v \mapsto L_1(v;\theta_1^0)$, and in particular $L_1(1;\theta_1^0) < L_1^*(\theta_1^0) = 1$.

From the definition (10.38), it follows that also $K_1(v_1;\theta_1^0) = 1 = K_1^*(\theta_1^0)$. By the scaling property (10.59), we also have $K_1(v_1^{-1};\theta_1^0) = K_1(v_1;\theta_1^0) = 1$. Since $L_1(1;\theta_1^0) < 1$, Lemma 10.11(a) implies that $K_1(1;\theta_1^0) < 1 = K_1^*(\theta_1^0)$. Thus, the corresponding solution of the system (10.8) satisfies the condition (10.9), and therefore $\nu(\theta_1^0, \alpha_1(\theta_1^0)) \geq 2$.

By continuity of $K_1(1;\theta)$ and $K_1^*(\theta)$, for small enough $\varepsilon > 0$ and all $\theta \in (\theta_1^0 - \varepsilon, \theta_1^0]$ we have that $K_1(1;\theta) < K_1^*(\theta)$, therefore the maximum is attained outside $v = 1$. Thus, by the same argument as before, the claim follows. □

The case $q = 3$ in the definition (10.48) and in Theorem 10.3 emerges because for $\theta \leq \tilde{\theta}_1$ and $\alpha = \alpha_1(\theta)$, there is a (hypothetically unique) solution $(u,v) = \big(\theta - \frac{\theta+2}{k}, 1\big)$ of the system (10.8), which is, however, not admissible due to the constraint (10.9) and, therefore, does not destroy the uniqueness of solution to (10.2). This hypothesis is conjectured below; if it is true then "if" in Theorem 10.3(b) (i.e., $q = 3$) can be enhanced to "if and only if" for all $k \geq 2$.

Conjecture 10.1. *If $q = 3$, $\theta \leq \tilde{\theta}_1$ and $\alpha = \alpha_1(\theta)$ then $(u,v) = \big(\theta - \frac{\theta+2}{k}, 1\big)$ is the sole solution of the system* (10.8).

Remark 10.2. Regardless of Conjecture 10.1, the inclusion of a proper part of the curve $\alpha = \alpha_1(\theta)$ in the uniqueness region in the case $q = 3$ is indeed necessary. Namely, it can be proved (see Proposition 10.1) that if $\varepsilon > 0$ is small enough then $\nu(\theta, \alpha_1(\theta)) = 1$ for $\theta_1 < \theta < \theta_1 + \varepsilon$ but $\nu(\theta, \alpha_1(\theta)) \geq 2$ for $\theta > \theta_1^0 - \varepsilon$, where θ_1^0 is the zero of $\alpha_1(\theta)$.

The following theorem describes some cases where it is possible to estimate the maximal number of solutions of the system (10.2).

Theorem 10.4. *Let $\alpha \neq 0$.*

(a) *If $q = 2$ then $\nu(\theta,\alpha) \leq 3$; moreover, $\nu(\theta,\alpha) = 3$ for all $\theta \geq 1$ large enough.*

(b) *If $k = 2$ then $\nu(\theta,\alpha) \leq 2^q - 1$ for all $\theta \geq 1$ and $\alpha \in \mathbb{R}$.*

Proof. (a) First, let $q = 2$. According to Lemma 10.1, the system (10.2) is reduced to the single equation (10.7), and by Theorem 10.1 the number of its solutions is not more than $3 = 2^2 - 1$; furthermore, for $\theta > \theta_{\mathrm{c}} = \frac{k+1}{k-1}$ and $\alpha_-(\theta) < \alpha < \alpha_+(\theta)$, there are exactly three solutions, so the upper bound is attained.

(b) Up to three solutions of the system (10.7) arising from the equation (10.5) are ensured by Lemma 10.1 (see also Theorem 10.1). Other solutions are determined by the system (10.6) indexed by $m = 1, \ldots, q-2$, which in turn depends on the solvability of the equation (10.37). In the case $k = 2$, the latter is a polynomial equation of degree 4, and therefore has at most four roots $v > 0$, for each m. The value $u > 0$ is then determined uniquely by formula (10.36), and it occupies the first place in the vector $\boldsymbol{z} = (z_1, \ldots, z_{q-1})$. As for the root $v > 0$, it occupies m out of the $q-2$ remaining places. Counting the total number of such permutations, we get the upper bound

$$3 + 4\sum_{m=1}^{q-2}\binom{q-1}{m} = 3 + 4\,(2^{q-2} - 1) = 2^q - 1,$$

as required. This completes the proof. □

10.1.3 *Properties of the critical curves*

In this section we list (without proofs) results from pages 46–67 of [28].

Properties of $\alpha_\pm(\theta)$

Lemma 10.4. *The quantities $a_\pm$, defined in* (10.15) *for $b \ge (k+1/k-1)^2$, satisfy the identity*

$$a_+ \times a_- = b^{-k-1}. \tag{10.49}$$

Lemma 10.5. *Suppose that $b \ge \left(\frac{k+1}{k-1}\right)^2$. Then the following inequalities hold,*

$$a_+ < b^{-1}, \qquad a_- > b^{-k}. \tag{10.50}$$

Lemma 10.6. *The functions $\alpha_\pm(\theta)$ satisfy the following identity,*

$$\alpha_-(\theta) + \alpha_+(\theta) = \frac{2\ln(q-1) + (k+1)\big(\ln b(\theta) - 2\big)}{\ln\theta}, \qquad \theta \ge \theta_{\mathrm{c}}, \tag{10.51}$$

where $b(\theta)$ is defined in (10.11). *In particular, if $q = 2$ then $\alpha_-(\theta) + \alpha_+(\theta) \equiv 0$ for all $\theta \ge \theta_{\mathrm{c}} = \frac{k+1}{k-1}$.*

Proposition 10.2. *The functions $\alpha_\pm(\theta)\colon [\theta_{\mathrm{c}}, \infty) \to \mathbb{R}$ defined in* (10.16) *have the following "boundary" values,*

$$\alpha_\pm(\theta_{\mathrm{c}}) = -(k+1) + \frac{1}{\ln\theta_{\mathrm{c}}}\left(\ln(q-1) + (k+1)\ln\frac{k+1}{k-1}\right), \tag{10.52}$$

$$\alpha_\pm(\theta) \to \pm(k-1) \qquad (\theta \to \infty). \tag{10.53}$$

In particular, $\alpha_\pm(\theta_{\mathrm{c}}) = 0$ if $q = 2$ and $\alpha_\pm(\theta_{\mathrm{c}}) > 0$ if $q > 2$.

Proposition 10.3. *The functions $\alpha_\pm(\theta)$ satisfy the following bounds,*

$$\alpha_-(\theta) > -(k-1), \qquad \theta \ge \theta_c, \tag{10.54}$$

$$\alpha_+(\theta) < k-1, \qquad \theta \ge \max\{\theta_c, \bar\theta\}, \tag{10.55}$$

where

$$\bar\theta = \bar\theta(k,q) := \begin{cases} \dfrac{q-2}{(q-1)^{(k-1)/k}-1}, & q>2,\\ 1, & q=2. \end{cases} \tag{10.56}$$

Remark 10.3. Note that $\bar\theta > 1$ for $q \ge 3$. Of course, if $\theta_c \ge \bar\theta$ then the upper bound (10.55) holds for all $\theta \ge \theta_c$; however, the ordering between θ_c and $\bar\theta$ depends on k and q. For example, for $k=2$ and $q=5$

$$\bar\theta = 3 < \theta_c \approx 4.6847,$$

whereas for $k=2$ and $q=50$

$$\bar\theta = 8 > \theta_c \approx 7.8904.$$

In fact, for large q the upper bound (10.55) fails near θ_c; indeed, using (10.10) we get

$$\theta_c = \frac{q-2}{2}\left(\sqrt{1+\frac{4(q-1)}{(q-2)^2}\left(\frac{k+1}{k-1}\right)^2}-1\right) \sim \left(\frac{k+1}{k-1}\right)^2, \qquad q\to\infty,$$

and, according to (10.52),

$$\alpha_\pm(\theta_c) \sim \frac{\ln q}{\ln\theta_c} \to \infty, \qquad q\to\infty.$$

Conjecture 10.2. *The function $\alpha_-(\theta)$ is monotone decreasing for all $\theta \ge \theta_c$, whereas $\alpha_+(\theta)$ is decreasing for $\theta \le \theta_0^+$ and increasing for $\theta \ge \theta_0^+$, with the unique minimum $\alpha_+(\theta_0^+) = 0$ at the critical point*

$$\theta_0^+ = \theta_0^+(k,q) := 1 + \frac{q}{k-1}. \tag{10.57}$$

In the case $q=2$, we have $\theta_c = \theta_0^+ = \frac{k+1}{k-1}$ and, by Lemma 10.6, $\alpha_+(\theta) \equiv -\alpha_-(\theta)$; hence, the function $\alpha_+(\theta)$ should be monotone increasing for all $\theta \ge \theta_c$.

This conjecture is supported by computer plots (see Fig. 10.4).

Properties of $L_m(v;\theta)$ and $K_m(v;\theta)$

Consider the case $q \ge 3$ and recall that $v_m = v_m(\theta)$ is the unique maximum of the function $v \mapsto L_m(v;\theta)$ (see Lemma 10.2(a)), and θ_m is defined by the relation (10.23), where $1 \le m \le q-2$.

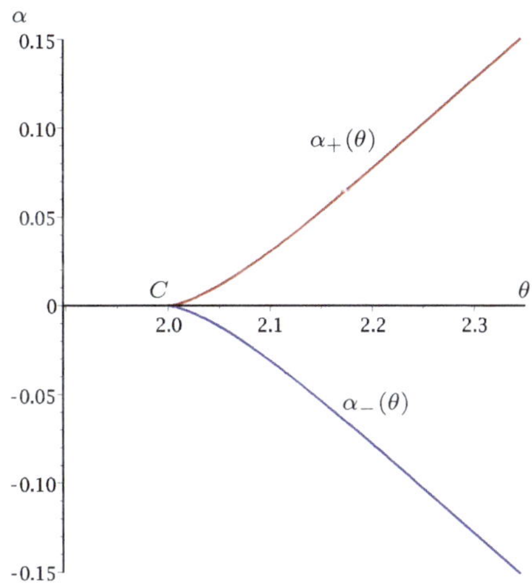

(a) $q = 2$ $(k = 3)$

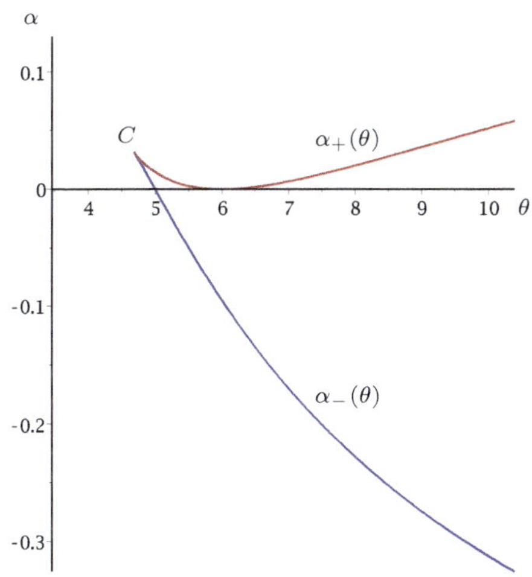

(b) $q = 5$ $(k = 2)$

Fig. 10.4 The graphs of the functions $\alpha_+(\theta)$ and $\alpha_-(\theta)$ (see (10.16)): (a) $q = 2$, $k = 3$; (b) $q = 5$, $k = 2$. The coordinates of the cusp point C (see (10.10) and (10.52)) are given numerically by: (a) $\theta_c = 2$, $\alpha_\pm(\theta_c) = 0$; (b) $\theta_c \approx 4.6847$, $\alpha_\pm(\theta_c) \approx 0.0319$. The axis $\alpha = 0$ is a tangent line for $\alpha_+(\theta)$ at $\theta_0^+ = 1 + q/(k-1)$ (see Proposition 10.8(b-i)): (a) $\theta_0^+ = 2$ $(= \theta_c)$; (b) $\theta_0^+ = 6$. In panel (b), the function $\alpha_-(\theta)$ has zero at $\theta_0^- = 5$ (see Proposition 10.8(b-ii) and Example 10.1), whereas in panel (a) we have $\theta_0^- = \theta_0^+ = 2$ (see Proposition 10.2). Note that the graphs display the monotone behavior as predicted by Conjecture 10.2.

The next lemma describes the useful scaling properties of the functions $L_m(v;\theta)$ and $K_m(v;\theta)$ under the conjugation $m \mapsto m'$.

Lemma 10.7. *For each $m = 1,\dots,q-2$ and $m' = q-1-m$, the following identities hold for all $v > 0$ and $\theta > \theta_m$,*

$$L_{m'}(v;\theta) = v^k L_m(v^{-1};\theta), \tag{10.58}$$

$$K_{m'}(v;\theta) = K_m(v^{-1};\theta). \tag{10.59}$$

For $\theta > \theta_m$, let $v_m = v_m(\theta)$ be the point where the function $v \mapsto L_m(v;\theta)$ attains its maximum value (see Lemma 10.2):

$$L_m(v_m;\theta) = \max_{v\in\mathcal{V}_m^+(\theta)} L_m(v;\theta).$$

The next result provides a strict lower bound for v_m (cf. Lemma 10.2(c) for $\theta = \theta_m$).

Lemma 10.8. *For all m in the range $1 \le m \le \frac{1}{2}(q-1)$, we have*

$$v_m(\theta) > 1, \qquad \theta > \theta_m. \tag{10.60}$$

For $\theta > \theta_m$, let $w_m = w_m(\theta)$ be the point where the function $v \mapsto K_m(v;\theta)$ attains its (positive) maximum value (see Lemma 10.2(c)):

$$K_m(w_m;\theta) = \max_{v\in\mathcal{V}_m^+(\theta)} K_m(v;\theta).$$

Note that $w_m(\theta_m) = v_m(\theta_m) = v_m^*$.

The next two lemmas provide useful bounds on $w_m = w_m(\theta)$. First, there is a simple uniform upper bound.

Lemma 10.9. *For all $m \in [1, q-2]$,*

$$w_m < \theta, \qquad \theta \ge \theta_m. \tag{10.61}$$

The important lower bound for $w_m = w_m(\theta)$ is established next.

Lemma 10.10.

(a) For all m in the range $1 \le m < \frac{1}{2}(q-1)$, we have

$$w_m > 1, \qquad \theta > \theta_m. \tag{10.62}$$

(b) If $m = \frac{1}{2}(q-1)$ then the maximum point w_m (which may not be unique) can be chosen so that $w_m \ge 1$.

Properties of θ_m and $\alpha_m(\theta)$

Proposition 10.4. *For each $m = 1, \dots, q-2$ and $m' = q-1-m$, we have*

$$\theta_m = \theta_{m'}. \tag{10.63}$$

Moreover, the functions $\theta \mapsto \alpha_m(\theta)$ (see (10.32)) satisfy the symmetry relation

$$\alpha_{m'}(\theta) \equiv \alpha_m(\theta), \qquad \theta > \theta_m. \tag{10.64}$$

Proposition 10.5. *Let $q \geq 5$, and set $m_0 := [\frac{1}{2}(q-1)]$. Then for $m = 1, \dots, m_0 - 1$*

$$\theta_m < \theta_{m+1}, \tag{10.65}$$

$$\alpha_m(\theta) > \alpha_{m+1}(\theta), \qquad \theta > \theta_{m+1}. \tag{10.66}$$

Proposition 10.6. *For all $m = 1, \dots, q-2$, the functions $\theta \mapsto \alpha_m(\theta)$ defined by formula (10.32) satisfy the upper bound*

$$\alpha_m(\theta) < k - 1, \qquad \theta > \theta_m. \tag{10.67}$$

Moreover, they have the following "boundary" values,

$$\lim_{\theta \downarrow \theta_m} \alpha_m(\theta) = -\infty, \qquad \lim_{\theta \uparrow \infty} \alpha_m(\theta) = k - 1. \tag{10.68}$$

Proposition 10.7. *For each $m = 1, \dots, q-2$, the function $\theta \mapsto K_m^*(\theta)$ is monotone increasing for $\theta > \theta_m$.*

Remark 10.4. The result of Proposition 10.7 is not trivial (unlike the similar statement for $L_m^*(\theta)$, see the footnote in the proof of Lemma 10.2(b)), because, for each $v > 0$, we have that $L_m(v;\theta) \to \infty$ and, therefore, $K_m(v;\theta) \to 0$ as $\theta \to \infty$ (see formula (10.38)).

Conjecture 10.3. *For each $m = 1, \dots, q-2$, the function $\theta \mapsto \alpha_m(\theta) = \ln K_m^*(\theta)/\ln\theta$ is monotone increasing.*

This conjecture is confirmed by computer plots (see Fig. 10.5) and is easy to prove at least for $\theta \leq \theta_m^0$, where θ_m^0 is the root of the equation $K_m^*(\theta) = 1$ (cf. Proposition 10.8(a)); that is, $\alpha_m(\theta) \leq 0$ for $\theta \leq \theta_m^0$. Indeed, for any $\theta \in (\theta_m, \theta_m^0]$, we have

$$\frac{\mathrm{d}\alpha_m(\theta)}{\mathrm{d}\theta} = \frac{\mathrm{d}K_m^*(\theta)}{\mathrm{d}\theta} \times \frac{1}{\ln\theta \cdot K_m^*(\theta)} - \frac{\ln K_m^*(\theta)}{\theta \ln^2\theta} > 0,$$

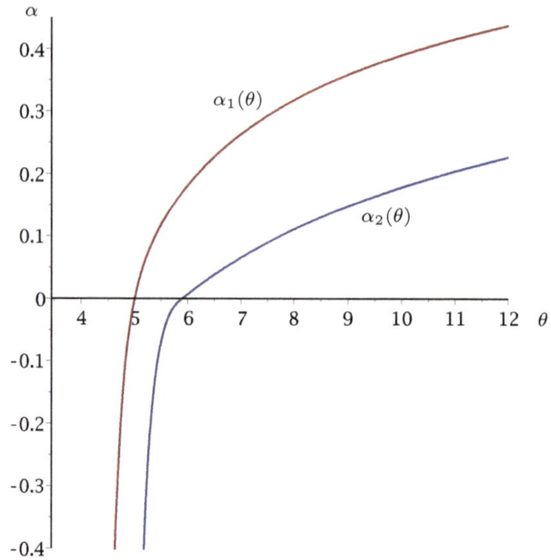

Fig. 10.5 The graphs of the functions $\alpha_m(\theta)$ $(\theta > \theta_m)$ for $k = 2$, $q = 5$ and $m = 1, 2$ (see the definition (10.32)). According to formula (10.33) (with $q = 5$), $\theta_1 = 1 + 2\sqrt{3} \approx 4.4641 < \theta_2 = 5$. Note that $\theta_1^0 = 5$ and $\theta_2^0 = 1+2\sqrt{6} \approx 5.8990$ are the zeros of $\alpha_1(\theta)$ and $\alpha_2(\theta)$, respectively (see Example 10.1). Note that the graphs are monotone increasing in line with Conjecture 10.3.

because $\mathrm{d}K_m^*/\mathrm{d}\theta > 0$ (Proposition 10.7), whereas $\ln K_m^*(\theta) \le \ln K_m^*(\theta_m^0) = 0$.

Zeros of $\alpha_\pm(\theta)$ and $\alpha_m(\theta)$. Recall that the functions $\alpha_\pm(\theta)$ and $\alpha_m(\theta)$ are defined in (10.16) and (10.32), respectively. As was observed in numerical examples (see Fig. 10.1 and also Figs. 10.4 and 10.5), the functions $\alpha_-(\theta)$ and $\alpha_1(\theta)$ have the same zero, $\theta_0^- = \theta_1^0$, whereas $\alpha_+(\theta) = 0$ at $\theta_0^+ = 1 + q/(k-1)$. Now we give results related to these observations.

Recall that

$$L_m^*(\theta) = \max_{v \in \mathcal{V}_m^+(\theta)} L_m(v;\theta),$$

$$K_m^*(\theta) = \max_{v \in \mathcal{V}_m^+(\theta)} K_m(v;\theta).$$

Lemma 10.11. *Let $q \geq 3$ and $k \geq 2$.*

(a) *For any $m \in [1, q-2]$, if $L_m(v;\theta) < 1$ for some $v \geq 1$ and $\theta > \theta_m$ then $K_m(v;\theta) < 1$.*

(b) *Let $1 \leq m \leq \frac{1}{2}(q-1)$. If $L_m^*(\theta) < 1$ for some $\theta > \theta_m$ then $K_m^*(\theta) < 1$. In particular, $L_m^*(\theta) = 1$ if and only if $K_m^*(\theta) = 1$.*

Proposition 10.8. *Let $q \geq 3$, and set $m_0 := \left[\frac{1}{2}(q-1)\right]$.*

(a) *For each m in the range $1 \leq m \leq m_0$, the function $\alpha_m(\theta)$ has a unique zero given by*

$$\theta_m^0 = \frac{m(v_m^0)^k + m' + 1}{p_k(v_m^0)}, \tag{10.69}$$

where $v_m^0 > 1$ is a sole positive root of the equation

$$m\sum_{i=1}^{k-1} iv^{k-i} - (m'+1)\sum_{i=1}^{k-1} iv^{i-k} = 0. \tag{10.70}$$

(b) (i) *The function $\alpha_+(\theta)$ has a unique zero given by $\theta_0^+ = 1 + \frac{q}{k-1}$. Moreover, $\alpha_+'(\theta_0^+) = 0$.*

(ii) *The function $\alpha_-(\theta)$ has a unique zero θ_0^-, which coincides with the zero θ_1^0 of the function $\alpha_1(\theta)$.*

(c) *The zeros $\theta_1^0, \dots, \theta_{m_0}^0$ follow in ascending order and are strictly below θ_0^+,*

$$\theta_1^0 < \cdots < \theta_{m_0}^0 < \theta_0^+ = 1 + \frac{q}{k-1}. \tag{10.71}$$

Example 10.1. Consider the case $k = 2$. Then the zero of the function $\alpha_+(\theta) = 0$ specializes to $\theta_0^+ = 1 + q$. Furthermore, the equation (10.70) is easily solved to yield $v_m^0 = \sqrt{(q-m)/m}$, and from (10.69) we readily find

$$\theta_m^0 = 1 + 2\sqrt{m(q-m)}, \qquad 1 \leq m \leq \tfrac{1}{2}(q-1). \tag{10.72}$$

Remark 10.5. In case $k = 2$, we note, by comparing (10.72) with (10.33), that $\theta_m^0 = \theta_{m+1}$ (cf.(8.28)). Moreover, taking $m = 1$ in (10.72) gives $\theta_0^- = \theta_1^0 = 1 + 2\sqrt{q-1}$.

10.2 An external field with non-zero coordinates

In this subsection following [255] we consider the (reduced) vector $\check{\boldsymbol{\xi}}^0$ (see Subsection 2.3.1) as a vector $\alpha = (\alpha_1, ..., \alpha_{q-1})$ and $h_x \equiv h = (h_1, ..., h_{q-1})$. Then the compatibility equations (2.18) is reduced to

$$h = \alpha + kF(h, \theta), \qquad (10.73)$$

where $F : h = (h_1, \ldots, h_{q-1}) \in \mathbb{R}^{q-1} \to F(h, \theta) = (F_1, \ldots, F_{q-1}) \in \mathbb{R}^{q-1}$ is defined as

$$F_i = \ln\left(\frac{(\theta - 1)e^{h_i} + \sum_{j=1}^{q-1} e^{h_j} + 1}{\theta + \sum_{j=1}^{q-1} e^{h_j}}\right),$$

and $\theta = \exp(J\beta)$.

Denote $z_i = \exp h_i, \lambda_i = \exp(\alpha_i), i = 1, ..., q-1$, then from (10.73) we get

$$z_j = \lambda_j \left(\frac{(\theta - 1)z_j + \sum_{i=1}^{q-1} z_i + 1}{\sum_{i=1}^{q-1} z_i + \theta}\right)^k. \qquad (10.74)$$

here $j = 1, 2, 3, .., q-1$.

Lemma 10.12. *Let* $\overrightarrow{z} = (z_1, \ldots, z_{q-1})$ *be a solution of (10.74). Then in this vector* $z_j = 1$ *(for some* j*) if and only if* $\lambda_j = 1$.

Proof. If $z_j = 1$, then from (10.74) we get $\lambda_j = 1$. If $\lambda_j = 1$ then (10.74) with respect to z_j may have several solutions, between them there exists $z_j = 1$ independently on values of parameters $k \geq 1$, $\theta > 0$. □

Thus the number of 1's in a solution of (10.74) coincides with the number of 1's in the vector $\overrightarrow{\lambda} = (\lambda_1, \ldots, \lambda_{q-1})$.

Consider the operator $U : \mathbb{R}_+^{q-1} \to \mathbb{R}_+^{q-1}$, defined by

$$U: \quad z_j' = \lambda_j \left(\frac{(\theta - 1)z_j + \sum_{i=1}^{q-1} z_i + 1}{\sum_{i=1}^{q-1} z_i + \theta}\right)^k, \quad j = 1, 2, 3, .., q-1. \qquad (10.75)$$

Then the equation (10.74) is the equation for fixed points of the operator U, that is, $z = U(z)$.

In general, finding all fixed points of this nonlinear operator is a difficult task. Therefore, we will look for fixed points on some invariant sets of the operator U.

For $m \in \{1, 2, ..., q-1\}$ and $a, b > 0$ denote

$$I_m = \{(x_1, x_2, ..., x_{q-1}) \in \mathbb{R}^{q-1} : x_1 = ... = x_m; \quad x_{m+1} = \cdots = x_{q-1}\}.$$

$$I_{m,a,b} = \{(x_1, x_2, ..., x_{q-1}) \in \mathbb{R}^{q-1} :$$

$$x_1 = ... = x_m = a; \quad x_{m+1} = \cdots = x_{q-1} = b\}.$$

The following lemma is straightforward

Lemma 10.13. *If $\overrightarrow{\lambda} = (\lambda_1, ..., \lambda_{q-1}) \in I_{m,a,b}$ for some $a, b > 0$. Then the set I_m is invariant with respect to the operator U, i.e., $U(I_m) \subseteq I_m$.*

Proof. Let $\lambda_i = a$, $i \leq m$ and $\lambda_i = b$, $i > m$. Then the operator (10.75) has the form

$$z'_j = a\left(\frac{(\theta-1)z_j + \sum_{i=1}^{q-1} z_i + 1}{\sum_{i=1}^{q-1} z_i + \theta}\right)^k, \quad j = 1, \ldots, m.$$

$$z'_j = b\left(\frac{(\theta-1)z_j + \sum_{i=1}^{q-1} z_i + 1}{\sum_{i=1}^{q-1} z_i + \theta}\right)^k, \quad j = m+1, \ldots, q-1. \tag{10.76}$$

Consequently, if $z \in I_m$, then $z' = U(z) \in I_m$. □

Let $\overrightarrow{\lambda} = (\lambda_1, ..., \lambda_{q-1}) \in I_{m,\lambda,\mu}$ and $\overrightarrow{z} = (z_1, ..., z_{q-1}) \in I_{m,u,v}$. Then the system (10.74) can be rewritten as

$$u = \lambda\left(\frac{(\theta+m-1)u + (q-m-1)v + 1}{mu + (q-m-1)v + \theta}\right)^k,$$

$$v = \mu\left(\frac{mu + (\theta+q-m-2)v + 1}{mu + (q-m-1)v + \theta}\right)^k. \tag{10.77}$$

For $\theta \neq 1$, $\lambda \neq 1$, $\mu \neq 1$ we have $u \neq \lambda$ and $v \neq \mu$. Therefore we introduce the following function

$$f_k(u, m, \theta, \lambda) = \frac{(mu+\theta)u^{1/k} - ((m+\theta-1)u + 1)\lambda^{1/k}}{(q-m-1)(\lambda^{1/k} - u^{1/k})}.$$

Find v from the first equation of (10.77), and find u from the second equation:

$$v = f_k(u, m, \theta, \lambda), \quad u = f_k(v, q-m-1, \theta, \mu). \tag{10.78}$$

From this system we get

$$u = f_k(f_k(u, m, \theta, \lambda), q-m-1, \theta, \mu). \tag{10.79}$$

For general parameter values, finding solutions of (10.79) seems impossible. But a computer analysis shows that for a specific choice of parameters, the number of positive solutions to this equation can be 1, 2, 3 and 4 (see table). Moreover, for each positive solution of the equation (10.79), due to the first equation of the system (10.78) corresponds a positive value of v.

k	θ	λ	μ	m	q	The number of positive solutions of (10.79)
3	0.5	0.3	0.5	1	7	1
3	20	0.3	1.5	1	7	2
3	10	0.3	1.5	1	7	3
3	10	0.3	1.01	1	7	4

In the system (10.77) for simplicity, put $\mu = 1$, then $v = 1$ is a solution to the second equation regardless of other parameters. For $v = 1$ from the first equation (10.77) we obtain

$$u = \lambda\left(\frac{(\theta + m - 1)u + q - m}{mu + \theta + q - m - 1}\right)^k. \tag{10.80}$$

Now by the change $u = \frac{q-m}{\theta+m-1}y$, the equation (10.80) can be rewritten as

$$ay = \left(\frac{1+y}{b+y}\right)^k, \quad y > 0, \tag{10.81}$$

where

$$a := \frac{(q-m)m^k}{\lambda(\theta + m - 1)^{k+1}}, \qquad b := \frac{(\theta + q - m - 1)(\theta + m - 1)}{q - m}. \tag{10.82}$$

By Proposition 3.1 for solutions of (10.81) we obtain

Lemma 10.14. *Let $b_0 \equiv b_0(k) := (\frac{k+1}{k-1})^2$ and $\nu \equiv \nu(a, b, k)$ denote the number of solutions to the equation (10.81) with $a, b > 0$ and $k \geq 2$. Then*

(i) If $b \leq b_0$, then for any $a > 0$ the equation (10.81) has a unique solution, i.e., $\nu = 1$.

(ii) For $b > b_0$ denote

$$a_\pm \equiv a_\pm(b, k) = \frac{1}{y_\pm}\left(\frac{1 + y_\pm}{b + y_\pm}\right)^k, \quad 0 < a_- < a_+,$$

here $y_\pm = y_\pm(b, k) > 0$ are two distinct solutions of the quadratic equation:

$$y^2 + (2 - (b-1)(k-1))y + b = 0,$$

i.e.,

$$y_\pm = \frac{(b-1)(k-1) - 2 \pm \sqrt{D(b,k)}}{2},$$

where $D(b, k) := (2 - (b-1)(k-1))^2 - 4b = (b-1)(k-1)^2(b - b_0)$.

Then

$$\nu = \begin{cases} 1, \text{ if } \ 0 < a < a_- \ \text{ or } \ a > a_+ \\ 2, \text{ if } \ a = a_- \ \text{ or } \ a = a_+ \\ 3, \text{ if } \ a_- < a < a_+. \end{cases} \tag{10.83}$$

Let $card(GM)_{TI}$ be the number of TISGMs for the Potts model with translation-invariant external field $\xi_x = \alpha \in \mathbb{R}^{q-1}$

By (10.82) one can see that the condition $b > b_0$ is equivalent to

$$\theta > \theta_c \equiv \theta_c(k, q, m)$$
$$= \frac{1}{2}\left(2 - q + \sqrt{(q-2)^2 + 4(q-1)b_0 + (q-m-1)(m-1)}\right). \tag{10.84}$$

When the condition (10.84) is satisfied, i.e., $b > b_0$, then by Lemma 10.14 the equation (10.81) (corresponding to the equation (10.80)) has three solutions for $\lambda^+ < \lambda < \lambda^-$, two solutions for $\lambda = \lambda^+$ or $\lambda = \lambda^-$ and one solution for $\lambda \notin [\lambda^+, \lambda^-]$, where

$$\frac{(q-m)m^k y_+}{(\theta+m-1)^k}\left(\frac{b+y_+}{1+y_+}\right)^k = \frac{(q-m)m^k}{(\theta+m-1)^k a_+} = \lambda^+,$$

$$\frac{(q-m)m^k y_-}{(\theta+m-1)^k}\left(\frac{b+y_-}{1+y_-}\right)^k = \frac{(q-m)m^k}{(\theta+m-1)^k a_-} = \lambda^-.$$

Conditions of the following theorem follows from the conditions of Lemma 10.14 for

$$\alpha_-(\theta, q, m) = \ln \lambda^-, \quad \alpha_+(\theta, q, m) = \ln \lambda^+.$$

Theorem 10.5. *Let $k \geq 2$. Then the following holds:*

$$card(GM)_{TI} \geq \begin{cases} 1, \text{ if } \ \theta \leq \theta_c \ \text{ or } \ \theta > \theta_c, \ \ \alpha \notin [\alpha_-(\theta,q,m), \alpha_+(\theta,q,m)] \\ 2, \text{ if } \ \theta > \theta_c \ \text{ and } \ \alpha = \alpha_-(\theta,q,m) \ \text{ or } \ \alpha = \alpha_+(\theta,q,m) \\ 3, \text{ if } \ \theta > \theta_c \ \text{ and } \ \alpha \in (\alpha_-(\theta,q,m), \alpha_+(\theta,q,m)). \end{cases} \tag{10.85}$$

Now consider the question of existence of a TISGM in case of a non-translation-invariant external field. Then $\overrightarrow{h}_x$ does not depend on x, i.e., $\overrightarrow{h}_x \equiv \overrightarrow{h} = (h_1, ..., h_{q-1})$, but $\overrightarrow{\alpha}_x$ depends on x.

Theorem 10.6. *For the Potts model with a non-translation-invariant external field, there is no a TISGM.*

Proof. Suppose the opposite, that is, for the Potts model with a non-TI external field, there is a TISGM. Since $\vec{\alpha}_x$ is not TI, there are vertices x and x' such that $\vec{\alpha}_x \neq \vec{\alpha}_{x'}$, but on the vertices we have $\vec{h}_x = \vec{h}_{x'} = h$. Then from (2.18) we get

$$\vec{h} = \vec{\alpha}_x + kF(\vec{h},\theta), \quad \vec{h} = \vec{\alpha}_{x'} + kF(\vec{h},\theta). \tag{10.86}$$

Consequently, $\vec{\alpha}_x - \vec{\alpha}_{x'} = 0$, i.e., $\vec{\alpha}_x \equiv \vec{\alpha}_{x'}$. This contradiction completes the proof. □

10.3 Potts model with competing interactions

In [243] several models on the Cayley tree were studied using the method of Markov random fields and recursive equations of this theory. When Hamiltonian of the system has interaction radius ≥ 2, then the method of Markov random fields is not always convenient. It is therefore natural to develop the Pirogov-Sinai theory (also called a contour method) on the Cayley tree. Note that the contour method on the lattice $\mathbb{Z}^d$, $d \geq 1$, is generally called the Pirogov-Sinai theory [267], [307].

10.3.1 *The model and its ground states*

In this section, for Potts model with competing interactions on trees, we are going to give the contour method of description of Gibbs measures.

For models given on a Cayley tree: the contour method was developed in [237] for a q-component model with nearest-neighbor interactions and in [239] for the Ising model with competing interactions of interaction radius $r = 2$. For more general case the method is developed in [240]. In this section we follow [33] to explain how this method used for the Potts model.

The Hamiltonian of the Potts model with competing interactions has the form

$$H(\sigma) = J_1 \sum_{\substack{\langle x,y\rangle,\\ x,y\in V}} \delta_{\sigma(x)\sigma(y)} + J_2 \sum_{\substack{x,y\in V:\\ d(x,y)=2}} \delta_{\sigma(x)\sigma(y)}, \tag{10.87}$$

where $J_1, J_2 \in \mathbb{R}$, $\sigma(x) \in \{1, 2, \ldots, q\}$ and $\langle x, y\rangle$ denotes edges of the tree.

For a pair of configurations σ and φ coinciding almost everywhere, i.e., everywhere except at a finite number of points, we consider the relative Hamiltonian $H(\sigma, \varphi)$ of the difference between the energies of the configurations σ and φ, i.e.,

$$H(\sigma,\varphi) = J_1 \sum_{\substack{\langle x,y\rangle,\\ x,y\in V}} (\delta_{\sigma(x)\sigma(y)} - \delta_{\varphi(x)\varphi(y)}) + J_2 \sum_{\substack{x,y\in V:\\ d(x,y)=2}} (\delta_{\sigma(x)\sigma(y)} - \delta_{\varphi(x)\varphi(y)}), \tag{10.88}$$

where $J = (J_1, J_2) \in \mathbb{R}^2$ is a fixed parameter.

Let M be the set of unit balls with vertices in V. The restriction of the configuration σ on the ball $b \in M$ is called *a bounded configuration* and denoted by σ_b. Let's define energy of the configuration σ_b on the ball b as follows:

$$U(\sigma_b) \equiv U(\sigma_b, J) = \frac{1}{2} J_1 \sum_{\substack{\langle x,y\rangle,\\ x,y\in b}} \delta_{\sigma(x)\sigma(y)} + J_2 \sum_{\substack{x,y\in b:\\ d(x,y)=2}} \delta_{\sigma(x)\sigma(y)}, \tag{10.89}$$

where $J = (J_1, J_2) \in \mathbb{R}^2$.

The following lemma can be easily proved.

Lemma 10.15. *Relative Hamiltonian (10.88) has the form*

$$H(\sigma,\varphi) = \sum_{b\in M} (U(\sigma_b) - U(\varphi_b)).$$

For simplicity of the illustration of the contour method we mainly consider the Cayley tree of order two (i.e., $k = 2$) and $q = 3$. It easy to see that $U(\sigma_b) \in \{U_1, \ldots, U_6\}$ for any σ_b, where

$$\begin{aligned} &U_1 = \frac{3}{2} J_1 + 3J_2, \quad U_2 = J_1 + J_2, \quad U_3 = 3J_2, \\ &U_4 = \frac{1}{2} J_1, \quad U_5 = J_2, \quad U_6 = \frac{1}{2} J_1 + J_2. \end{aligned} \tag{10.90}$$

Definition 10.1. A configuration φ is called a ground state of the relative Hamiltonian H if

$$U(\varphi_b) = min\{U_1, U_2, \ldots, U_6\},$$

for any $b \in M$.

Denote

$$C_i = \{\sigma_b : U(\sigma_b) = U_i\}$$

$$U_i(J) = U(\sigma_b, J), \quad \text{if} \quad \sigma_b \in C_i \quad i = 1, 2, \ldots, 6.$$

Recall that there exists a one-to-one correspondence between the set V of vertices of the Cayley tree of order $k \geq 1$ and the group G_k of free products of a $k+1$ cyclic groups of order two with the generators $a_1, a_2, ..., a_{k+1}$.

Let $F_k \subset G_k$ be a finite-index subgroup. We define a F_k-periodic configuration as a configuration $\sigma(x)$ that is invariant under F_k, i.e., $\sigma(yx) = \sigma(x)$ for any $x \in G_k$ and $y \in F_k$. For a given periodic configuration, the index of the subgroup is called the configuration's period. A configuration which is invariant under all translations is said to be translation invariant.

Theorem 10.7.

1. *For any class C_i, $i = 1, 2, 3, 4$, and any bounded configuration $\sigma_b \in C_i$, there exists a periodic configuration φ (on the Cayley tree) with a period $p \leq 6$ such that $\varphi_{b'} \in C_i$ for any $b' \in M$ and $\varphi_b = \sigma_b$.*
2. *For any $\sigma_b \in C_5$ there exists a configuration φ (generally not periodic) on the Cayley tree such that $\varphi_{b'} \in C_5$ for any $b' \in M$ and $\varphi_b = \sigma_b$.*

Proof. 1. For any class C_i, $i = 1, 2, 3, 4$ and $\sigma_b \in C_i$ we construct a configuration φ as follows. Without loss of generality, we can take the ball b centered at $e \in G_2$ (where e is the unit element of G_2), i.e., $b = \{e, a_1, a_2, a_3\}$.

Let us consider several cases:

Case C_1. In this case we have $\sigma(x) = i$, $i = 1, 2, 3$, for any $x \in b$, therefore we can take φ as a translation invariant configuration, i.e., $\varphi^{(i)} = \{\varphi^{(i)}(x) = i\}$, where $i = 1, 2, 3$.

Case C_2. Consider

$$H_2 = \{x \in G_2 : \omega_2(x) \, is \, even\},$$

where $\omega_2(x)$ is the number of letters a_2 in the word $x \in G_2$. Note that H_2 is a normal subgroup of index 2 of G_2. For the initial bounded configuration σ_b we have

$$\sigma_b(x) = \begin{cases} i \text{ for any } x \in b \bigcap H_2 \\ j \text{ for any } x \in b \bigcap (G_2 \backslash H_2), \end{cases}$$

where $i \neq j$, $i, j = 1, 2, 3$.

We extend the configuration $\sigma_b \in C_i$ to the entire tree Γ^2 (which is denoted by $\varphi^{(ij)}$) as follows:

$$\varphi^{(ij)}(x) = \begin{cases} i, \text{ if } x \in H_2, \\ j, \text{ if } x \in G_2 \setminus H_2, \end{cases}$$

where $i \neq j$, $i, j = 1, 2, 3$. This configuration is H_2-periodic, and its period is $p = 2$ (=the index of the subgroup).

Case C_3. In this case consider

$$H_3 = \{x \in G_2 : |x| \, is \, even\},$$

where $|x|$ is the length of the word x. Note that H_3 is also a normal subgroup of the group G_2 of index two. In this case, the configuration $\psi^{(ij)}$ can be determined as

$$\psi^{(ij)}(x) = \begin{cases} i, & \text{if } x \in H_3, \\ j, & \text{if } x \in G_2 \setminus H_3, \end{cases}$$

where $i \neq j$, $i, j = 1, 2, 3$.

We thus obtain a H_3-periodic configuration $\psi^{(ij)}$ with period two; it is clear that $\psi_b^{(ij)} = \sigma_b$.

Case C_4. Let S_3 be the group of three-order permutations. Choose $\pi_0, \pi_1, \pi_2 \in S_3$ such that

$$\pi_0 = \begin{pmatrix} 1\,2\,3 \\ 1\,2\,3 \end{pmatrix}, \; \pi_1 = \begin{pmatrix} 1\,2\,3 \\ 2\,1\,3 \end{pmatrix}, \; \pi_2 = \begin{pmatrix} 1\,2\,3 \\ 1\,3\,2 \end{pmatrix}.$$

It is easy to see that $\pi_0 = \pi_1^2 = \pi_2^2$. Consider the map $u : \{a_1, a_2, a_3\} \mapsto \{\pi_0, \pi_1, \pi_2\}$:

$$u(a_i) = \begin{cases} \pi_0, \text{ if } & i = 3, \\ \pi_i, \text{ if } & i = 1 \text{ or } 2, \end{cases}$$

and let $f : G_2 \mapsto S_3$ is defined as

$$f(x) = f(a_{i_1} a_{i_2} \dots a_{i_n}) = u(a_{i_1}) \dots u(a_{i_n}).$$

Note that (see [91], and Chapter 1 of [243]),

$$H_4 = \{x \in G_2 : \; f(x) = \pi_0\},$$

is a normal subgroup of index six. Let $G_2/H_4 = \{\aleph_0, \dots, \; \aleph_5\}$ be the quotient group.

In this case we define the following configuration

$$\varphi^{(ijp)}(x) = \begin{cases} i, & \text{if } x \in \aleph_0 \bigcup \aleph_3, \\ j, & \text{if } x \in \aleph_1 \bigcup \aleph_5, \\ p, & \text{if } x \in \aleph_2 \bigcup \aleph_4. \end{cases}$$

We thus obtain a H_4-periodic configuration $\varphi^{(ijp)}$ with the period $p = 6$ such that $\varphi_b^{(ijp)} = \sigma_b$.

2. *Case* C_5. For $x \in G_2$ we denote $l(x) = l(a_{i_1} \dots a_{i_n}) = i_1$, $r(x) = r(a_{i_1} \dots a_{i_n}) = i_n$.

In this case $\psi^{(ijp)}$ is defined as

$$\psi^{(ijp)}(x) = \begin{cases} i, & \text{if } r(x) = 1, \ l(x) \neq 3 \text{ or } r(x) = 3, \ l(x) = 3, \\ j, & \text{if } r(x) = 2, \\ p, & \text{if } r(x) = 3, \ l(x) \neq 3 \text{ or } r(x) = 1, \ l(x) = 3, \\ p, & \text{if } x = e. \end{cases}$$

For each configuration φ we have $\varphi_b \in C_i$, $i = 1,2,3,4,5$, for any $b \in M$. □

For each $m = 1,2,3,4,5,6$ denote by A_m the set

$$A_m = \{J \in \mathbb{R}^2 : \min\{U_1(J), U_2(J), \dots, U_6(J)\} = U_m\}.$$

It is easy to see that

$$\begin{aligned} A_1 &= \{J \in R^2 : J_1 \leq 0, J_1 + 4J_2 \leq 0\}, \\ A_2 &= \{J \in R^2 : J_2 \geq 0, -4J_2 \leq J_1 \leq -2J_2\}, \\ A_3 &= \{J \in R^2 : J_1 \geq 0, J_2 \leq 0\}, \\ A_4 &= \{J \in R^2 : J_2 \geq 0 \ -2J_2 \leq J_1 \leq 2J_2\}, \\ A_5 &= \{J \in R^2 : J_2 \geq 0, J_1 - 2J_2 \geq 0\}, \\ A_6 &= \{J \in R^2 : J_1 = 0, J_2 = 0\}, \end{aligned}$$

and $\mathbb{R}^2 = \bigcup_{i=1}^{6} A_i$.

Since there is no interaction in the case of A_6, we do not consider the class C_6 in what follows. For the same reason, we also did not include this class in Theorem 10.7.

Now we introduce the following sets:

$$B = A_1 \cap A_3, \quad B_0 = A_1 \cap A_2,$$

$$B_1 = A_2 \cap A_4, \quad B_2 = A_4 \cap A_5, \quad B_3 = A_5 \cap A_3.$$

$$\tilde{A}_1 = A_1 \setminus (B \cup B_0), \quad \tilde{A}_2 = A_2 \setminus (B_0 \cup B_1),$$

$$\tilde{A}_3 = A_3 \setminus (B \cup B_2), \quad \tilde{A}_4 = A_4 \setminus (B_1 \cup B_2), \quad \tilde{A}_5 = A_5 \setminus (B_2 \cup B_3).$$

Let $GS(H)$ be the set of all ground states, and let $GS_p(H)$ be the set of all periodic ground states.

Theorem 10.8.

(i) *If* $J = (0, 0)$ *then* $GS(H) = \Omega$.

(ii) *1. If* $J \in \tilde{A}_1$, *then*

$$GS_p(H) = \{\varphi^{(i)} : i = 1, 2, 3\}.$$

2. If $J \in \tilde{A}_2$, *then*

$$GS_p(H) = \{\varphi^{(ij)} : i, j = 1, 2, 3, \ i \neq j\},$$

$| GS_p(H) |= 3$.

3. If $J \in \tilde{A}_3$, *then*

$$GS_p(H) = \{\psi^{(ij)} : i, j = 1, 2, 3, \ i \neq j\},$$

$| GS_p(H) |= 3$.

4. If $J \in \tilde{A}_4$, *then*

$$GS_p(H) = \{\varphi^{(ijp)} : i, j, p = 1, 2, 3, \ i \neq j \neq p\},$$

$| GS_p(H) |= 3$.

5. If $J \in \tilde{A}_5$, *then*

$$GS(H) \supset \{\psi^{(ijp)} : i, j, p = 1, 2, 3, \ i \neq j \neq p\},$$

The configurations $\varphi^{(i)}$, $\varphi^{(ij)}$, $\psi^{(ij)}$, $\varphi^{(ijp)}$, $\psi^{(ijp)}$ *for each case are defined in the proof of Theorem 10.7.*

(iii) *(1) If* $J \in B_i \backslash \{(0, 0)\}$, $i = 0, 1, 2, 3$ *then* $| GS_p(H) |= 6$.

(2) If $J \in B \backslash \{(0, 0)\}$, *then* $| GS_p(H) |= 6$.

Moreover, if $J \in \mathbb{R}^2 \backslash B$, *then there is at least a countable set of nonperiodic ground states.*

Proof. Case (i) is trivial. In cases (ii) and (iii), for a given configuration σ_b for which the energy $U(\sigma_b)$ is minimal, by Theorem 10.7 one can construct the periodic configurations $\varphi^{(ij)}$, $\varphi^{(ijp)}$ (with a period $p \leq 6$). In each case, the exact number of such ground states coincides with the number of different configurations φ_b such that the energy $U(\varphi_b)$ is minimal for any $b \in M$. □

Definition 10.2. A ball $b \in M$ is called a degenerate ball of a configuration σ if $\sigma_b \neq \varphi_b$ for all $\varphi \in GS(H)$. The set of all degenerate balls of a configuration σ is called the boundary configuration and is denoted by $\partial(\sigma)$.

Definition 10.3. The relative Hamiltonian H with the set of ground states $GS(H)$ satisfies the Peierls condition if for any $\varphi \in GS(H)$ and all configurations σ, coinciding almost everywhere with φ, the following inequality holds:

$$H(\sigma, \varphi) \geq \lambda |\partial(\sigma)|,$$

where λ is a positive constant independent of σ, and $|\partial(\sigma)|$ is the number of unit balls in $\partial(\sigma)$.

Theorem 10.9. *If $J \neq (0,0)$, then the Peierls condition is satisfied.*

Proof. Denote $\mathbf{U} = \{U_1, ..., U_6\}$, $U^{\min} = \min\{U_1, ..., U_6\}$ and

$$\lambda_0 = \min\left\{\mathbf{U} \setminus \{U_j : U_j = U^{\min}\}\right\} - U^{\min}. \tag{10.91}$$

Note that $U_1 = ... = U_6$ if and only if $J = (0,0)$, consequently $\lambda_0 > 0$ if $J \neq (0,0)$.

Suppose σ coincides almost everywhere with a ground state $\varphi \in GS(H)$ then we have $U(\sigma_b) - U(\varphi_b) \geq \lambda_0$ for any $b \in \partial(\sigma)$ since φ is a ground state. Thus

$$H(\sigma, \varphi) = \sum_{b \in M} (U(\sigma_b) - U(\varphi_b)) = \sum_{b \in \partial(\sigma)} (U(\sigma_b) - U(\varphi_b)) \geq \lambda_0 |\partial(\sigma)|.$$

Therefore, the Peierls condition is satisfied for $\lambda = \lambda_0$. □

10.3.2 *Contours on trees*

In this subsection we consider the case $J \in \tilde{A}_1$, i.e., the case where only constant configurations are ground states of the Hamiltonian (see Theorem 10.8).

Let $\Lambda \subset V$ be a finite subset, $\Lambda' = V \setminus \Lambda$ and $\omega_{\Lambda'} = \{\omega(x), x \in \Lambda'\}$, $\sigma_\Lambda = \{\sigma(x), x \in \Lambda\}$ be given configurations. The energy of the configuration σ_Λ is

$$\begin{aligned} H_\Lambda(\sigma_\Lambda|\omega_{\Lambda'}) = J_1 &\sum_{\langle x,y\rangle : x,y \in \Lambda} \delta_{\sigma(x)\sigma(y)} + J_1 \sum_{\langle x,y\rangle : x \in \Lambda, y \in \Lambda'} \delta_{\sigma(x)\omega(y)} \\ &+ J_2 \sum_{x,y \in \Lambda : d(x,y)=2} \delta_{\sigma(x)\sigma(y)} + J_2 \sum_{x \in \Lambda, y \in \Lambda' d(x,y)=2} \delta_{\sigma(x)\sigma(y)}. \end{aligned} \tag{10.92}$$

Let $\omega_{\Lambda'}^{(i)} \equiv i$, $i = 1, \ldots, q$ be constant configurations outside Λ. For each configuration σ_Λ in the interior of Λ, we extend the configuration to the entire tree using the ith constant configuration; we let $\sigma_\Lambda^{(i)}$ denote this configuration and set $\Omega_\Lambda^{(i)} = \{\sigma_\Lambda^{(i)}\}$.

Consider the ball V_n (with the center at e and radius is n) and for a given configuration $\sigma_\Lambda^{(i)} \in \Omega_\Lambda^{(i)}$, $\Lambda \subset V_n$, we introduce

$$V_n^{(j)} \equiv V_n^{(j)}(\sigma_\Lambda^{(i)}) = \{t \in V_n : \sigma_\Lambda^{(i)}(t) = j\}, \quad j = 1, \dots, q, j \neq i.$$

Let $G^{n,j} = (V_n^{(j)}, L_n^{(j)})$ be a graph such that

$$L_n^{(j)} = \{l = \langle x, y\rangle \in L : x, y \in V_n^{(j)}\}, \quad j = 1, \dots, q.$$

It is easy to see that for each fixed n the graph $G^{n,j}$ contains finitely many (their number is denoted by m) maximal connected subgraphs $G_r^{n,j}$, i.e.,

$$G^{n,j} = \{G_1^{n,j}, \dots, G_m^{n,j}\}, \quad G_r^{n,j} = (V_{n,r}^{(j)}, L_{n,r}^{(j)}), \quad r = 1, \dots, n.$$

Here $V_{n,r}^{(j)}$ is the set of vertices and $L_{n,r}^{(j)}$ is the set of edges of $G_r^{n,j}$.

Two edges $l_1, l_2 \in L$ $(l_1 \neq l_2)$ are called nearest neighboring edges (write it as $< l_1, l_2 >_1$) if $|i(l_1) \cap i(l_2)| = 1$.

For any connected component $K \subset \Gamma^k$ denote by $E(K)$ the set of edges in K and

$$b(K) = \{l \in L \setminus E(K) : \exists l_1 \in E(K) \text{ such that } < l, l_1 >_1\}.$$

Definition 10.4. An edge $l = \langle x, y\rangle \in L_{n+1}$ is called a boundary edge of the configuration $\sigma_{V_n}^{(i)}$, if $\sigma_{V_n}^{(i)}(x) \neq \sigma_{V_n}^{(i)}(y)$. The set of boundary edges of a configuration is called the edge boundary $\Gamma(\sigma_{V_n}^{(i)}) \equiv \Gamma$ of this configuration.

Note that the boundary Γ consists $\frac{q(q-1)}{2}$ parts

$$\Gamma_\varepsilon(\sigma_{V_n}^{(i)}) \equiv \Gamma_\varepsilon, \quad \varepsilon \in \{ij : i < j; i, j = 1, \dots, q\} \equiv Q_q,$$

where, for example, Γ_{12} is the set of edges $l = \langle x, y\rangle$ with $\sigma(x) = 1$ and $\sigma(y) = 2$.

The finite set $b(G_r^{n,j})$, $j = 1, \dots, q, j \neq i; r = 1, \dots, m$ (for each edge from this set together with the information about the boundary part $\Gamma_\varepsilon, \varepsilon \in Q_q$ containing this edge), is called a *subcontour* of the boundary Γ.

The sets $V_{n,r}^{(j)}$, $j = 1, \dots, q, j \neq i; r = 1, \dots, m$ are called the interior Int $b(G_r^{n,j})$ of the subcontour $b(G_r^{n,j})$.

For a graph G we denote by $V(G)$ the set of its vertices and by $E(G)$ the set of edges.

For any two subcontours T_1 and T_2, the distance $\mathrm{dist}(T_1, T_2)$ is defined as

$$\mathrm{dist}(T_1, T_2) = \min_{\substack{x \in V(T_1)\\ y \in V(T_2)}} d(x, y),$$

where $d(x, y)$ is the distance between $x, y \in V$ on the Cayley tree.

Definition 10.5. The subcontours T_1 and T_2 are called adjacent if $d(T_1, T_2) \leq 2$. The set of subcontours A is said to be connected if for any two subcontours $T_1, T_2 \in A$, there are subcontours $T_1 = \widetilde{T}_1, \widetilde{T}_2, \ldots, \widetilde{T}_l = T_2$ in the set A such that subcontours $\widetilde{T}_i$ and $\widetilde{T}_{i+1}$ are adjacent for each $i = 1, \ldots, l-1$.

Definition 10.6. Any maximal connected set (component) of subcontours is called the contour of the boundary Γ (see Fig. 10.6).

Let $\gamma = \{\gamma_r, r = 1, 2, \ldots\}$ be a contour of the boundary Γ and γ_r be a subcontour. Denote

$$\text{Int}\gamma = \cup_j \text{Int}\gamma_j; \quad \text{imp}\gamma = \{b \in \partial : b \cap \gamma \neq \emptyset\}; \quad |\gamma| = |\text{imp}\gamma|.$$

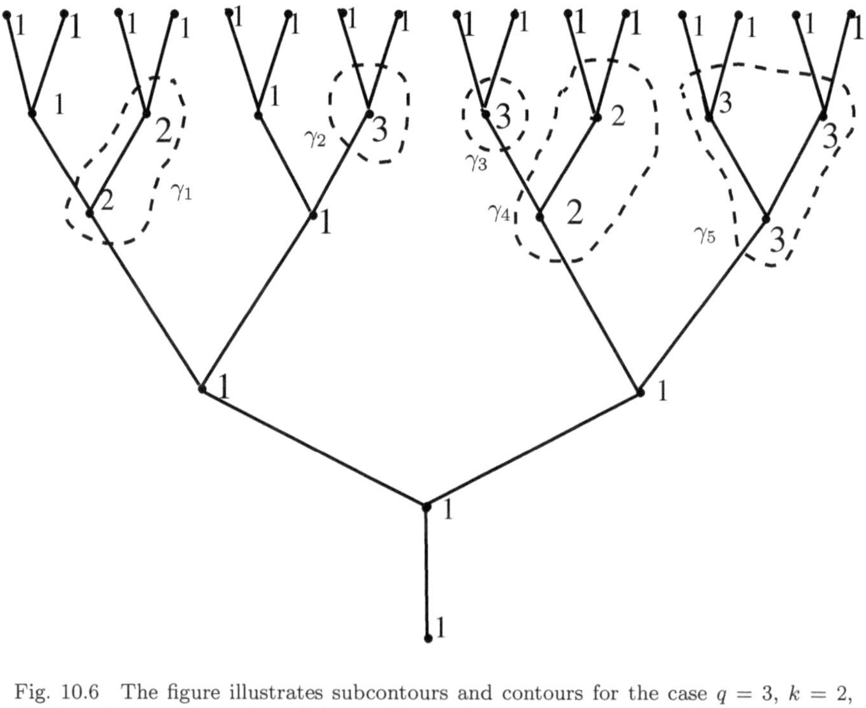

Fig. 10.6 The figure illustrates subcontours and contours for the case $q = 3$, $k = 2$, the boundary configuration $\sigma(x) \equiv 1$. Here subcontours: γ_i, $i = 1, 2, 3, 4, 5$; Contours: $\Upsilon_1 = \{\gamma_1\}$, $\Upsilon_2 = \{\gamma_2\}$, $\Upsilon_3 = \{\gamma_3, \gamma_4\}$, $\Upsilon_4 = \{\gamma_5\}$.

For $A \subset V$, we denote

$$B(A) = \{b \in M : b \cap A \neq \emptyset\};$$

$$D_{int}(A) = \{x \in A : \exists y \in V \backslash A, \text{such that } \langle x, y \rangle\};$$

$$D(A) = \{x \in V \backslash A : \exists y \in A, \text{such that } \langle x, y \rangle\}.$$

Lemma 10.16. *Let K be a connected subgraph of the Cayley tree Γ^k, $k \geq 2$ such that $|V(K)| = n$. Then*
(a) $|D(V(K))| = (k-1)n + 2$,
(b) $|B(V(K))| = kn + 2$.

Proof. (a) We proceed by induction on n. For $n = 1$, 2 the statement is trivial. Assume that the statement is true for $n = m$, that is $|V(K)| = m$ implies $|D(V(K))| = (k-1)m+2$. We prove the statement for $n = m+1$, i.e., for $\tilde{K} = K \cup \{x\}$. Since $\tilde{K}$ is a connected graph, we see that $x \in D(V(K))$ and there is a unique vertex $y \in S_1(x) = \{u \in V : d(x, u) = 1\}$ such that $y \in V(K)$. Thus

$$D(V(\tilde{K})) = (D(V(K)) \setminus \{x\}) \cup (S_1(x) \setminus \{y\}).$$

Consequently,

$$|D(V(\tilde{K}))| = |D(V(K))| - 1 + k = (k-1)(m+1) + 2.$$

(b) It is clear that $|B(V(K))| = |V(K)| + |D(V(K))|$. By part a) of the lemma we get $|B(V(K))| = n + (k-1)n + 2 = kn + 2$. □

Lemma 10.17. *[30] Let G be a countable graph of maximal degree $k+1$ (i.e., each $x \in V(G)$ has at most $k+1$ neighbors) and let $\tilde{N}_{n,G}(x)$ be the number of connected subgraphs $G' \subset G$ with $x \in V(G')$ and $|E(G')| = n$. Then*

$$\tilde{N}_{n,G}(x) \leq (e \cdot k)^n.$$

Let γ be a contour. For $x \in V$ we write $x \in \gamma$, if there exists an edge $l \in \gamma$ such that $x \in i(l)$.

Denote

$$N_r(x) = |\{\gamma : x \in \gamma, |\gamma| = r\}|.$$

Lemma 10.18. *For any $k \geq 2$ we have*

$$N_r(x) \leq \alpha^r, \tag{10.93}$$

where $\alpha = (2ke)^{4k+3}$.

Proof. Denote by K_γ the minimal connected subgraph of the tree Γ^k, which contains the contour $\gamma = \{\gamma_1, \dots, \gamma_m\}$, where γ_i are subcontours and $m \geq 1$.

By Lemma 10.16 we get

$$|B(\mathrm{Int}\gamma_i)| = k|\mathrm{Int}\gamma_i| + 2 = k\{|D_{\mathrm{int}}(\mathrm{Int}\gamma_i)| + |\mathrm{Int}\gamma_i \backslash D_{\mathrm{int}}(\mathrm{Int}\gamma_i)|\} + 2.$$

Using (see Lemma 10.16)

$$|\mathrm{Int}\gamma_i \backslash D_{\mathrm{int}}(\mathrm{Int}\gamma_i)| = \frac{|D_{\mathrm{int}}(\mathrm{Int}\gamma_i)| - 2}{k-1},$$

we get

$$|B(\mathrm{Int}\gamma_i)| = k \cdot \frac{k|D_{\mathrm{int}}(\mathrm{Int}\gamma_i)| - 2}{k-1} + 2 \leq 2k|D_{\mathrm{int}}(\mathrm{Int}\gamma_i)|. \tag{10.94}$$

Note that

$$|D_{\mathrm{int}}(\mathrm{Int}\gamma_i)| \leq |\mathrm{imp}\gamma_i| = |\gamma_i|. \tag{10.95}$$

By (10.95), from (10.94) we get

$$|B(\mathrm{Int}\gamma_i)| \leq 2k|\gamma_i|.$$

Moreover, we have

$$\sum_{i=1}^{m} |\gamma_i| \leq |\gamma| + m, \quad |B(V(K_\gamma))| \leq \sum_{i=1}^{m} |B(\mathrm{Int}\gamma_i)| + 2m - 2. \tag{10.96}$$

By (10.95), from (10.96) we get

$$B(V(K_\gamma))| \leq 2k \sum_{i=1}^{m} |\gamma_i| + 2m - 2 \leq 2k|\gamma| + 2(k+1)m - 2. \tag{10.97}$$

Since γ contains m subcontours, we have $m \leq |\gamma|$, and (10.97) implies

$$N_r(x) \leq \sum_{m=1}^{r} \binom{2kr + 2(k+1)m - 2}{r} \tilde{N}_{2kr+2(k+1)m-2,\Gamma^k}(x). \tag{10.98}$$

By Lemma 10.17 and $\binom{n}{r} \leq 2^{n-1}$ from (10.98) we get (10.93). □

10.3.3 *Non uniqueness of Gibbs measure*

For a given volume Λ and the boundary condition $\omega_\Lambda^{(i)}$ we define $Z_\Lambda^{(i)}$ as

$$Z_\Lambda^{(i)} = \sum_{\sigma\in\Omega_\Lambda} \exp(-\beta H(\sigma|\omega_{\Lambda'}^{(i)})),$$

where $\beta = \frac{1}{T}$ is the inverse temperature. Then the conditional Gibbs measure $\mu_\Lambda^{(i)}$ in the volume Λ under the boundary condition $\omega_\Lambda^{(i)}$ is defined as

$$\mu_{\Lambda,\beta}^{(i)}(\sigma) = \frac{\exp(-\beta H(\sigma|\omega_{\Lambda'}^{(i)}))}{Z_\Lambda^{(i)}}, \quad \sigma \in \Omega_\Lambda.$$

The following lemma gives the contour representation of the Hamiltonian.

Lemma 10.19. *The Hamiltonian* $H_n(\sigma_n) = H_{V_n}(\sigma_n|\ \omega_{V_n'}^{(1)} \equiv 1)$ *(see (10.92)) has the following form*

$$H_n(\sigma_n) = \sum_{i=2}^{6}(U_i - U_1)|\partial_i| + |M_n|U_1, \tag{10.99}$$

where

$$\sigma_n = \sigma_{V_n}^{(1)},\ \sigma_{n,b} = (\sigma_n)_b,\ |\partial_i| = \sum_{\gamma\in\Gamma} |\{b \in \mathrm{imp}\gamma : \sigma_{n,b} \in C_i\}|,$$

$$M_n = \{b \in M : b \cap V_n \neq \emptyset\}.$$

Proof. We have

$$H_n(\sigma_n) = \sum_{b\in M_n} U(\sigma_{n,b}) = \sum_{j=2}^{6} U_j|\partial_j| + (|M_n| - |\partial|)U_1.$$

We obtain (10.99) by using the equality $|\partial| = \sum_{j=2}^{6} |\partial_j|$. □

Lemma 10.20. *Let* $J \in \tilde{A}_1$, γ *is a fixed contour and*

$$P(\gamma) = \mu_{n,\beta}^{(1)}\{\sigma_n : \gamma \in \Gamma(\sigma_n)\}.$$

Then

$$P(\gamma) \le \exp\{-\beta\lambda_0|\gamma|\}, \tag{10.100}$$

where

$$\lambda_0 = \min\{U\backslash\{U_j : U_j = U^{\min}\}\} - U^{\min},$$

$$U = \{U_1, \dots, U_6\},\ \ U^{\min} = \min\{U_1, \dots, U_6\}.$$

Proof. Denote $\Omega_\gamma = \{\sigma_n : \gamma \subset \Gamma(\sigma_n)\}$ and $\Omega_\gamma^0 = \{\sigma_n : \gamma \cap \Gamma(\sigma_n) = \emptyset\}$. Define the map $\chi_\gamma : \Omega_\gamma \to \Omega_\gamma^0$ as

$$\chi_\gamma(\sigma_n)(x) = \begin{cases} 1, & \text{if } x \in \text{Int}\gamma, \\ \sigma_n(x), & \text{if } x \notin \text{Int}\gamma. \end{cases}$$

The map χ_γ is bijective for a given γ.

For $\sigma_n \in \Omega_{V_n}$ and $j = 2, \dots, 6$ we have $|\partial_j(\sigma_n)| = |\partial_j(\chi_\gamma(\sigma_n)| + |\gamma_j|$, where $\gamma_j = \{b \in \text{imp}\gamma : \sigma_{\mathrm{b}} \in \mathrm{C_j}\}$.

Consequently, by Lemma 10.19 we get

$$P(\gamma) = \frac{\sum\limits_{\sigma_n \in \Omega_\gamma} \exp\{-\beta \sum\limits_{j=2}^{6} (U_j - U_1)|\partial_j(\sigma_n)|\}}{\sum\limits_{\widetilde{\sigma}_n} \exp\{-\beta \sum\limits_{j=2}^{6} (U_j - U_1)|\partial_j(\widetilde{\sigma}_n)|\}}$$

$$\le \frac{\sum\limits_{\sigma_n \in \Omega_\gamma} \exp\{-\beta \sum\limits_{j=2}^{6} (U_j - U_1)|\partial_j(\sigma_n)|\}}{\sum\limits_{\widetilde{\sigma}_n \in \Omega_\gamma} \exp\{-\beta \sum\limits_{j=2}^{6} (U_j - U_1)|\partial_j(\chi_\gamma(\widetilde{\sigma}_n))|\}}. \tag{10.101}$$

Since $J \in \widetilde{A}_1$, by Theorem 10.8 we have $\mathrm{GS_p(H)} = \{\varphi^{(\mathrm{i})},\ \mathrm{i} = 1, 2, 3\}$, and hence $U_j - U_1 \ge \lambda_0 > 0$ for all $j = 2, 3, \dots, 6$. Thus the relation (10.101) implies (10.100). □

Lemma 10.21. *For any $k \ge 2$ and a sufficiently large β, there exists a constant $C = C(k, \beta) > 0$ such that*

$$\mu_\beta^{(1)}\{\sigma_\Lambda : |\gamma| > C\ln|\Lambda| \text{ for some } \gamma \subset \Gamma(\sigma_\Lambda)\} \to 0 \text{ as } |\Lambda| \to \infty.$$

Proof. Assume the condition $\beta > \lambda_0^{-1}\ln\alpha$ is satisfied. Using Lemma 10.18 and Lemma 10.20 we get

$$\mu_\beta^{(1)}\{\sigma_\Lambda, t \in \gamma,\ |\gamma| = r\} \le N_r(t)e^{-\beta r \lambda_0} \le \alpha^r e^{-\beta r \lambda_0},$$

where $\alpha = (2ke)^{4k+3}$,

$$\mu_\beta^{(1)}\{\sigma_\Lambda, t \in \gamma,\ |\gamma| > C_1\ln|\Lambda|\} \le \sum_{r \ge C_1 \ln|\Lambda|} (\alpha e^{-\beta\lambda_0})^r$$

$$= \frac{(\alpha e^{-\beta\lambda_0})^{C_1\ln|\Lambda|}}{1 - \alpha e^{\beta\lambda_0}} = \frac{|\Lambda|^{C_1(\ln\alpha - \beta\lambda_0)}}{1 - \alpha e^{-\beta\lambda_0}},$$

and the constant C_1 is defined subsequently. Thus we have

$$\mu_\beta^{(1)}\{\sigma_\Lambda,\ |\gamma| > C_1 \ln|\Lambda|\} \le \frac{|\Lambda|^{C_1(\ln\alpha - \beta\lambda_0)+1}}{1-\alpha e^{-\beta\lambda_0}}.$$

The last expression tends to zero if $|\Lambda| \to \infty$ and $C_1 > \frac{1}{\lambda_0\beta - \ln\alpha}$. □

Lemma 10.22. *For all $k \ge 2$, $x \in \Lambda$ we have uniform convergence in $x \in \Lambda$*

$$\mu_\beta^{(1)}\{\sigma(\Lambda) : \sigma(x) = j\} \to 0, \quad j \ne 1 \quad \text{as} \quad \beta \to \infty.$$

Proof. If $\sigma(x) = j$, then x is a vertex in the interior of some contour. We write this as $x \in \text{Int}_j\gamma$. Take $t \in \gamma$ and $x \in \text{Int}_j\gamma$, then we have $|\gamma| \ge |t| + 2$.

Consequently,

$$\mu_\beta^{(1)}\{\sigma_\Lambda : x \in \text{Int}_j\gamma,\quad t \in \gamma,\ |\gamma| < C_1\ln|\Lambda|\} \le \frac{(\alpha e^{-\beta\lambda_0})^{|t|+2}}{1-\alpha e^{-\beta\lambda_0}}.$$

$$\begin{aligned}
\mu_\beta^{(1)}\{\sigma(x) = j\} &\le \mu_\beta^{(1)}\{\sigma_\Lambda : x \in \text{Int}_j\gamma,\quad \gamma \subset \Gamma(\sigma_\Lambda)\} \\
&\le \sum_{|t|=1}^{C_1\ln|\Lambda|} \frac{(\alpha e^{-\beta\lambda_0})^{|t|+2}}{1-\alpha e^{-\beta\lambda_0}} \\
&\quad + \mu_\beta^{(1)}\ \{\sigma_\Lambda :\ \exists\gamma \subset \Gamma(\sigma_\Lambda),\ \text{such that},\ |\gamma| \ge C_1\ln|\Lambda|\} \\
&\le \frac{(\alpha e^{-\beta\lambda_0})^3}{(1-\alpha e^{-\beta\lambda_0})^2} + \frac{|\Lambda|^{C_1(\ln\alpha-\beta\lambda_0)+1}}{1-\alpha e^{-\beta\lambda_0}}.
\end{aligned}$$

For $|\Lambda| \to \infty$ and $\beta \to \infty$ we obtain $\mu_\beta^{(1)}\{\sigma(x) = j\} \to 0, \quad j \ne 1.$ □

Theorem 10.10. *For all sufficiently large β, there exist at least three Gibbs measures for the 3-state (i.e., $q = 3$) Potts model with competing interactions on the Cayley tree of order $k \ge 2$.*

Proof. We note that an analogue of Lemma 10.22 holds for $i = 2, 3$, i.e., for $\mu_\beta^{(i)}$. By Lemma 10.22, we therefore obtain

$$\lim_{\beta\to\infty} \mu_\beta^{(i)}\{\sigma(x) = j\} = \begin{cases} 0, & i \ne j, \\ 1, & i = j. \end{cases}$$

By this equality, for any $i, j \in \{1, 2, 3\}, i \ne j$, and sufficiently large β we get

$$\mu_\beta^{(i)}\{\sigma(x) = i\} \ne \mu_\beta^{(j)}\{\sigma(x) = i\}.$$

□

Remark 10.6. In this subsection we have presented the contour method on trees for one model (Potts model with competing interaction). Interested reader can see Chapter 10 of [243] for more general contour argument used to study finite range lattice models on a Cayley tree with two basic properties: the existence of only a finitely many ground states and with Peierls type condition. By a contour argument it is shown the existence of s different (where s is the number of ground states) Gibbs measures.

Chapter 11

Periodic Gibbs measures

In this chapter, we consider the q-state Potts model on a Cayley tree of degree $k \geq 2$ in an external (may be zero) field. The chapter is devoted to periodic and weakly periodic Gibbs measures. It is shown that any periodic measure is ether translation-invariant or 2-periodic (known as a chessboard periodic). For this model with a zero external field, we find conditions (in particular, ferromagnetic case) on parameters under which there are only translation-invariant Gibbs measures. We also studied the periodic Gibbs measures for the Potts model with a periodic external field. It is proved that under certain conditions the number of such measures maybe at least three. In case of the anti-ferromagnetic Potts model for some values of q and k we show that there are several periodic (non-translation-invariant) Gibbs measures. For anti-ferromagnetic case on the Cayley tree of order $k \geq 6$ it is proved non-uniqueness of weakly periodic (non-periodic) Gibbs measures. Under some condition on parameters of the model we show that there are at last $2^q - 2$ weakly periodic measures.

11.1 Definitions

Let $\Gamma^k = (V, L)$ be a Cayley tree of order $k \geq 2$.

Consider q-state Potts model with an external field:

$$H(\sigma) = -J \sum_{\langle x,y\rangle \in L} \delta_{\sigma(x)\sigma(y)} - \alpha \sum_{x \in V} \delta_{1\sigma(x)}, \tag{11.1}$$

where $J, \alpha \in \mathbb{R}$.

By Theorem 2.2 we have that to each splitting Gibbs measures of this model corresponds a set of vectors $h_x \in \mathbb{R}^{q-1}$, $x \in V$ which satisfies

$$h_x = \sum_{y \in S(x)} F(h_y, \theta, \alpha), \tag{11.2}$$

where $F : h = (h_1, \ldots, h_{q-1}) \in \mathbb{R}^{q-1} \to F(h, \theta, \alpha) = (F_1, \ldots, F_{q-1}) \in \mathbb{R}^{q-1}$ defined as:

$$F_i = \alpha\beta\delta_{1i} + \ln\left(\frac{(\theta - 1)e^{h_i} + \sum_{j=1}^{q-1} e^{h_j} + 1}{\theta + \sum_{j=1}^{q-1} e^{h_j}}\right),$$

$\theta = \exp(J\beta)$, and $S(x)$ is the set of all direct successors of $x \in V$ (see (2.2)).

Let $G_k/G_k^* = \{H_1, ..., H_r\}$ be the quotient group, where G_k^* is a normal subgroup of index $r \geq 1$ (see Subsection 2.1.1).

Definition 11.1. A set of vectors $h = \{h_x, x \in G_k\}$ is said to be G_k^* — periodic, if $h_{yx} = h_x$ for any $x \in G_k$ and $y \in G_k^*$. A G_k — periodic set is called translation-invariant.

Definition 11.2. A set of vectors $h = \{h_x, x \in G_k\}$ is said to be G_k^*-weakly periodic, if $h_x = h_{ij}$ for $x \in H_i$ and $x_\downarrow \in H_j$ for any $x \in G_k$.

Definition 11.3. A measure μ is said to be G_k^*-periodic (weakly periodic), if it corresponds to the G_k^*-periodic (weakly periodic) set of vectors h. The G_k−periodic measure is said to be translation-invariant.

In this chapter we give results of [123]–[126], [223]–[226], [244] related to (weakly) periodic Gibbs measures of the Potts model.

11.2 Characterization of periodic Gibbs measures

Recall that an one-to-one function is a function that maps distinct elements of its domain to distinct elements of its codomain.

Lemma 11.1. *Let $\theta \neq 1$. The function $F(h, \theta, \alpha)$, $h \in \mathbb{R}^{q-1}$ is one-to-one.*

Proof. It is clear that from $h = u$ it follows $F(h, \theta, \alpha) = F(u, \theta, \alpha)$. We show that from $F(h, \theta, \alpha) = F(u, \theta, \alpha)$ it follows that $h = u$.

Denoting

$$e^{h_i} = z_i, \quad e^{u_i} = t_i, \quad e^{F_i(h)} = K_i(h),$$

we shall show that from $K_i(h)-K_i(u)=0$ it follows that $h-u=0$. Indeed, consider

$$K_i(h)-K_i(u)=e^{\alpha\beta\delta_{1i}}\left(\theta+\sum_{j=1}^{q-1}z_j\right)^{-1}\left(\theta+\sum_{j=1}^{q-1}t_j\right)^{-1}$$
$$\times\left[(\theta-1)z_i+\sum_{j=1}^{q-1}z_j+1\right]\left(\theta+\sum_{j=1}^{q-1}t_j\right)$$
$$-\left[(\theta-1)t_i+\sum_{j=1}^{q-1}t_j+1\right]\left(\theta+\sum_{j=1}^{q-1}z_j\right)=0.$$

After some simplifications, we obtain

$$K_i(h)-K_i(u)$$
$$=M_i\left[\left(\theta+\sum_{j=1}^{q-1}t_j\right)(z_i-t_i)+(1-t_i)\sum_{j=1}^{q-1}(z_j-t_j)\right]=0, \qquad (11.3)$$

where

$$M_i=\frac{e^{\alpha\beta\delta_{1i}}(\theta-1)}{\left(\theta+\sum_{j=1}^{q-1}z_j\right)\left(\theta+\sum_{j=1}^{q-1}t_j\right)}\neq 0.$$

Consequently, denoting $a_i=z_i-t_i$, $i=1,\dots,q-1$, from (11.3), we get the following system of equations (with respect to a_i):

$$\left(\theta+\sum_{j=1}^{q-1}t_j\right)a_i+(1-t_i)\sum_{p=1}^{q-1}a_p=0,$$

which has only zero solution if the determinant $\det\mathbf{A}=$

$$\begin{vmatrix}\theta+\sum_{j=1}^{q-1}t_j+1-t_1 & 1-t_1 & \dots. & 1-t_1\\ 1-t_2 & \theta+\sum_{j=1}^{q-1}t_j+1-t_2 & \dots. & 1-t_2\\ \dots & \dots & \dots & \dots\\ 1-t_{q-1} & 1-t_{q-1} & \dots & \theta+\sum_{j=1}^{q-1}t_j+1-t_{q-1}\end{vmatrix}$$

is non-zero, where $\mathbf{A}$ is the matrix of the system. It is easy to see that

$$\det\mathbf{A}=\left(\theta+\sum_{j=1}^{q-1}t_j\right)^{q-2}(\theta+q-1).$$

Since $\theta>0$, $t_j>0$, $j=1,\dots,q-1$ and $q\geq 2$ we have $\det\mathbf{A}\neq 0$. Thus, $a_i=z_i-t_i=0$, i.e., $h=u$. □

Let $G_k^{(2)}$ be the subgroup in G_k consisting of all words of even length. Clearly, $G_k^{(2)}$ is a subgroup of index 2.

The following theorem characterizes all periodic Gibbs measures.

Theorem 11.1. *Let K be a normal subgroup of finite index in G_k. Then each K-periodic Gibbs measure for the Potts model (11.1) is either translation-invariant or $G_2^{(2)}$-periodic.*

Proof. Follows from the property that F is one-to-one (Lemma 11.1), similarly as proof of Theorem 2.3 of [243]. □

Since translation-invariant SGMs are already studied in Chapter 9, by Theorem 11.1 it suffices to study $G_k^{(2)}$-periodic Gibbs measure. Such measures correspond to the set of vectors $h = \{h_x \in \mathbb{R}^{q-1} : x \in G_k\}$ of the form (see Fig. 11.1)

$$h_x = \begin{cases} \mathbf{h}^1, & \text{if } |x| \text{ is even,} \\ \mathbf{h}^2, & \text{if } |x| \text{ is odd,} \end{cases}$$

where $|x|$ denotes the length of the word x and the vectors $\mathbf{h}^1 = (h_1, h_2, ..., h_{q-1})$, $\mathbf{h}^2 = (l_1, l_2, ..., l_{q-1})$, by (11.2), should satisfy

$$\begin{cases} h_i = k \ln \dfrac{(\theta - 1)\exp(l_i) + \sum_{j=1}^{q-1} \exp(l_j) + 1}{\sum_{j=1}^{q-1} \exp(l_j) + \theta}, \\ \qquad\qquad\qquad\qquad\qquad\qquad\qquad\qquad\qquad i = 1, \ldots, q-1. \\ l_i = k \ln \dfrac{(\theta - 1)\exp(h_i) + \sum_{j=1}^{q-1} \exp(h_j) + 1}{\sum_{j=1}^{q-1} \exp(h_j) + \theta} \end{cases}$$

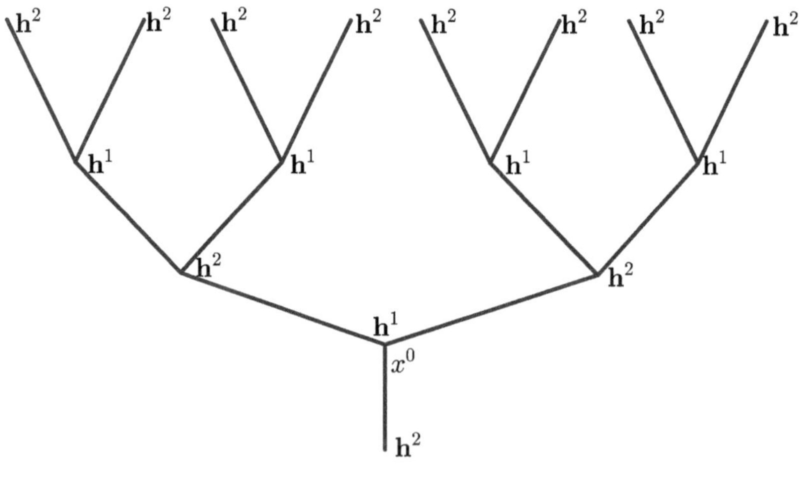

Fig. 11.1 Periodic vector-valued function h_x on tree.

Denoting $\exp(h_i) = x_i$, $\exp(l_i) = y_i$ the last system can be written as

$$x_i = \left(\frac{(\theta-1)y_i + \sum_{j=1}^{q-1} y_j + 1}{\sum_{j=1}^{q-1} y_j + \theta}\right)^k,$$

$$i = 1, \dots, q-1. \qquad (11.4)$$

$$y_i = \left(\frac{(\theta-1)x_i + \sum_{j=1}^{q-1} x_j + 1}{\sum_{j=1}^{q-1} x_j + \theta}\right)^k$$

Consider mapping $W : \mathbb{R}_+^{q-1} \times \mathbb{R}_+^{q-1} \to \mathbb{R}_+^{q-1} \times \mathbb{R}_+^{q-1}$, defined by

$$x_i' = \left(\frac{(\theta-1)y_i + \sum_{j=1}^{q-1} y_j + 1}{\sum_{j=1}^{q-1} y_j + \theta}\right)^k,$$

$$i = 1, \dots, q-1. \qquad (11.5)$$

$$y_i' = \left(\frac{(\theta-1)x_i + \sum_{j=1}^{q-1} x_j + 1}{\sum_{j=1}^{q-1} x_j + \theta}\right)^k$$

Note that (11.4) is the equation $z = W(z)$, with

$$z = (x_1, \dots, x_{q-1}, y_1, \dots, y_{q-1}),$$

that is equation for fixed points of the mapping W.

For $m \leq q-1$ introduce the set

$$I_m = \{x = (x_1, \dots, x_{q-1}) \in \mathbb{R}_+^{q-1} : x_1 = \dots = x_m,\ x_{m+1} = \dots = x_{q-1} = 1\}.$$

Let S_{q-1} be the group of permutations on $\{1, \dots, q-1\}$.

For $\pi = (\pi(1), \dots, \pi(q-1)) \in S_{q-1}$ and $x \in \mathbb{R}^{q-1}$ define

$$\pi x = (x_{\pi(1)}, \dots, x_{\pi(q-1)}),$$

and for a subset $I \subset \mathbb{R}^{q-1}$ define

$$\pi I = \{\pi x : x \in I\}.$$

The following lemma is useful to find special solutions of (11.4).

Lemma 11.2. *For any $\pi \in S_{q-1}$ and $m \leq q-1$ the set $\pi I_m \times \pi I_m$ is invariant with respect to the mapping W.*

Proof. For each $i = 1, 2, \dots, q-1$, from ith equation of the system (11.5) we note that $x_i = 1$ if and only if $y_i' = 1$. Assume for a given $M \subset \{1, \dots, q-1\}$ we have $x_i = y_i = 1$ for any $i \notin M$. Let $\emptyset \neq M \subset \{1, \dots, q-1\}$, without loss of generality we can take $M = \{1, 2, \dots, m\}$, $m \leq q-1$, i.e., $x_i = y_i = 1$, $i = m+1, \dots, q$. Taking $x_1 = \dots = x_m$ and $y_1 = \dots = y_m$ one can see that $x_1' = \dots = x_m'$ and $y_1' = \dots = y_m'$. Therefore, the permutation symmetry of the variables of the mapping W completes the proof. □

To find solutions of (11.4) which belongs to the invariant set $I_m \times I_m$, reduce W on this set. Then (11.4) becomes (denoting $x_1 = \cdots = x_m = x$, $y_1 = \cdots = y_m = y$)

$$x = f_m(y), \quad y = f_m(x)$$
$$\text{with } f_m(z) \equiv \left(\frac{(\theta + m - 1)z + q - m}{mz + q - m - 1 + \theta}\right)^k, \tag{11.6}$$

for some $m = 1, \dots, q-1$.

Remark 11.1. Again by symmetry of solution, if a solution (x, y) of (11.6) is known then the vector $(\pi x, \pi y)$ will be solution of (11.4) for any $\pi \in S_{q-1}$.

A $G_k^{(2)}$-periodic (non-translation-invariant) Gibbs measure corresponds to a solution (x, y) of (11.6) iff $x \neq y$. The following lemma gives conditions of existence of such solutions.

Lemma 11.3. *For any $k \geq 2$, $0 < \theta < 1$ we have*

1. *If $q \geq k+1$ then (11.6) has only solutions with $x = y$.*
2. *Let $2 \leq q < k+1$. Then system of equations (11.6) has at least three solutions for $0 < \theta < \theta_{cr}$ (where $\theta_{cr} = \frac{k-q+1}{k+1}$), it has at least one solution for $\theta = \theta_{cr}$ and it has only one solution for $\theta > \theta_{cr}$.*

Proof. By (11.6) we obtain

$$x = g(x) \equiv f_m(f_m(x)).$$

We have

$$f_m'(x) = k\left(\frac{(\theta + m - 1)x + q - m}{mx + q - m - 1 + \theta}\right)^{k-1} \frac{(\theta - 1)(\theta + q - 1)}{(mx + q - m - 1 + \theta)^2};$$
$$g'(x) = f_m'(f_m(x)) f_m'(x).$$

Consequently, for $0 < \theta < 1$ and $x > 0$ the function $f_m(x)$ decreases monotonically and the equation $f_m(x) = x$ has a unique solution $x = 1$, with $f_m'(1) = k(\theta - 1/\theta + q - 1)$. Since derivative of $g(x)$ is positive (for $0 < \theta < 1$ and $x > 0$), it is an increasing function and $x = 1$ is a solution to $g(x) = x$. If $g'(1) = k^2(f_m'(1))^2 = (k(\theta - 1/\theta + q - 1))^2 > 1$ then this solution is not unique, because in this case for $x > 1$ the graph of the function g lies above the bisector and the function g is bounded. Thus a critical value for θ can be found by the equation $(k(\theta - 1/\theta + q - 1))^2 = 1$ which for $\theta < 1$ gives

$$\theta_{cr} = \frac{k - q + 1}{k + 1} = 1 - \frac{q}{k + 1}.$$

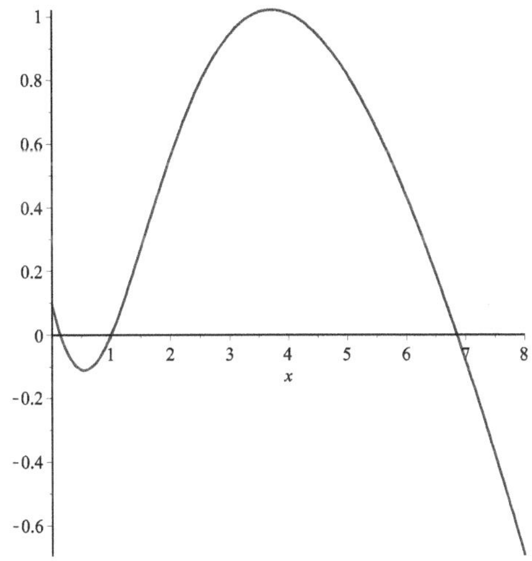

Fig. 11.2 Non-uniqueness: The graph of function $g(x) - x$ for $k = 2$, $q = 2$, $m = 1$, $\theta = 0.25$.

The condition $0 < 1 - \frac{q}{k+1}$ then gives $q < k + 1$, which for $k = 2$, has solution $q = 2$ only. In case $k = 3$, the parameter q may be $q = 2, 3$ and so on.

Hence it follows that for $0 < \theta < \theta_{cr}$ the equation $g(x) = x$ has at least three solutions $x_0^* < x_1^* = 1 < x_2^*$, i.e. the equation $g(x) = x$ has at least two roots, which are distinct from roots of the equation $f(x) = x$. For $\theta = \theta_{cr}$ the graph of the function g tangents to the bisector in $x = 1$. This means that in this condition the equation $g(x) = x$ has at least one solution. Besides it is clear that for $\theta > \theta_{cr}$ the equation $g(x) = x$ has a unique solution $x_1^* = 1$, which is solution to equation $f_m(x) = x$ (see Figs. 11.2–11.5). □

11.3 Zero external field

In this section consider the case $\alpha = 0$.

11.3.1 *Ferromagnetic case*

In this subsection, following [125], we show that in the ferromagnetic case (i.e., $J > 0$) each periodic measure is translation-invariant.

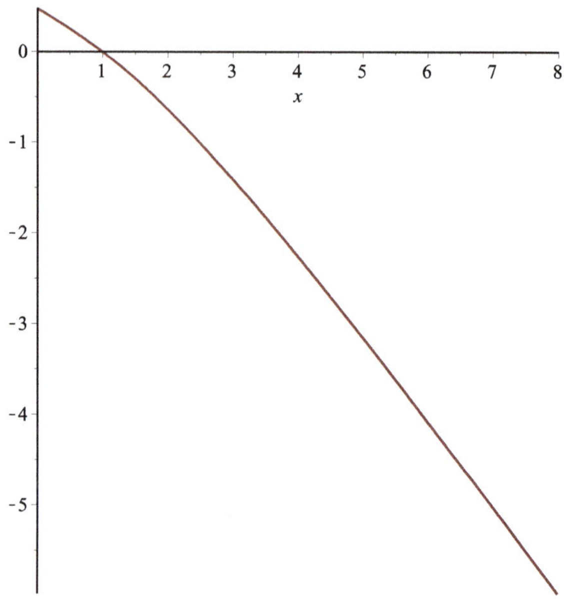

Fig. 11.3 Uniqueness: The graph of function $g(x)-x$ for $k=2$, $q=3$, $m=1$, $\theta=0.25$.

Theorem 11.2. *Let $k \geq 2,\ q \geq 2,\ J>0,\ \alpha=0$. Then any periodic Gibbs measure of the model (11.1) is translation-invariant.*

Proof. Find $x_i - y_i,\ i=1,\dots,q-1$ from (11.4):

$$x_i - y_i = \left(\frac{(\theta-1)y_i+\sum_{j=1}^{q-1} y_j+1}{\sum_{j=1}^{q-1} y_j+\theta} - \frac{(\theta-1)x_i+\sum_{j=1}^{q-1} x_j+1}{\sum_{j=1}^{q-1} x_j+\theta}\right)\cdot A_i$$

$$= \frac{A_i}{XY}\cdot\left((\theta^2-\theta)(y_i-x_i)+(\theta-1)\left(y_i\sum_{j=1}^{q-1}x_j - x_i\sum_{j=1}^{q-1}y_j\right)\right.$$

$$\left. +(\theta-1)\left(\sum_{j=1}^{q-1}y_j-\sum_{j=1}^{q-1}x_j\right)\right),$$

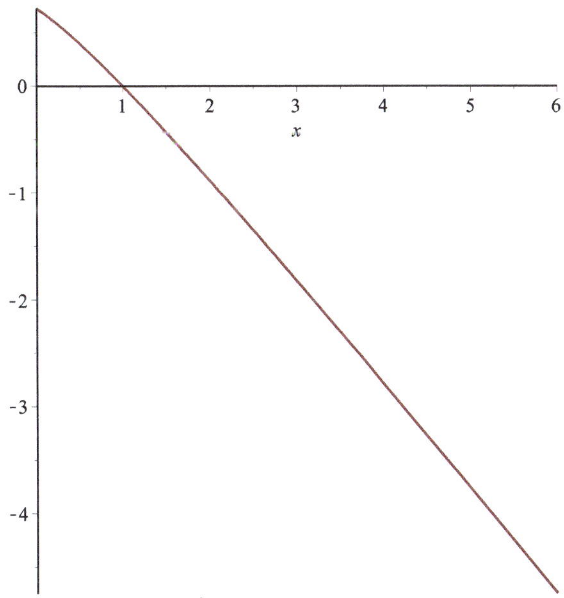

Fig. 11.4 Uniqueness: The graph of function $g(x)-x$ for $k=5$, $q=4$, $m=2$, $\theta=0.7$.

where

$$A_i=\left(\frac{(\theta-1)y_i+\sum_{j=1}^{q-1}y_j+1}{\sum_{j=1}^{q-1}y_j+\theta}\right)^{k-1}+\ldots+\left(\frac{(\theta-1)x_i+\sum_{j=1}^{q-1}x_j+1}{\sum_{j=1}^{q-1}x_j+\theta}\right)^{k-1},$$

$$X=\sum_{j=1}^{q-1}x_j+\theta,\ Y=\sum_{j=1}^{q-1}y_j+\theta.$$

Simplifying we get

$$x_i-y_i=\frac{A_i(\theta-1)}{XY}\cdot\left(\theta(y_i-x_i)+y_i\sum_{j=1}^{q-1}(x_j-y_j)+(y_i-x_i)\right.$$
$$\left.\times\sum_{j=1}^{q-1}y_j+\sum_{j=1}^{q-1}y_j-\sum_{j=1}^{q-1}x_j\right).$$

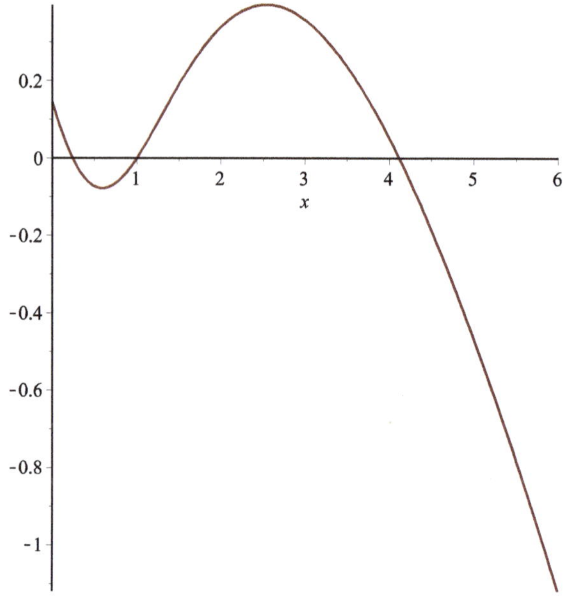

Fig. 11.5 Non-uniqueness: The graph of function $g(x) - x$ for $k = 5$, $q = 4$, $m = 2$, $\theta = 0.25$.

Consequently, we obtain the following system

$$\begin{cases} a_{11}(x_1 - y_1) + a_{12}(x_2 - y_2) + ... + a_{1q-1}(x_{q-1} - y_{q-1}) = 0 \\ a_{21}(x_1 - y_1) + a_{22}(x_2 - y_2) + ... + a_{2q-1}(x_{q-1} - y_{q-1}) = 0 \\ \cdots\cdots\cdots\cdots\cdots\cdots \\ a_{q-11}(x_1 - y_1) + a_{q-12}(x_2 - y_2) + ... + a_{q-1q-1}(x_{q-1} - y_{q-1}) = 0, \end{cases} \tag{11.7}$$

with

$$a_{ii} = 1 + \frac{A_i(\theta - 1)(\theta + 1 + \sum_{j=1}^{q-1} y_j - y_i)}{XY},$$

$$a_{il} = \frac{(\theta - 1)(1 - y_i)A_i}{XY}, \ i \neq l, \ i, l = 1, \dots, q-1.$$

This system has only zero solution if $\det \mathbf{A} =$

$$\begin{vmatrix} 1+\frac{A_1(\theta-1)(1+Y-y_1)}{XY} & \frac{(\theta-1)(1-y_1)A_1}{XY} & .... & \frac{(\theta-1)(1-y_1)A_1}{XY} \\ \frac{(\theta-1)(1-y_2)A_2}{XY} & 1+\frac{A_2(\theta-1)(1+Y-y_2)}{XY} & .... & \frac{(\theta-1)(1-y_2)A_2}{XY} \\ ... & ... & ... & ... \\ \frac{(\theta-1)(1-y_{q-1})A_{q-1}}{XY} & \frac{(\theta-1)(1-y_{q-1})A_{q-1}}{XY} & ... & 1+\frac{A_{q-1}(\theta-1)(1+Y-y_{q-1})}{XY} \end{vmatrix}$$

is non-zero. Here $\mathbf{A}$ is the matrix of the system. Rewrite the determinant of $\mathbf{A}$ as

$$\det \mathbf{A} = C \begin{vmatrix} 1+B_1 & 1 & .... & 1 \\ 1 & 1+B_2 & .... & 1 \\ ... & ... & ... & ... \\ 1 & 1 & ... & 1+B_{q-1} \end{vmatrix},$$

where

$$C = \frac{A_1(\theta-1)(1-y_1)}{XY} \cdot \frac{A_2(\theta-1)(1-y_2)}{XY} \cdot \dots \cdot \frac{A_{q-1}(\theta-1)(1-y_{q-1})}{XY},$$

$$B_i = \frac{XY + A_i(\theta-1)Y}{A_i(\theta-1)(1-y_i)}.$$

Let us show that $\det \mathbf{A} \neq 0$.

Using the following known (see [70]) formula

$$\begin{vmatrix} 1+a_1 & 1 & ... & 1 & 1 \\ 1 & 1+a_2 & ... & 1 & 1 \\ 1 & 1 & ... & 1 & 1 \\ ... & ... & ... & ... & ... \\ 1 & 1 & ... & 1 & 1+a_n \end{vmatrix} = a_1a_2a_3...a_n \cdot \left(1 + \frac{1}{a_1} + \frac{1}{a_2} + ... + \frac{1}{a_n}\right),$$

we obtain

$$\det \mathbf{A} = \frac{1}{X^{q-1}Y^{q-1}} \cdot (XY + A_1(\theta-1)Y) \cdot \dots \cdot (XY + A_{q-1}(\theta-1)Y)$$
$$\times \left(1 + \frac{A_1(\theta-1)(1-y_1)}{XY + A_1(\theta-1)Y} + \dots + \frac{A_{q-1}(\theta-1)(1-y_{q-1})}{XY + A_{q-1}(\theta-1)Y}\right).$$

Consequently,

$$\det \mathbf{A} = \frac{1}{X^{q-1}Y^{q-1}} \cdot P,$$

where

$$P = (XY + A_1(\theta - 1)Y) \cdot (XY + A_2(\theta - 1)Y) \cdot \ldots \cdot (XY + A_{q-1}(\theta - 1)Y)$$
$$+ (XY + A_2(\theta - 1)Y) \cdot (XY + A_3(\theta - 1)Y) \cdot \ldots \cdot (XY + A_{q-1}(\theta - 1)Y)$$
$$\times A_1(\theta - 1)(1 - y_1) + (XY + A_1(\theta - 1)Y) \cdot (XY + A_3(\theta - 1)Y) \cdot \ldots \cdot$$
$$\times (XY + A_{q-1}(\theta - 1)Y) \cdot A_2(\theta - 1)(1 - y_2) + \ldots + (XY + A_1(\theta - 1)Y)$$
$$\times (XY + A_2(\theta - 1)Y) \cdot \ldots \cdot (XY + A_{q-2}(\theta - 1)Y)\, A_{q-1}(\theta - 1)(1 - y_{q-1}).$$

Regrouping P into powers of XY we get

$$P = (XY)^{q-1} + (XY)^{q-2}(\theta - 1) \cdot \left[Y \sum_{j=1}^{q-1} A_j + \sum_{j=1}^{q-1} (1 - y_j) A_j \right]$$
$$+ (XY)^{q-3}(\theta - 1)^2 Y \cdot \left[Y \sum_{i=1}^{q-2} A_i A_{i+1} + \sum_{i=1}^{q-1} \left(\sum_{j=1}^{q-1} A_j - A_i \right) A_i (1 - y_i) \right]$$
$$+ \ldots + (XY)^{q-1-i}(\theta - 1)^i Y^{i-1} \cdot [Y(A_1 A_2 \ldots A_i + \ldots + A_{q-i} A_{q-i+1} \ldots A_{q-1})$$
$$+ (A_2 A_3 \ldots A_i + \ldots + A_{q-i+1} \ldots A_{q-1}) A_1 (1 - y_1)$$
$$+ (A_1 A_3 \ldots A_i + \ldots + A_{q-i+1} \ldots A_{q-1}) A_2 (1 - y_2)$$
$$+ \ldots + (A_1 A_2 \ldots A_{i-1} + \ldots + A_{q-i} A_{q-i+1} \ldots A_{q-2}) A_{q-1} (1 - y_{q-1})]$$
$$+ \ldots + (\theta - 1)^{q-1} Y^{q-2} \cdot [Y(A_1 A_2 \ldots A_{q-1}) + A_1 A_2 \ldots A_{q-1}(1 - y_1)$$
$$+ \ldots + A_1 A_2 \ldots A_{q-1}(1 - y_{q-1})].$$

Now we shall prove that $P > 0$. Since $\theta > 1,\ X > 0,\ Y > 0,\ A_i > 0$, it suffices to show that the expressions in the square brackets are positive.

We have that the expression in the first bracket is positive:

$$Y \sum_{j=1}^{q-1} A_j + \sum_{j=1}^{q-1} (1 - y_j) A_j = (\theta + 1) \sum_{j=1}^{q-1} A_j + \sum_{j \neq k} y_j A_k > 0,$$

in the second bracket is

$$Y(A_1A_2 + \ldots + A_{q-2}A_{q-1}) + \left(\sum_{j=1}^{q-1} A_j - A_1\right) A_1(1 - y_1)$$
$$+ \ldots + \left(\sum_{j=1}^{q-1} A_j - A_{q-1}\right) A_{q-1}(1 - y_{q-1})$$
$$= (\theta + 1)(A_1A_2 + \ldots + A_{q-2}A_{q-1})$$
$$+ y_1 \sum_{j\neq k\neq 1} A_jA_k + \ldots + y_{q-1} \sum_{j\neq k\neq q-1} A_jA_k > 0,$$

and in i-th bracket we have

$$Y(A_1A_2 \cdot \ldots \cdot A_i + \ldots + A_{q-i}A_{q-i+1} \cdot \ldots \cdot A_{q-1})$$
$$+ (A_2A_3 \cdot \ldots \cdot A_i + \ldots + A_{q-i+1} \ldots A_{q-1})A_1(1 - y_1)$$
$$+ (A_1A_3 \cdot \ldots \cdot A_i + \ldots + A_{q-i+1} \cdot \ldots \cdot A_{q-1})A_2(1 - y_2) + \ldots$$
$$+ (A_1A_2 \cdot \ldots \cdot A_{i-1} + \ldots + A_{q-i}A_{q-i+1} \cdot \ldots \cdot A_{q-2})A_{q-1}(1 - y_{q-1}))$$
$$= (\theta + 1)(A_1 \cdot \ldots \cdot A_i + \ldots + A_{q-i} \cdot \ldots \cdot A_{q-1})$$
$$+ y_1(A_2 \cdot \ldots \cdot A_{i+1} + \ldots + A_{q-i} \cdot \ldots \cdot A_{q-1})$$
$$+ \ldots + y_{q-1}(A_1 \cdot \ldots \cdot A_i + \ldots + A_{q-i-1} \cdot \ldots \cdot A_{q-2}) > 0.$$

Finally, the last bracket has the expression:

$$YA_1A_2 \ldots A_{q-1} + A_1A_2 \ldots A_{q-1}(1 - y_1) + \ldots + A_1A_2 \ldots A_{q-1}(1 - y_{q-1})$$
$$= A_1A_2 \ldots A_{q-1}(\theta + 1) > 0.$$

Thus, $P > 0$. Consequently, the system (11.7) has solutions only in the form $x_i = y_i$. Then $h = l$ and $h_x = h$ (independent on x). This completes the proof. □

For $q = 3$ this theorem was proved in [244].

11.3.2 *Anti-ferromagnetic case*

Note that the case $q = 2$ corresponds to anti-ferromagnetic Ising model, which has periodic measures for any $k \geq 2$ ([98], [243]).

In this subsection we consider the anti-ferromagnetic case (i.e., $J < 0$) and following [125] we show that for $k = 2$, $q \geq 3$ there is no periodic (except translation-invariant) Gibbs measures. But for $k \geq 3$, $q \geq 2$ there are several periodic (non-translation-invariant) Gibbs measures.

11.3.2.1 *Case:* $k = 2$.

Theorem 11.3. *Let $k = 2$, $q \geq 3$, $J < 0$, and $\alpha = 0$. Then the $G_k^{(2)}$ — periodic Gibbs measure for the Potts model is unique. Moreover, this measure coincides with the unique translation-invariant Gibbs measure.*

Proof. We show that (11.4) has unique solution $x_i = y_i = 1$, $i = 1, \dots, q-1$. Note that this solution is unique in class of translation-invariant solutions of (11.2) and the Gibbs measure corresponding to this solution is translation-invariant (see Theorem 8.2).

Substitute the expressions for y_i in the right-hand sides of the first $q-1$ equations in (11.4). We then obtain the equalities

$$\sqrt{x_i} = \frac{\gamma + (\theta - 1)\tilde{x}_i + \sum_{j=1}^{q-1} \tilde{x}_j}{\theta\gamma + \sum_{j=1}^{q-1} \tilde{x}_j}, \quad i = 1, \dots, q-1, \tag{11.8}$$

where

$$\tilde{x}_i = (1 + (\theta - 1)x_i + \sum_{j=1}^{q-1} x_j)^2, \quad i = 1, \dots, q-1,$$

$$\gamma = \left(\theta + \sum_{j=1}^{q-1} x_j\right)^2.$$

Subtracting 1 from both sides of each quality in (11.8) we get

$$\sqrt{x_i} - 1 = \frac{(\theta - 1)(\tilde{x}_i - \gamma)}{\theta\gamma + \sum_{j=1}^{q-1} \tilde{x}_j}, \quad i = 1, \dots, q-1.$$

Denote

$$L = \frac{\theta - 1}{\theta\gamma + \sum_{j=1}^{q-1} \tilde{x}_j},$$

and then the last system of equations can be written as

$$\sqrt{x_i} - 1 = L(\tilde{x}_i - \gamma), \ i = 1, 2, \dots q-1. \tag{11.9}$$

Using

$$\tilde{x}_i - \gamma = (\theta - 1) \cdot (x_i - 1) \cdot \left((\theta + 1)(x_i + 1) + 2\left(\sum_{j=1}^{q-1} x_j - x_i\right)\right),$$

from (11.9) we obtain

$$(\sqrt{x_i}-1)\left[1-L(\theta-1)(\sqrt{x_i}+1)\left((\theta+1)(x_i+1)+2\left(\sum_{j=1}^{q-1}x_j-x_i\right)\right)\right]=0.$$

Consequently, for $i=1,2,\dots q-1$ we get $x_i=1$ or

$$1-L(\theta-1)\cdot(\sqrt{x_i}+1)\cdot\left((\theta+1)(x_i+1)+2\left(\sum_{j=1}^{q-1}x_j-x_i\right)\right)=0. \quad (11.10)$$

As mentioned above $x_i=1,\ i=1,2,\dots q-1$ corresponds to translation-invariant measure. Therefore, we assume $x_i\neq 1$. Rewrite (11.10) as

$$\frac{1}{L}=(\theta-1)\cdot(\sqrt{x_i}+1)\cdot\left((\theta+1)(x_i+1)+2\left(\sum_{j=1}^{q-1}x_j-x_i\right)\right),\quad i=1,\dots q-1.$$

Then for $i\neq j,\ i,j=1,2,\dots q-1$ we obtain

$$\begin{aligned}&(\theta+1)\cdot(\sqrt{x_i}+1)\cdot(x_i+1)+2\cdot(\sqrt{x_i}+1)\cdot\left(\sum_{p=1}^{q-1}x_p-x_i\right)\\&\quad=(\theta+1)\cdot(\sqrt{x_j}+1)\cdot(x_j+1)+2\cdot(\sqrt{x_j}+1)\cdot\left(\sum_{p=1}^{q-1}x_p-x_j\right).\end{aligned}$$

Consequently,

$$\begin{aligned}&(\sqrt{x_i}-\sqrt{x_j})[(\theta+1)(x_i+x_j+1)+(\theta-1)(\sqrt{x_i}+\sqrt{x_j}+\sqrt{x_ix_j})\\&\quad+2\left(\sum_{p=1}^{q-1}x_p-x_i-x_j\right)]=0.\end{aligned}$$

Then $x_i=x_j$ or

$$(\theta+1)(x_i+x_j+1)+(\theta-1)(\sqrt{x_i}+\sqrt{x_j}+\sqrt{x_ix_j})+2\left(\sum_{p=1}^{q-1}x_p-x_i-x_j\right)=0. \quad (11.11)$$

The case $x_i=x_j$ gives $x_i=x_j=1$, therefore, we assume $x_i\neq x_j$. Then from (11.11) we get

$$\begin{aligned}&\theta(x_i+x_j+1+\sqrt{x_i}+\sqrt{x_j}+\sqrt{x_ix_j})+2(\sum_{p=1}^{q-1}x_p-x_i-x_j)\\&\quad+x_i+x_j+1-(\sqrt{x_i}+\sqrt{x_j}+\sqrt{x_ix_j})=0. \quad (11.12)\end{aligned}$$

Now we show that (11.12) does not have positive solutions. For this it suffices to show the following inequality

$$x_i + x_j + 1 > \sqrt{x_i} + \sqrt{x_j} + \sqrt{x_i x_j}.$$

Denoting $\sqrt{x_i} = s$, $\sqrt{x_j} = t$, we get the inequality (with respect to s)

$$s^2 - (t+1)s + t^2 - t + 1 > 0,$$

discriminant of which $D = -3(t-1)^2 < 0$ for all $t \neq 1$. Hence, the system (11.4) has only solutions with $x_i = x_j$, consequently $y_i = y_j$. That is the solution $(x, y) \in I_m \times I_m$, therefore Lemma 11.3 completes the proof. □

11.3.2.2 *Case:* $k \geq 3$.

In case $k \geq 3$, we have (see [124], [126]) the following:

Theorem 11.4. *Let $k \geq 3$, $q \geq 3$, $J < 0$ and $\theta_{cr} = \frac{k-q+1}{k+1}$. Then for the Potts model for $0 < \theta < \theta_{cr}$ there are exactly three $G_k^{(2)}$– periodic Gibbs measures, corresponding to solutions on $I_m \times I_m$ (i.e. the solutions of system (11.6)). Moreover, only one of these measures is translation-invariant.*

Proof. Under conditions of theorem we show that the system (11.6) has exactly two solutions with $x \neq y$. We have $f_m(0) = \left(\frac{q-m}{\theta+q-m-1}\right)^k > 0$, and the function $f_m(x)$ strictly decreasing for $0 < \theta < 1$ and $x > 0$. From (11.6) we get $f_m(f_m(x)) = x$. Since $f_m(x)$ is invertible for $x > 0$, we can rewrite the last equation as

$$f(x) = f^{-1}(x) = g(x),$$

$$\text{where } \; g(x) = f^{-1}(x) = \frac{q - m - (\theta + q - m - 1)\sqrt[k]{x}}{m\sqrt[k]{x} - \theta - m + 1}.$$

By $f(x) > 0$, we have $g(x) > 0$, therefore

$$\theta_1 = \left(\frac{\theta + m - 1}{m}\right)^k < x < \left(\frac{q - m}{\theta + q - m - 1}\right)^k = \theta_2.$$

Let us denote $h(x) = \ln \frac{f(x)}{g(x)} = \ln f(x) - \ln g(x)$ and solve $h(x) = 0$. It is clear that $x = 1$ is a solution to this equation, i.e., $h(1) = 0$. Using

$$f'(x) = \frac{k(\theta - 1)(\theta + q - 1)f(x)}{[(\theta + m - 1)x + q - m](mx + \theta + q - m - 1)},$$

and

$$g'(x) = \frac{(\theta - 1)(\theta + q - 1)g(x)}{k\sqrt[k]{x^{k-1}}(m\sqrt[k]{x} - \theta - m + 1)[q - m - (\theta + q - m - 1)\sqrt[k]{x}]},$$

we obtain

$$h'(x) = \frac{f'(x)}{f(x)} - \frac{g'(x)}{g(x)}$$
$$= \frac{(\theta-1)(\theta+q-1)}{k} \cdot \frac{k^2}{[(\theta+m-1)x+q-m](mx+\theta+q-m-1)}$$
$$- \frac{(\theta-1)(\theta+q-1)}{k}$$
$$\times \frac{1}{\sqrt[k]{x^{k-1}}(m\sqrt[k]{x}-\theta-m+1)(q-m-(\theta+q-m-1)\sqrt[k]{x})}.$$

Denoting $\sqrt[k]{x} = t$, rewrite $h'(x)$ as

$$v(t) = \frac{(\theta-1)(\theta+q-1)p(t)}{s(t)},$$

where

$$p(t) = m(\theta+m-1)t^{2k} + k^2 m(\theta+q-m-1)t^{k+1} - (k^2-1)(\theta^2+(q-2)\theta+2mq$$
$$- 2m^2 - q + 1)t^k + k^2(\theta+m-1)(q-m)t^{k-1} + (\theta+q-m-1)(q-m).$$

$$s(t) = kt^{k-1}[(\theta+m-1)t^k+q-m](mt^k+\theta+q-m-1)(mt-\theta-m+1)$$
$$\times [(\theta+q-m-1)t-q+m].$$

Using the fact (see [221]) that the number of positive roots of a polynomial does not exceed the number of sign changes of its coefficients, we see that the polynomial $p(t)$ and $h'(x)$ has up to two positive solutions.

It is easy to see that

$$\lim_{x\to\theta_1} h(x) = -\infty, \ h(1) = 0, \ \lim_{x\to\theta_2} h(x) = +\infty.$$

Consequently, the equation $h(x) = 0$ has at least one solution x_0 for $x < 1$ and at least one solution x_2 for $x > 1$, if $h'(1) < 0$. From condition

$$h'(1) = \frac{(\theta-1)(\theta+q-1)}{k} \cdot \left[\frac{k^2}{(\theta+q-1)^2} - \frac{1}{(\theta-1)^2}\right] < 0,$$

it follows that $\theta < \frac{k-q+1}{k+1} = \theta_{cr}$, i.e., $\frac{(\theta-1)(\theta+q-1)}{k} < 0$.

Moreover,

$$\lim_{x\to\theta_1} h'(x) = +\infty, \ \lim_{x\to\theta_2} h'(x) = +\infty$$

for $\theta < \theta_{cr}$. Then from $h'(1) < 0$ it follows that the function $h(x)$ has exactly two critical points ξ_1, ξ_2: $\theta_1 < \xi_1 < 1$ and $1 < \xi_2 < \theta_2$.

Thus $h(x)$ is an increasing function for $\theta_1 < x < \xi_1$, $x > \xi_2$ and a decreasing function for $\xi_1 < x < \xi_2$. Therefore, the equation $h(x) = 0$ has exactly three solutions $x_0 < x_1 = 1 < x_2$. It remains to verify the condition $x \neq y$. Since $f_m(x)$ is strictly decreasing, from the second equation in (11.6), we obtain $f(x_0) = y_0 > f(x_1) = y_1 = 1 > f(x_2) = y_2$. Therefore, for the Potts model, under conditions of theorem, there exist exactly three $G_k^{(2)}$−periodic Gibbs measures corresponding to the solutions $x_0 < x_1 = 1 < x_2$. Moreover, the measure corresponding to $x_1 = 1$ is translation-invariant. Theorem is proved. □

Using symmetry of solutions, similarly to arguments of Chapter 9, one can prove the following:

Theorem 11.5. *For the Potts model on the Cayley tree of order $k \geq 3$, if $3 \leq q < k+1$ and $0 < \theta < \theta_{cr}$ then the number of $G_k^{(2)}$−periodic Gibbs measures is at least $2 \cdot (2^q - 1)$.*

11.4 Non-zero external field

In case of non-zero external field the system of equations for periodic solutions becomes more complicated to solve. Because in this case the symmetry of solutions will be lost.

11.4.1 *The case $k = 2$, $q = 3$*

In this section following [244] we consider a particular case of parameters and show that under some conditions on parameters the model has periodic (non-translation-invariant) Gibbs measures.

Let $k = 2$, $q = 3$ and $e^{\alpha\beta} = \lambda$, where $\alpha \in \mathbb{R}$ is the external field, and as before, $\beta = \frac{1}{T}$, $T > 0$.

We are interested to solutions of the form $((z,1),(t,1))$, which satisfies

$$\begin{cases} z = f(t) \\ t = f(z), \end{cases} \text{ where } f(z) = \lambda\left(\frac{\theta z + 2}{z + \theta + 1}\right)^2. \tag{11.13}$$

Denoting $\sqrt{z} = x$, $\sqrt{t} = y$, $\sqrt{\lambda} = \zeta$ we obtain:

$$\begin{cases} x = \zeta \cdot \dfrac{\theta y^2 + 2}{y^2 + \theta + 1} \\ y = \zeta \cdot \dfrac{\theta x^2 + 2}{x^2 + \theta + 1}. \end{cases} \tag{11.14}$$

This system, in case $x = y$, can be written as

$$x^3 - \zeta\theta x^2 + (\theta + 1)x - 2\zeta = 0, \tag{11.15}$$

which gives translation-invariant measures.

Substituting the expression for y from the second equation in the first equation, we obtain a fifth-degree equation in x. Dividing the fifth-degree polynomial from the obtained equation by the third-degree polynomial from (11.15), we obtain the quadratic equation

$$(\zeta^2\theta^2 + \theta + 1)x^2 + (-2\zeta + \zeta\theta(\theta + 1))x + 2\zeta^2\theta + (\theta + 1)^2 = 0, \tag{11.16}$$

which describes periodic Gibbs measures. We now study solutions of this quadratic equation. For this, we calculate its discriminant taking into account that $\sqrt{\lambda} = \zeta$:

$$D = -8\theta^3\lambda^2 - [3\theta^2(\theta + 1)^2 + 12\theta(\theta + 1) - 4]\lambda - 4(\theta + 1)^3.$$

It is known that equation (11.16) has two positive real roots if $D > 0$ and if $-2\zeta + \zeta\theta(\theta + 1) = \zeta(\theta^2 + \theta - 2) < 0$. Solving these inequalities, we find that equation (11.16) has two real solutions if $0 < \theta < \frac{\sqrt{17}-3}{6} = \theta^*$ and $\lambda_1(\theta) < \lambda < \lambda_2(\theta)$, where $\lambda_1(\theta)$, $\lambda_2(\theta)$ has the following form

$$\begin{aligned} \lambda_1(\theta) &= \frac{-3[3\theta^2(\theta + 1)^2 + 12\theta(\theta + 1) - 4] - (\theta - 1)(\theta + 2)\sqrt{K}}{48\theta^3}, \\ \lambda_2(\theta) &= \frac{-3[3\theta^2(\theta + 1)^2 + 12\theta(\theta + 1) - 4] + (\theta - 1)(\theta + 2)\sqrt{K}}{48\theta^3}, \end{aligned} \tag{11.17}$$

where

$$K = (\theta - 1)(\theta + 2)(9\theta^2 + 9\theta - 2).$$

Note that for $0 < \theta < \theta^*$ both $\lambda_1(\theta)$ and $\lambda_2(\theta)$ are positive.

If $\theta^* < \theta < 1$, then equation (11.16) has no real solutions. Moreover, equation (11.16) also has no real solutions for $\theta > 1$.

Let now $\theta = \theta^*$. Then for $\lambda = \lambda^*$, where

$$\lambda^* = -\frac{3(\theta^*)^2(\theta^* + 1)^2 + 12\theta^*(\theta^* + 1) - 4}{16(\theta^*)^3}, \tag{11.18}$$

the discriminant of the equation (11.16) is $D = 0$. Consequently, the equation (11.16) has unique solution:

$$x^* = \frac{\zeta^*(2 - \theta^* - (\theta^*)^2)}{2((\zeta^*)^2(\theta^*)^2 + \theta^* + 1)},$$

where $\zeta^* = \sqrt{\lambda^*}$.

Hence, the system (11.13) has solutions $(z^*,\ t^*)$, where

$$z^* = \frac{\lambda^*}{4}\left[\frac{2-\theta^*-(\theta^*)^2}{(\zeta^*)^2(\theta^*)^2+\theta^*+1}\right]^2,\ t^* = \lambda^*\left(\frac{\theta^* z^*+2}{z^*+\theta^*+1}\right)^2.$$

Thus we have the following theorem:

Theorem 11.6. *For the Potts model, in case $k=2, q=3, \alpha \neq 0$ we have*

1. *If $\lambda \in (\lambda_1(\theta),\ \lambda_2(\theta))$ and $\theta \in (0,\ \theta^*)$, then there are at least two $G_k^{(2)}-$ periodic (non-translation-invariant) Gibbs measures, where $\theta^* = \frac{\sqrt{17}-3}{6}$, and $\lambda_1,\ \lambda_2$ defined by (11.17);*
2. *For $\lambda = \lambda^*$ and $\theta = \theta^*$, there are at least one $G_k^{(2)}-$ periodic (non-translation-invariant) Gibbs measure, where λ^* is defined in (11.18).*

11.4.2 *The case of translation-invariant external field*

Here we follow [255] and use the notations of the Section 10.4.

Consider the Potts model with an external field. We note that for any normal subgroup H of finite index each H-periodic Gibbs measure of this model is either $G_k^{(2)}$ - periodic or translation invariant.

Therefore we need only to study $G_k^{(2)}$-periodic Gibbs measures. Note that $G_k^{(2)}$-periodic set of vectors $\overrightarrow{h} = \{h_x \in \mathbb{R}^{q-1} : x \in G_k\}$ has the following form

$$\overrightarrow{h}_x = \begin{cases} \overrightarrow{h}_1, \text{ if } \ x \in G_k^{(2)} \\ \overrightarrow{h}_2, \text{ if } \ x \in G_k \setminus G_k^{(2)}, \end{cases}$$

and a translation-invariant external field has the form:

$$\overrightarrow{\alpha}_x = \overrightarrow{\alpha}, \forall x \in G_k.$$

here $\overrightarrow{h}_i = (h_{i1}, h_{i2}, ..., h_{iq-1}),\ i = 1, 2,\ \overrightarrow{\alpha} = (\alpha_1, ..., \alpha_{q-1})$.

Then by the consistency equation we get

$$\begin{cases} \overrightarrow{h}_1 = \overrightarrow{\alpha} + kF(\overrightarrow{h}_2, \theta) \\ \overrightarrow{h}_2 = \overrightarrow{\alpha} + kF(\overrightarrow{h}_1, \theta). \end{cases} \tag{11.19}$$

Denote $z_{ij} = \exp h_{ij}, \lambda_i = \exp(\alpha_i), i = 1, 2, j = 1, 2, ..., q-1$, then from (11.19) we get

$$\begin{cases} z_{1j} = \lambda_j\left(\frac{(\theta-1)z_{2j}+\sum_{r=1}^{q-1} z_{2j}+1}{\sum_{r=1}^{q-1} z_{2j}+\theta}\right)^k \\ z_{2j} = \lambda_j\left(\frac{(\theta-1)z_{1j}+\sum_{r=1}^{q-1} z_{1j}+1}{\sum_{r=1}^{q-1} z_{1j}+\theta}\right)^k, \end{cases} \tag{11.20}$$

where $j = 1, 2, ..., q-1$.

Let $\overrightarrow{z}_r = (z_{r1}, z_{r2}, ..., z_{rq-1}) \in I_m$ for $r = 1,2$ and $\overrightarrow{\lambda} = (\lambda_1, \lambda_2, ..., \lambda_{q-1}) \in I_m$. For $r = 1,2$ denote by z_r the coordinates of $\overrightarrow{z}_r$, which are different from 1, and by λ we denote coordinates of $\overrightarrow{\lambda}$, which are different from 1. Then from the system (11.20) we obtain

$$\begin{cases} z_1 = \lambda \left(\frac{(\theta+m-1)z_2+q-m}{mz_2+\theta+q-m-1} \right)^k \\ z_2 = \lambda \left(\frac{(\theta+m-1)z_1+q-m}{mz_1+\theta+q-m-1} \right)^k. \end{cases} \tag{11.21}$$

Introduce $\sqrt[k]{z_1} = x$, $\sqrt[k]{z_2} = y$, $\sqrt[k]{\lambda} = \zeta$, then we have

$$\begin{cases} x = \zeta \frac{(\theta+m-1)y^k+q-m}{my^k+\theta+q-m-1} \\ y = \zeta \frac{(\theta+m-1)x^k+q-m}{mx^k+\theta+q-m-1}. \end{cases} \tag{11.22}$$

Consider the function $g : (0, +\infty) \to (0, +\infty)$, defined by

$$g(x) \equiv g(x, \zeta, \theta, q, m, k) = \zeta \frac{(\theta + m - 1)x^k + q - m}{mx^k + \theta + q - m - 1}.$$

Lemma 11.4. *The function $g(x)$, $x > 0$ has the following properties*

1. *$g(x)$ is an increasing for $\theta > 1$ and a decreasing function for $\theta < 1$.*
2. *$g(x)$ is a bounded function:*

$$\zeta \cdot \min M < g(x) < \zeta \cdot \max M.$$

where

$$M = \left\{ \frac{q - m}{\theta + q - m - 1}, \frac{\theta + m - 1}{m} \right\}$$

3. *Strictly convex at $\theta > 1$, $x \in (0, \hat{x})$ and $\theta < 1$, $x \in (\hat{x}, +\infty)$, (resp. strictly concave at $\theta > 1$, $x \in (\hat{x}, +\infty)$ and $\theta < 1$, $x \in (0, \hat{x})$) where*

$$\hat{x} = \left(\frac{(k - 1)(\theta + q - m - 1)}{m(k + 1)} \right)^{1/k}.$$

Proof. Follows from the following qualities

$$g'(x) = \frac{\zeta k(\theta - 1)(\theta - 1 + q)x^{k-1}}{(mx^k + \theta + q - m - 1)^2}.$$

$$g''(x) = \frac{\zeta k(\theta - 1)(\theta - 1 + q)x^{k-2}}{(mx^k + \theta + q - m - 1)^3} \cdot \left[-m(k + 1)x^k + (k - 1)(\theta + q - m - 1) \right].$$

□

Let $\overrightarrow{\alpha} \in I_1$ and α be the non-zero coordinate of the vector $\overrightarrow{\alpha}$.

The following theorem gives conditions on parameters of the system (11.22) describing $G_k^{(2)}$-periodic Gibbs measures.

Theorem 11.7. *For the q-state Potts model with a translation-invariant external field on a Cayley tree the following assertions hold*

1. *If $k \geq 2$, $\theta > 1$ then all $G_k^{(2)}$-periodic Gibbs measures are transnational-invariant.*
2. *If $k = 2$, $\theta < 1$, $m = 2, \ldots, q-2$, $q \geq 4$, then all $G_k^{(2)}$-periodic Gibbs measures are transnational-invariant.*
3. *If $k = 2$, $\theta < 1$, $m = 1$, then*
   3.1. *for $\theta \in (0, \theta^*)$ and $\alpha \in (\ln \lambda_1(\theta, q), \ln \lambda_2(\theta, q))$, where*
   $$\theta^* = (\sqrt{32 - 32q + 9q^2} + 6 - 3q)/6, \tag{11.23}$$
   there are two $G_k^{(2)}$-periodic (non-translation-invariant) Gibbs measures.
   3.2. *for $\theta = \theta^*$ and $\alpha = \ln \lambda^*(\theta, q)$ there exists one $G_k^{(2)}$-periodic (non-translation-invariant) Gibbs measure.*
   3.3. *for $\theta > \theta^*$ and $\theta \neq 1$ there is no $G_k^{(2)}$-periodic (non-translation-invariant) Gibbs measure, where*

$$\lambda_{1,2}(\theta, q) = \frac{-[3\theta^2(\theta + q - 2)^2 + 6\theta(q-1)(\theta + q - 2) - (q-1)^2] \pm \sqrt{D'}}{8(q-1)\theta^3},$$

$$\lambda^*(\theta, q) = -\frac{(3\theta^*)^2(\theta^* + q - 2)^2 + 6\theta^*(q-1)(\theta^* + q - 2) - (q-1)^2}{8(q-1)(\theta^*)^3}$$

and

$$D' = (3\theta^2(\theta+q-2)^2+6\theta(q-1)(\theta+q-2)-(q-1)^2)^2-64\theta^3(\theta+q-2)^3(q-1).$$

4. *If $k = 2$, $\theta < 1$, $m = q-1$, then*
   4.1. *for $\theta \in (0, \theta^*)$ and $\alpha \in (\ln \lambda_3(\theta, q), \ln \lambda_4(\theta, q))$, where θ^* is defined in (11.23), there are two $G_k^{(2)}$-periodic (non-translation-invariant) Gibbs measure,*
   4.2. *for $\theta = \theta^*$ and $\alpha = \ln \lambda_3(\theta, q)$ there exists one $G_k^{(2)}$-periodic (non-translation-invariant) Gibbs measure.*
   4.3. *for $\theta > \theta^*$ and $\theta \neq 1$ there is no $G_k^{(2)}$-periodic (non-translation-invariant) Gibbs measure, where $\lambda_3(\theta, q) \leq \lambda_4(\theta, q)$ are solutions of the equation*
   $$[4(q+\theta-2)((\theta-1)^2+(q-1)(2\theta+q-3))]\lambda^2 - [(\theta-1)^2((\theta+q-1)^2 -4\theta^2)-(4(q-1))\theta(\theta+q-2+(2\theta+q-3)\theta)]\lambda+4(q-1)\theta^3 = 0. \tag{11.24}$$

Proof. 1. For $x = y$ the system (11.22) is reduced to the equation $g(x) = x$, solutions of which determine TISGMs. For $\theta > 1$ we show that $g(g(x)) = x$ does not have solutions different from the solutions of $g(x) = x$. Assume that there is a solution x with $g(x) \neq x$. Then $g(x) > x$ (or $g(x) < x$). In case $g(x) > x$ (similarly in case $g(x) < x$) by Lemma 11.4 we have $g(g(x)) > g(x) > x$, i.e., x can not be a solution for $g(g(x)) = x$.

2. Let $\theta < 1$. For $k = 2$ the equation $g(x) = x$ has the form

$$mx^3 - \zeta(\theta + m - 1)x^2 + (\theta + q - m - 1)x - \zeta(q - m) = 0, \tag{11.25}$$

whose solutions describe TISGMs. For $k = 2$ in the system of equations (11.22), substituting the expression for y from the second equation into the first, we obtain an equation of the fifth degree with respect to x. Dividing the polynomial of the fifth degree by the polynomial of the third degree from (11.25), we obtain the quadratic equation

$$\begin{aligned}&[m(\theta + q - m - 1) + \zeta^2((\theta - 1)^2 + m(m + 2\theta - 2))]x^2\\ &\quad +[\zeta(\theta - 1)(\theta + q - 1)]x + \zeta^2(q - m)(\theta + m - 1) + (\theta + q - m - 1)^2 = 0,\end{aligned} \tag{11.26}$$

which describes periodic Gibbs measures. Let's explore the solutions to this quadratic equation. To do this, we calculate its discriminant, taking into account that $\sqrt{\lambda} = \zeta$:

$$\begin{aligned}D = &-4\lambda^2(q - m)(\theta + m - 1)[(\theta - 1)^2 + m(m + 2\theta - 2)]\\ &+\lambda[(\theta - 1)^2\{(\theta + q - 1)^2 - 4(\theta + q - m - 1)^2\}\\ &-\{(q - m)(\theta + m - 1) + (m + 2\theta - 2)(\theta + q - m - 1)\}\\ &\times 4m(\theta + q - m - 1)] - 4m(\theta + q - m - 1)^3.\end{aligned} \tag{11.27}$$

It is known that (11.26) has two positive real solutions if $D > 0$ and $\zeta(\theta - 1)(\theta + q - 1) < 0$. For $\theta < 1$ it is clear that $\zeta(\theta - 1)(\theta + q - 1) < 0$. To solve the inequality $D > 0$, we multiply its both sides to -1, and get

$$A\lambda^2 - B\lambda + C < 0, \tag{11.28}$$

where

$$A = 4(q - m)(\theta + m - 1)[(\theta - 1)^2 + m(m + 2\theta - 2)],$$

$$B = (\theta - 1)^2\{(\theta + q - 1)^2 - 4(\theta + q - m - 1)^2\}$$

$$-4m(\theta + q - m - 1)\{(q - m)(\theta + m - 1) + (m + 2\theta - 2)(\theta + q - m - 1)\},$$

$$C = 4m(\theta + q - m - 1)^3.$$

Note that under the conditions of the part 2 of the theorem, we have $A > 0$, $C > 0$. Therefore, for $\lambda > 0$, in order for the inequality (11.28) to have a solution, it is necessary to fulfill $B > 0$, i.e.,

$$B \equiv B(\theta) = (\theta - 1)^2\{(\theta + q - 1)^2 - 4(\theta + q - m - 1)^2\}$$

$$-4m(\theta+q-m-1)\{(q-m)(\theta+m-1)+(m+2\theta-2)(\theta+q-m-1)\} > 0. \tag{11.29}$$

We are interested to solutions of (11.29) with respect to $\theta \in (0, 1)$. Note that $B(\theta)$ is a polynomial of degree 4, i.e.,

$$\begin{aligned} B(\theta) = &-3\theta^4 - 6(q-2)\theta^3 + 3(4m^2 - 4mq - q^2 + 6q - 6)\theta^2 \\ &+ 6(2m^2q - 2mq^2 - 4m^2 + 4mq + q^2 - 3q + 2)\theta - 8m^4 + 16m^3q \\ &- 8m^2q^2 - 12m^2q + 12mq^2 + 12m^2 - 12qm - 3q^2 + 6q - 3. \end{aligned}$$

For $m = 2, 3, \ldots, q-2, \quad q \geq 4$ denote

$$E(m,q) = -8m^4 + 16m^3q - 8m^2q^2 - 12m^2q + 12mq^2 + 12m^2 - 12qm - 3q^2 + 6q - 3.$$

We note that for all $m = 2, 3, \ldots, q-2, \quad q \geq 4$ it holds that $E(m,q) < 0$. Indeed, the equation $E(m,q) = 0$ has the following solutions (with respect to m):

$$m_{1,2} = \frac{q}{2} \pm \frac{1}{2}\sqrt{q^2 - (3 - \sqrt{3})(q-1)},$$

$$m_{3,4} = \frac{q}{2} \pm \frac{1}{2}\sqrt{q^2 - (3 + \sqrt{3})(q-1)}.$$

For $q \geq 4$ we have

$$m_1 > q - 2, \quad m_2 < 2, \quad m_3 > q - 2, \quad m_4 < 2. \tag{11.30}$$

Since the coefficient at m^4 in $E(m,q)$ is negative, from (11.30) we get $E(m,q) < 0$.

By a computer one can find all solutions of $B(\theta) = 0$:

$$\theta_1 = 1 - \frac{q}{2} + \frac{1}{6}\sqrt{9q^2 - (72 - 24\sqrt{3})m(q-m)},$$

$$\theta_2 = 1 - \frac{q}{2} - \frac{1}{6}\sqrt{9q^2 - (72 - 24\sqrt{3})m(q-m)},$$

$$\theta_3 = 1 - \frac{q}{2} + \frac{1}{6}\sqrt{9q^2 - (72 + 24\sqrt{3})m(q-m)},$$

$$\theta_4 = 1 - \frac{q}{2} - \frac{1}{6}\sqrt{9q^2 - (72 + 24\sqrt{3})m(q-m)}.$$

Note that

$$m(q-m) \in \{q-1, 2(q-2), ..., [q/2](q-[q/2])\}. \tag{11.31}$$

Therefore,

$$9q^2-(72-24\sqrt{3})m(q-m) \geq 9q^2-(72-24\sqrt{3})\frac{q}{2}(q-\frac{q}{2}) = 3\sqrt{3}q^2(2-\sqrt{3}) > 0,$$

for all $1 \leq m < q$ and

$$9q^2 - (72+24\sqrt{3})m(q-m) > 0,$$

for sufficiently large q, with

$$m \in \left(1, \frac{q}{4}(2-\sqrt{1+\sqrt{3}})\right) \cup \left(\frac{q}{4}(2+\sqrt{1+\sqrt{3}}), q-1\right).$$

Moreover, $\theta_i < 0$, $i = 2, 3, 4$. But $\theta_1 > 0$ iff $m = 1$ and $m = q-1$. Indeed, $\theta_1 > 0$ is reduced to

$$3(q-1) > (6-2\sqrt{3})m(q-m). \tag{11.32}$$

It is easy to see that if m is a solution to the last inequality, then $q-m$ is also a solution. Obviously, (11.32) is true for $m = 1$ and $m = q-1$. But for $m \geq 2$, i.e., $m(q-m) \in \{2(q-2), ..., [q/2](q-[q/2])\}$ from (11.32) we get

$$q < \varphi(m) = \frac{(6-2\sqrt{3})m^2-3}{(6-2\sqrt{3})m-3}. \tag{11.33}$$

Now we show that (11.33) does not hold if $q \geq 4$ and $m = 2, \ldots, q-2$. Note that $\varphi(m)$ is an increasing function. Therefore it suffices to show that (11.33) does not hold for $q \geq 4$ and $m = q-2$:

$$q < \varphi(q-2) \quad \Leftrightarrow \quad q < \frac{21-8\sqrt{3}}{9-4\sqrt{3}} \approx 3.448,$$

that is, (11.33) does not hold for all $q \geq 4$.

Thus for all $m = 2, \ldots, q-2$ we have $\theta_i < 0$, $i = 1, 2, 3, 4$. Since $E(m, q) < 0$ and the coefficient at θ^4 in $B(\theta)$ is negative it follows that $B(\theta) < 0$ for all $\theta \in (0, 1)$. Consequently, the inequality (11.28) does not have solution under conditions of the part 2.

3. If $k = 2$, $m = 1$, then from (11.26) we get

$$[\theta^2\zeta^2 + \theta + q - 2]x^2 + (\theta-1)(\theta+q-1)\zeta x + (q-1)\theta\zeta^2 + (\theta+q-2)^2 = 0. \tag{11.34}$$

In this case, the inequality (11.29) for $0 < \theta < 1$ has a solution of the form

$$0 < \theta < \theta_1,$$

where

$$\theta_1 = 1 - \frac{q}{2} + \frac{1}{6}\sqrt{9q^2 - (72 - 24\sqrt{3})(q-1)}. \tag{11.35}$$

For such values of θ and $m = 1$, inequality (11.28) has solutions if

$$D' := (3\theta^2(\theta+q-2)^2 + 6\theta(q-1)(\theta+q-2) - (q-1)^2)^2 - 64\theta^3(\theta+q-2)^3(q-1) \geq 0. \tag{11.36}$$

From (11.36) we get

$$(\theta-1)^3(\theta+q-1)^3\left(\theta - (1 - \frac{q}{2} + \frac{1}{6}\sqrt{9q^2 - 32(q-1)})\right)$$
$$\times\left(\theta - (1 - \frac{q}{2} - \frac{1}{6}\sqrt{9q^2 - 32(q-1)})\right) \geq 0.$$

Introduce

$$\theta^* = 1 - \frac{q}{2} + \frac{1}{6}\sqrt{9q^2 - 32(q-1)}.$$

Then for $0 < \theta < 1$ the inequality (11.36) has the form

$$0 < \theta \leq \theta^*.$$

It is easy to see that if $q \geq 2$ then $\theta^* < \theta_1$, consequently, for $0 < \theta \leq \theta^*$ we have (11.29) and (11.36). Hence for $\theta \in (0, \theta^*)$ the inequality (11.28) has solutions of the form (see Fig. 11.6)

$$\lambda_1(\theta, q) < \lambda < \lambda_2(\theta, q),$$

where $\lambda_1(\theta, q), \lambda_2(\theta, q)$ are solutions to the equation

$$4(q-1)\theta^3\lambda^2 + [3\theta^2(\theta+q-2)^2 + 6\theta(q-1)(\theta+q-2) - (q-1)^2]\lambda + 4(\theta+q-2)^3 = 0,$$

i.e.,

$$\lambda_{1,2}(\theta, q) = \frac{-[3\theta^2(\theta+q-2)^2 + 6\theta(q-1)(\theta+q-2) - (q-1)^2] \pm \sqrt{D'}}{8(q-1)\theta^3}. \tag{11.37}$$

Note that for $0 < \theta \leq \theta^*$ both $\lambda_1(\theta, q)$ and $\lambda_2(\theta, q)$ are positive. Since $m = 1$, we have $\vec{\lambda} \in (\vec{\lambda}_1(\theta, q), \vec{\lambda}_2(\theta, q))$, where $\vec{\lambda}_i(\theta, q) = (1, ..1, \lambda_i^{\{j\}}(\theta, q), 1, ..., 1) \in I_1, i = 1, 2, j = 1, 2, ..., q-1$ and $\lambda_i^{\{j\}}(\theta, q)$

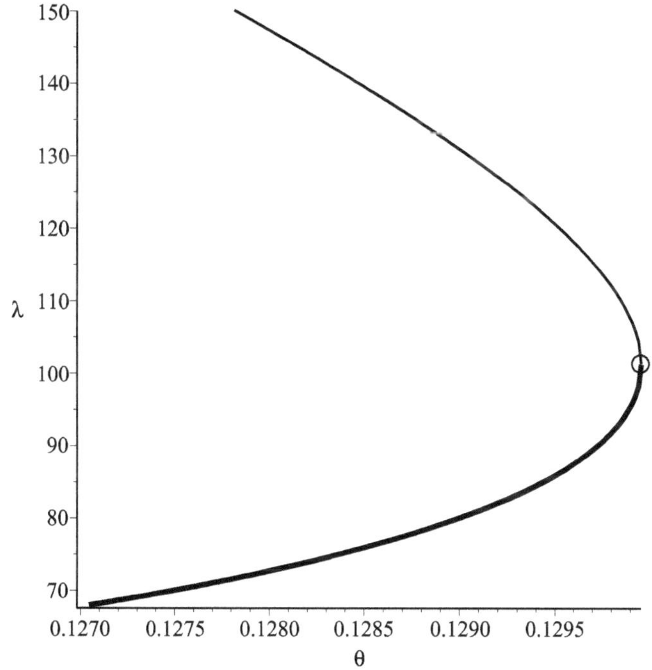

Fig. 11.6 *The graphs of* λ_1 *(bold) and* λ_2 *for* $q = 7$.

is j-th coordinate of the vector $\overrightarrow{\lambda}_i(\theta, q)$. Consequently, we get $\overrightarrow{\alpha} \in (\overrightarrow{\alpha}_1(\theta, q), \overrightarrow{\alpha}_2(\theta, q))$, where

$$\overrightarrow{\alpha}_i(\theta, q) = (0, ..0, (\ln \lambda_i(\theta, q))^{\{j\}}, 0, ..., 0), i = 1, 2. \tag{11.38}$$

If $D' < 0$, i.e. $\theta^* < \theta < 1$, then the quadratic equation (11.34) does not have real solutions.

Let now $\theta = \theta^*$, i.e., $D' = 0$. Then for

$$\lambda = \lambda^*(\theta, q) = -\frac{(3\theta^*)^2(\theta^* + q - 2)^2 + 6\theta^*(q-1)(\theta^* + q - 2) - (q-1)^2}{8(q-1)(\theta^*)^3},$$

i.e.,

$$\overrightarrow{\alpha} = \overrightarrow{\alpha}^*(\theta, q) := (0, ...0, (\ln \lambda^*(\theta, q))^{\{j\}}, 0, ..., 0), \tag{11.39}$$

we have $D = 0$, here $(\ln \lambda^*(\theta, q))^{\{j\}}$ j-th coordinate of the vector $\overrightarrow{\alpha}^*(\theta, q)$. This means that in this case the equation (11.34) has one solution.

4. If $k=2$, $m=q-1$, then from (11.26) we get

$$[((\theta-1)^2+(q-1)(2\theta+q-3))\zeta^2+(q-1)\theta]x^2 + (\theta-1)(\theta+q-1)\zeta x+(q+\theta-2)\zeta^2+\theta^2=0. \quad (11.40)$$

In this case, similarly to case 3, the inequality (11.29) has solutions $0<\theta<\theta_1$, where θ_1 is defined in (11.35).

As in case 3, it is easy to show that for $m=q-1$ the inequality (11.28) has solutions if

$$0<\theta\le\theta^*,$$

consequently, for $m=q-1$ and $0<\theta\le\theta^*$ the inequality (11.29) holds. Hence for $m=q-1$, $\theta\in(0,\theta^*)$ the inequality (11.28) has solutions of the form

$$\lambda_3(\theta,q)<\lambda<\lambda_4(\theta,q),$$

where $\lambda_3(\theta,q),\lambda_4(\theta,q)$ is a solution of (11.24) (see Fig. 11.7).

Thus, we have two solutions $\vec{\alpha}\in(\vec{\alpha}_3(\theta,q),\vec{\alpha}_4(\theta,q))$, where

$$\vec{\alpha}_i(\theta,q)=(0,..0,(\ln\lambda_i(\theta,q))^{\{j\}},0,...,0), i=3,4. \quad (11.41)$$

Let $\theta=\theta^*$. In this case the equation (11.40) has one solution. □

Let $\vec{z_i}\in I_m$, $\vec{\lambda}\in I_n, i=1,2$. If $m<n$, then we have from (11.20) the following

$$\begin{cases} z_1=\lambda f^k(z_2)\\ z_2=\lambda f^k(z_1)\\ 1=\lambda f^k(1), \end{cases} \quad (11.42)$$

where

$$f(z)=\left(\frac{(\theta+m-1)z+q-m}{mz+\theta+q-m-1}\right).$$

Since $f(1)=1$, from (11.42) we have $\lambda=1$. Thus, the system of equations (11.42) is reduced to the following system of equations

$$\begin{cases} z_1=f^k(z_2)\\ z_2=f^k(z_1), \end{cases} \quad (11.43)$$

whose solutions describe periodic Gibbs measures for the Potts model with zero external field.

If $m>n$, in a similar way we obtain a system of equations for periodic Gibbs measures for the Potts model with a zero external field.

If $m=n$ and the coordinates of the vectors $\vec{\lambda}$ and $\vec{z}$, different from 1 do not coincide, i.e., if there is i_j such that $\lambda_{i_j}=1, z_{i_j}\ne1$ or $\lambda_{i_j}\ne1, z_{i_j}=1$, then by a similar method we obtain the system of equations (11.43).

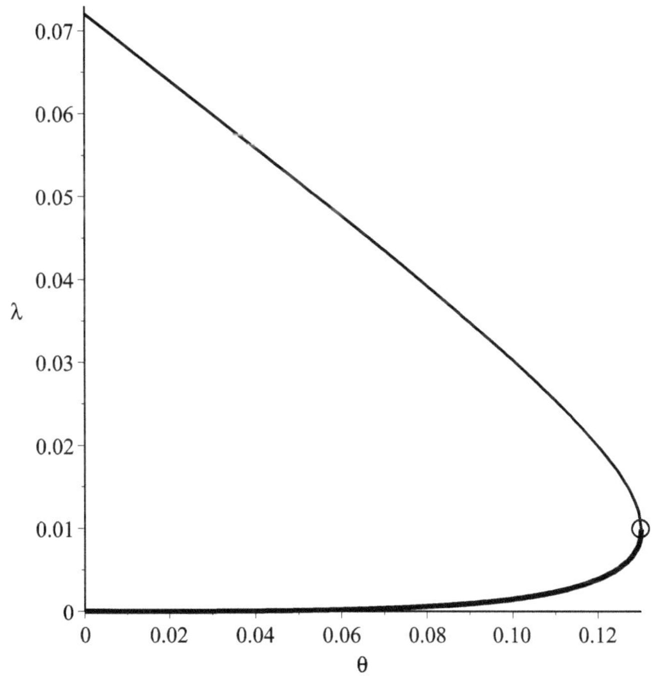

Fig. 11.7 *The graphs of functions λ_3 (bold) and λ_4 for $q = 7$.*

11.4.3 *A periodic external field*

Consider periodic Gibbs measures with a periodic external field.

Recall

$$G_k^{(2)} = \{x \in G_k : |x| \text{ is even}\},$$

then $G_k/G_k^{(2)} = \{G_k^{(2)}, G_k \setminus G_k^{(2)}\}$ is the factor group. Note that $G_k^{(2)}$ — periodic set of vectors $h = \{h_x \in \mathbb{R}^{q-1} : x \in G_k\}$ has the form

$$h_x = \begin{cases} \overrightarrow{h}_1, \text{ if } x \in G_k^{(2)}; \\ \overrightarrow{h}_2, \text{ if } x \in G_k \setminus G_k^{(2)}. \end{cases}$$

Let the external field be also $G_k^{(2)}$ - periodic, i.e.,

$$\alpha_x = \begin{cases} \overrightarrow{\alpha}_1, \text{ if } x \in G_k^{(2)}; \\ \overrightarrow{\alpha}_2, \text{ if } x \in G_k \setminus G_k^{(2)}. \end{cases}$$

Here $\overrightarrow{\alpha}_1 \neq \overrightarrow{\alpha}_2$. Note that $\overrightarrow{\alpha}_1 = \overrightarrow{\alpha}_2$ the Potts model with a periodic external field coincides with the Potts model with a TI external field, which was studied in the previous subsection.

Then by virtue of (2.18) we have

$$\begin{cases} \overrightarrow{h}_1 = \overrightarrow{\alpha}_1 + kF(\overrightarrow{h}_2, \theta), \\ \overrightarrow{h}_2 = \overrightarrow{\alpha}_2 + kF(\overrightarrow{h}_1, \theta). \end{cases} \tag{11.44}$$

Here $\overrightarrow{h}_i = (h_{i1}, h_{i2}, ..., h_{iq-1})$, $\overrightarrow{\alpha}_i = (\alpha_{i1}, \alpha_{i2}, ..., \alpha_{iq-1})$, $i = 1, 2$.

The following theorem is true.

Theorem 11.8. *Let $k \geq 2$. Then, for the Potts model with a periodic external field, for any value of $\theta > 0$, there is at least one $G_k^{(2)}$ — periodic (not translation-invariant) Gibbs measure.*

Proof. Introduce the notation $z_{ij} = e^{h_{ij}}, \lambda_{ij} = e^{\alpha_{ij}}, i = 1, 2, j = 1, 2, ..., q-1$. Let $\overrightarrow{z}_i = (z_{i1}, z_{i2}, ..., z_{iq-1}) \in I_m$, $\overrightarrow{\lambda}_i = (\lambda_{i1}, \lambda_{i2}, ..., \lambda_{iq-1}) \in I_m$. Then the system of equations (11.44) has the following form

$$\begin{cases} z_{11} = \lambda_{11} f(z_{21}) \\ z_{21} = \lambda_{21} f(z_{11}), \end{cases} \tag{11.45}$$

where

$$f(z) = \left(\frac{(\theta + m - 1)z + q - m}{mz + \theta + q - m - 1}\right)^k.$$

From (11.45) we get

$$z_{11} = \lambda_{11} f(\lambda_{21} f(z_{11})). \tag{11.46}$$

The function $f(z)$ has the following properties

(1) Since $m \leq q - 1$ and $\theta > 0$, we have $f(0) = \left(\frac{q-m}{\theta+q-m-1}\right)^k > 0$.

(2) Consider

$$f'(z) = k\left(\frac{(\theta + m - 1)z + q - m}{mz + \theta + q - m - 1}\right)^{k-1} \frac{(\theta - 1)(\theta + q - 1)}{(mx + \theta + q - m - 1)^2}.$$

It follows from the equality that $f(z)$ an increasing for $\theta > 1$ and a decreasing function for $\theta < 1$.

(3) The function $f(z)$ is bounded. Because of $\lim_{z \to +\infty} f(z) = \left(\frac{\theta+m-1}{m}\right)^k$ and properties 1), 2).

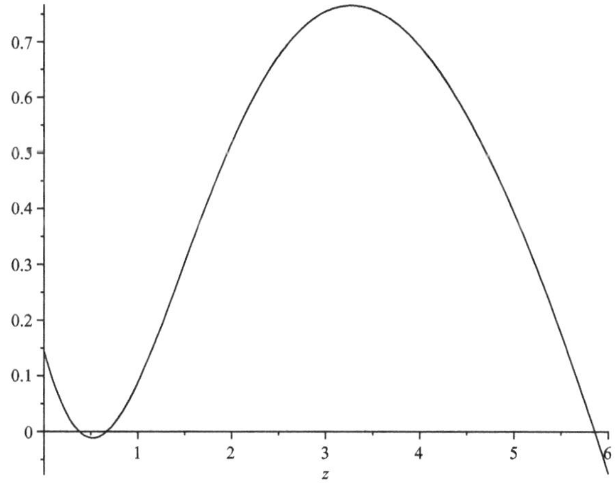

Fig. 11.8 *The graph of the function* $\varphi(x) - x$ *for* $q = 3; m = 1; \lambda_{11} = \frac{1}{\lambda_{21}} = 0.53$; $\theta = 4.7$.

Denote $\varphi(z) := \lambda_{11} f(\lambda_{21} f(z))$. Since $\lambda_{11}, \lambda_{21}$ are positive, from properties (1)–(3) we have $\varphi(0) > 0$ and $\varphi(z)$ — is bounded.

Consequently, for any $\theta > 0, \theta \neq 1$ and $\lambda_{11}, \lambda_{21}$ the equation (11.46) has at least one solution $z_{11} := x_*$. Therefore, the system of equations (11.45) has a solution of the form $(z_{11}, z_{21}) := (x_*, y_*)$. Thus this solution corresponds to a $G_k^{(2)}$ — periodic Gibbs measure. Since the considered external field is not a TI, then by Theorem 10.6 we have that the resulting measure is not a TI. □

Denote

$$A(\lambda_{11}, \lambda_{21}, \theta, q, m) = \lambda_{11}\lambda_{21}(\theta - 1)^2(\theta + q - 1)^2(mx_* + \theta + q - m - 1)$$
$$\times \frac{[\lambda_{21}(\theta + m - 1)((\theta + m - 1)x_* + q - m)^2 + (q - m)(mx_* + \theta + q - m - 1)^2]}{(\lambda_{21}m((\theta + m - 1)x_* + q - m)^2 + (\theta + q - m - 1)(mx_* + \theta + q - m - 1)^2)^3}.$$

Theorem 11.9. *Let* $k = 2$. *If* $A(\lambda_{11}, \lambda_{21}, \theta, q, m) < 1/4$, *then for the Potts model with a periodic external field, there is at least one* $G_k^{(2)}$ — *periodic Gibbs measure; if* $A(\lambda_{11}, \lambda_{21}, \theta, q, m) = 1/4$, *then there are at least two* $G_k^{(2)}$ — *periodic Gibbs measures; if* $A(\lambda_{11}, \lambda_{21}, \theta, q, m) > 1/4$, *then there are at least three* $G_k^{(2)}$ - *periodic Gibbs measures.*

Proof. Denote $z_{ij} = e^{h_{ij}}, \lambda_{ij} = e^{\alpha_{ij}}, i = 1, 2, j = 1, 2, ..., q - 1$. Let $k = 2$ and $\overrightarrow{z}_i \in I_m, \overrightarrow{\lambda}_i \in I_m, i = 1, 2$. Then the system (11.44) has the following form

$$\begin{cases} z_{11} = \lambda_{11} f(z_{21}) \\ z_{21} = \lambda_{21} f(z_{11}), \end{cases} \tag{11.47}$$

where

$$f(z) = \left(\frac{(\theta + m - 1)z + q - m}{mz + \theta + q - m - 1} \right)^2 .$$

Consider $\varphi(z) := \lambda_{11} f(\lambda_{21} f(z))$. It is known that $\varphi(0) > 0$, $\varphi(z)-$ bounded and has a fixed point $z = x_*$. It is clear that if $\varphi'(x_*) > 1$, then the equation

$$z_{11} = \lambda_{11} f(\lambda_{21} f(z_{11})), \tag{11.48}$$

has at least three solutions (see Fig. 11.8), for $\varphi'(x_*) = 1$ has at least two solutions, and for $\varphi'(x_*) < 1$ has at least one solution. Calculating $\varphi'(x_*) = 4A(\lambda_{11}, \lambda_{21}, \theta, q, m)$ one completes the proof. □

11.5 Weakly periodic measures

In this section following [223]-[226] we give some weakly periodic Gibbs measures (see Definition 11.3) of the Potts model.

The level of difficulty in describing of weakly periodic Gibbs measures is related to the structure and index of the normal subgroup relative to which the periodicity condition is imposed.

From Chapter 1 of [243] (see also [205]) we know that in the group G_k, there is no normal subgroup of odd index different from one. Therefore, we consider normal subgroups of even indices. In this section we restrict our self to the case of indices two.

It is known (see [90] and [243]) that any normal subgroup of index two of the group G_k has the form

$$H_A = \left\{ x \in G_k : \sum_{i \in A} \omega_x(a_i) \text{—even} \right\},$$

where $\emptyset \neq A \subseteq N_k = \{1, 2, \ldots, k+1\}$, and $\omega_x(a_i)$ is the number of letters a_i in a word $x \in G_k$.

Let $A \subseteq N_k$ and H_A be the corresponding normal subgroup of index two. We note that in the case $|A| = k + 1$, i.e., in the case $A = N_k$, weak periodicity coincides with ordinary periodicity.

Consider $G_k/H_A = \{H_A, G_k \setminus H_A\}$ the quotient group.

For simplicity of notations, we denote $H_0 = H_A$, $H_1 = G_k \setminus H_A$.

The H_0 -weakly periodic collections of vectors $h = \{h_x \in \mathbb{R}^{q-1} : x \in G_k\}$ have the form:

$$h_x = \begin{cases} h_1, \text{ if } x_\downarrow \in H_0, \ x \in H_0 \\ h_2, \text{ if } x_\downarrow \in H_0, \ x \in H_1 \\ h_3, \text{ if } x_\downarrow \in H_1, \ x \in H_0 \\ h_4, \text{ if } x_\downarrow \in H_1, \ x \in H_1. \end{cases} \tag{11.49}$$

Here $h_i = (h_{i1}, h_{i2}, ..., h_{iq-1})$, $i = 1, 2, 3, 4$.

Remark 11.2. The definition of $x_\downarrow$ depends on the (fixed) root e of the Cayley tree. Therefore the function h_x given in (11.49) is defined for $x \neq e$.

By (11.2) we then have

$$\begin{cases} h_1 = (k - |A|)F(h_1, \theta, \alpha) + |A|F(h_2, \theta, \alpha) \\ h_2 = (|A| - 1)F(h_3, \theta, \alpha) + (k + 1 - |A|)F(h_4, \theta, \alpha) \\ h_3 = (|A| - 1)F(h_2, \theta, \alpha) + (k + 1 - |A|)F(h_1, \theta, \alpha) \\ h_4 = (k - |A|)F(h_4, \theta, \alpha) + |A|F(h_3, \theta, \alpha). \end{cases} \tag{11.50}$$

Before going to solutions of this system let us give examples geometrically presenting the vector-values of h_x on the Cayley tree.

Example 11.1. Let $A = \{1\}$. Then

$$H_A = H_{\{1\}} = \{x \in G_k : \ \text{number of} \ \ a_1 \ \ \text{in} \ \ x \ \ \text{is even}\},$$

In this case $H_0 = H_{\{1\}}$ (red points in Fig. 11.9) and $H_1 = G_k \setminus H_0$ is the set of words with an odd number of a_1 (green points in the Fig. 11.9).

In the Fig. 11.10 we showed the distribution of the four values of function (11.49).

Denote $z_{ij} = \exp h_{ij}, \lambda_j = \exp(\alpha_j), i = 1, 2, 3, 4, j = 1, 2, ...q - 1$. Then the last system of equations can be rewritten as

$$\begin{cases} z_{1j} = \lambda_j \left(\frac{(\theta-1)z_{1j}+\sum_{i=1}^{q-1} z_{1i}+1}{\sum_{i=1}^{q-1} z_{1i}+\theta}\right)^{k-|A|} \left(\frac{(\theta-1)z_{2j}+\sum_{i=1}^{q-1} z_{2i}+1}{\sum_{i=1}^{q-1} z_{2i}+\theta}\right)^{|A|} \\ z_{2j} = \lambda_j \left(\frac{(\theta-1)z_{3j}+\sum_{i=1}^{q-1} z_{3i}+1}{\sum_{i=1}^{q-1} z_{3i}+\theta}\right)^{|A|-1} \left(\frac{(\theta-1)z_{4j}+\sum_{i=1}^{q-1} z_{4i}+1}{\sum_{i=1}^{q-1} z_{4i}+\theta}\right)^{k+1-|A|} \\ z_{3j} = \lambda_j \left(\frac{(\theta-1)z_{2j}+\sum_{i=1}^{q-1} z_{2i}+1}{\sum_{i=1}^{q-1} z_{2i}+\theta}\right)^{|A|-1} \left(\frac{(\theta-1)z_{1j}+\sum_{i=1}^{q-1} z_{1i}+1}{\sum_{i=1}^{q-1} z_{1i}+\theta}\right)^{k+1-|A|} \\ z_{4j} = \lambda_j \left(\frac{(\theta-1)z_{4j}+\sum_{i=1}^{q-1} z_{4i}+1}{\sum_{i=1}^{q-1} z_{4i}+\theta}\right)^{k-|A|} \left(\frac{(\theta-1)z_{3j}+\sum_{i=1}^{q-1} z_{3i}+1}{\sum_{i=1}^{q-1} z_{3i}+\theta}\right)^{|A|}, \end{cases} \tag{11.51}$$

here $j = 1, 2, 3, .., q - 1$.

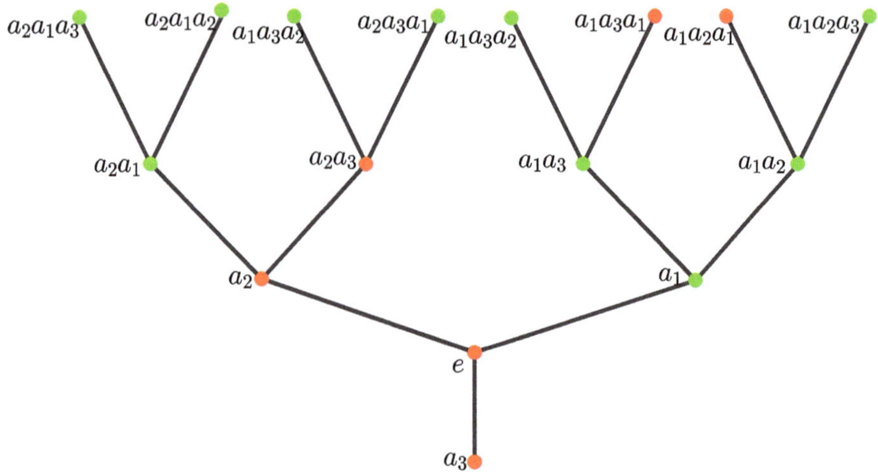

Fig. 11.9 Partition (red and green) of the Cayley tree with respect to subgroup $H_{\{1\}}$ of index two.

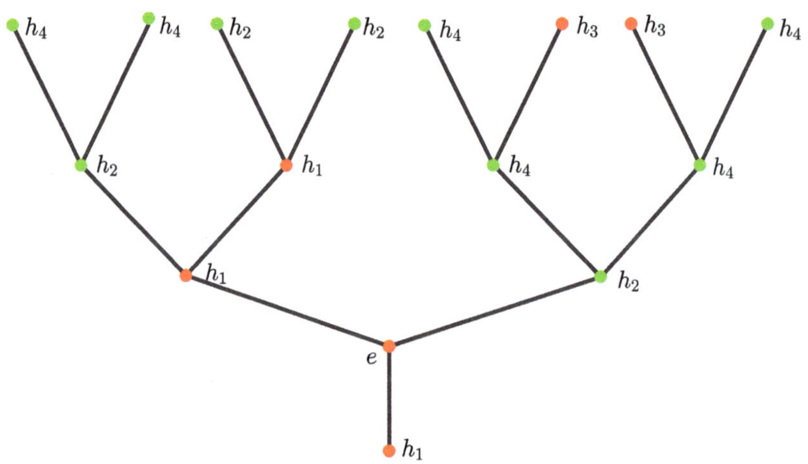

Fig. 11.10 The values of the weakly-periodic function (11.49) in case $A = \{1\}$, i.e., subgroup $H_{\{1\}}$ of index two.

Define the map $K : \mathbb{R}^{4(q-1)} \to \mathbb{R}^{4(q-1)}$, as

$$\begin{cases} z'_{1j} = \lambda_j \left(\frac{(\theta-1)z_{1j}+\sum_{i=1}^{q-1} z_{1i}+1}{\sum_{i=1}^{q-1} z_{1i}+\theta}\right)^{k-|A|} \left(\frac{(\theta-1)z_{2j}+\sum_{i=1}^{q-1} z_{2i}+1}{\sum_{i=1}^{q-1} z_{2i}+\theta}\right)^{|A|} \\ z'_{2j} = \lambda_j \left(\frac{(\theta-1)z_{3j}+\sum_{i=1}^{q-1} z_{3i}+1}{\sum_{i=1}^{q-1} z_{3i}+\theta}\right)^{|A|-1} \left(\frac{(\theta-1)z_{4j}+\sum_{i=1}^{q-1} z_{4i}+1}{\sum_{i=1}^{q-1} z_{4i}+\theta}\right)^{k+1-|A|} \\ z'_{3j} = \lambda_j \left(\frac{(\theta-1)z_{2j}+\sum_{i=1}^{q-1} z_{2i}+1}{\sum_{i=1}^{q-1} z_{2i}+\theta}\right)^{|A|-1} \left(\frac{(\theta-1)z_{1j}+\sum_{i=1}^{q-1} z_{1i}+1}{\sum_{i=1}^{q-1} z_{1i}+\theta}\right)^{k+1-|A|} \\ z'_{4j} = \lambda_j \left(\frac{(\theta-1)z_{4j}+\sum_{i=1}^{q-1} z_{4i}+1}{\sum_{i=1}^{q-1} z_{4i}+\theta}\right)^{k-|A|} \left(\frac{(\theta-1)z_{3j}+\sum_{i=1}^{q-1} z_{3i}+1}{\sum_{i=1}^{q-1} z_{3i}+\theta}\right)^{|A|}, \end{cases} \tag{11.52}$$

here $j = 1, 2, 3, .., q-1$.

Denote

$$I_m = \{(z_1, ..., z_{q-1}) \in \mathbb{R}^{q-1} : z_1 = ... = z_m, z_{m+1} = ... = z_{q-1} = 1\}, \tag{11.53}$$

$$M_m = \{(z^{(1)}, z^{(2)}, z^{(3)}, z^{(4)}) \in \mathbb{R}^{4(q-1)} : z^{(i)} \in I_m, i = 1, 2, 3, 4\}, \tag{11.54}$$

here $m = 1, 2, ..., q-1$.

Lemma 11.5. *We have*

(1) *For any fixed $m \geq 1$, $\lambda > 0$ and*

$$\lambda_i = \begin{cases} \lambda & if \ \ 1 \leq i \leq m \\ 1 & if \ \ m < i \leq q-1 \end{cases} \tag{11.55}$$

the set M_m is an invariant set with respect to the map K, i.e. $K(M_m) \subset M_m$.

(2) *For $\alpha = 0$ the sets M_m are invariant with respect to the map K for all $m = 1, 2, ..., q-1$.*

Proof.

(1) For the fixed m, λ_i given by (11.55) and $\mathbf{z} = (z^{(1)}, z^{(2)}, z^{(3)}, z^{(4)}) \in M_m$ it is clear, that $z^{(i)} \in I_m, i = 1, 2, 3, 4$. From (11.53) we have, that $z^{(i)} = (z_i, ..., z_i, 1, ..., 1)$, where $z_i \neq 1, i = 1, 2, 3, 4$. Using (11.52) for any $j = m, m+1, ...q-1$ we obtain

$$\begin{cases} z'_{1j} = \left(\frac{\theta+q-2+z_1}{\theta+q-2+z_1}\right)^{k-|A|} \left(\frac{\theta+q-2+z_2}{\theta+q-2+z_2}\right)^{|A|} = 1, \\ z'_{2j} = \left(\frac{\theta+q-2+z_3}{\theta+q-2+z_3}\right)^{|A|-1} \left(\frac{\theta+q-2+z_4}{\theta+q-2+z_4}\right)^{k+1-|A|} = 1, \\ z'_{3j} = \left(\frac{\theta+q-2+z_2}{\theta+q-2+z_2}\right)^{|A|-1} \left(\frac{\theta+q-2+z_4}{\theta+q-2+z_4}\right)^{k+1-|A|} = 1, \\ z'_{4j} = \left(\frac{\theta+q-2+z_4}{\theta+q-2+z_4}\right)^{k-|A|} \left(\frac{\theta+q-2+z_3}{\theta+q-2+z_3}\right)^{|A|} = 1. \end{cases}$$

Consequently, $K(\mathbf{z}) \in M_m$.

The proof of the second part of lemma is similar. □

Case $\alpha \neq 0$. Let us consider the case $\alpha \neq 0$ and λ_i given by (11.55). For $\mathbf{z} \in M_m$ we denote $z_i = z_{ij}, i = 1,2,3,4; j = 1,2,...,m$.. Then on the invariant set M_m the system of equations (11.51) has the following form

$$\begin{cases} z_1 = \lambda\left(\frac{(\theta+m-1)z_1+q-m}{mz_1+\theta+q-m-1}\right)^{k-|A|}\left(\frac{(\theta+m-1)z_2+q-m}{mz_2+\theta+q-m-1}\right)^{|A|} \\ z_2 = \lambda\left(\frac{(\theta+m-1)z_3+q-m}{mz_3+\theta+q-m-1}\right)^{|A|-1}\left(\frac{(\theta+m-1)z_4+q-m}{mz_4+\theta+q-m-1}\right)^{k+1-|A|} \\ z_3 = \lambda\left(\frac{(\theta+m-1)z_2+q-m}{mz_2+\theta+q-m-1}\right)^{|A|-1}\left(\frac{(\theta+m-1)z_1+q-m}{mz_1+\theta+q-m-1}\right)^{k+1-|A|} \\ z_4 = \lambda\left(\frac{(\theta+m-1)z_4+q-m}{mz_4+\theta+q-m-1}\right)^{k-|A|}\left(\frac{(\theta+m-1)z_3+q-m}{mz_3+\theta+q-m-1}\right)^{|A|}. \end{cases} \tag{11.56}$$

We introduce the notation

$$g_m(z) = \frac{(\theta+m-1)z+q-m}{mz+\theta+q-m-1}.$$

It is easy to prove the following:

Lemma 11.6. *The function $g_m(z)$ is strictly decreasing for $0 < \theta < 1$, $1 \leq m \leq q-1$ and strictly increasing for $1 < \theta$.*

Proposition 11.1. *Let $\mathbf{z} = (z_1, z_2, z_3, z_4)$ is a solution of the system of equations (11.56). If $z_i = z_j$ for some $i \neq j$, then $z_1 = z_2 = z_3 = z_4$.*

Proof.

From the system of equations (11.56), we obtain

$$\frac{z_1}{z_2} = \left(\frac{g_m(z_1)}{g_m(z_4)}\right)^{k-|A|}\left(\frac{g_m(z_2)}{g_m(z_3)}\right)^{|A|-1}\left(\frac{g_m(z_2)}{g_m(z_4)}\right), \tag{a_1}$$

$$\frac{z_1}{z_3} = \left(\frac{g_m(z_2)}{g_m(z_1)}\right), \tag{a_2}$$

$$\frac{z_1}{z_4} = \left(\frac{g_m(z_1)}{g_m(z_4)}\right)^{k-|A|}\left(\frac{g_m(z_2)}{g_m(z_3)}\right)^{|A|}, \tag{a_3}$$

$$\frac{z_2}{z_3} = \left(\frac{g_m(z_3)}{g_m(z_2)}\right)^{|A|-1}\left(\frac{g_m(z_4)}{g_m(z_1)}\right)^{k-|A|+1}, \tag{a_4}$$

$$\frac{z_2}{z_4} = \left(\frac{g_m(z_4)}{g_m(z_3)}\right), \tag{a_5}$$

$$\frac{z_3}{z_4} = \left(\frac{g_m(z_1)}{g_m(z_4)}\right)^{k-|A|} \left(\frac{g_m(z_2)}{g_m(z_3)}\right)^{|A|-1} \left(\frac{g_m(z_1)}{g_m(z_3)}\right). \qquad (a_6)$$

Let $\mathbf{z} = \{z_1, z_2, z_3, z_4\}$ be a solution of system of equations (11.56) and $z_1 = z_2$. Then from strictly monotonicity of $g_m(z)$ and from equality (a_2) we obtain, that $z_1 = z_2 = z_3$. In this case from (a_4) we obtain $z_1 = z_4$, consequently $z_1 = z_2 = z_3 = z_4$.

Let $z_1 = z_3$. Then from strictly monotonicity of $g_m(z)$ and from equality (a_2) we obtain, that $z_1 = z_2 = z_3$. In this case from (a_4) we obtain $z_1 = z_4$, consequently, $z_1 = z_2 = z_3 = z_4$.

Let $z_1 = z_4$. Then from strictly monotonicity of $g_m(z)$ and from equality (a_3) we obtain, that $z_2 = z_3$. Then from (a_5) we obtain the following inequality

$$z_2 g_m(z_2) = z_4 g_m(z_4). \qquad (11.57)$$

We consider the function

$$\phi(z) = z g_m(z) = z\frac{(\theta + m - 1)z + q - m - 1}{mz + \theta + q - m - 1},$$

and calculate its derivative

$$\phi'(z) = [(\theta - 1 + m)mz^2 + 2(\theta - 1 + m)(\theta + q - m - 1)z \\ + (q - m)(\theta + q - m - 1)] \times [(mz + \theta + q - m - 1)^2]^{-1}.$$

From $\theta > 0,\ \ z > 0,\ \ 1 \le m \le q - 1$ and $q \ge 2$ we obtain, that $\phi(z)$ is a strictly increasing function. Consequently (11.57) is true only when $z_2 = z_4$.

The other cases are proved similarly. □

Consider now the anti-ferromagnetic Potts model (i.e. $0 < \theta < 1$).

Let $|A| = k$. Then system of equation (11.56) has the form

$$\begin{cases} z_1 = \lambda\left(g_m(z_2)\right)^k \\ z_2 = \lambda\left(g_m(z_3)\right)^{k-1}\left(g_m(z_4)\right) \\ z_3 = \lambda\left(g_m(z_2)\right)^{k-1}\left(g_m(z_1)\right) \\ z_4 = \lambda\left(g_m(z_3)\right)^k. \end{cases} \qquad (11.58)$$

Solving the system (11.58) can be reduced to analyzing of the following system of equations

$$\begin{cases} z_2 = \lambda\left(g_m(z_3)\right)^{k-1} \cdot g_m(\lambda(g_m(z_3))^k) \\ z_3 = \lambda\left(g_m(z_2)\right)^{k-1} \cdot g_m(\lambda(g_m(z_2))^k). \end{cases} \qquad (11.59)$$

Introducing the notation

$$\psi(z) = \lambda\left(g_m(z)\right)^{k-1} \cdot g_m(\lambda(g_m(z))^k). \tag{11.60}$$

Then we reduce the system of equations (11.59) to the form

$$\begin{cases} z_2 = \psi(z_3) \\ z_3 = \psi(z_2). \end{cases} \tag{11.61}$$

The number of solutions of the system (11.61) coincides with the number of solutions of the equation $\psi(\psi(z)) = z$.

The following lemma is useful (see [121]):

Lemma 11.7. *Let $\gamma : [0,1] \to [0,1]$ be a continuous function with a fixed point $\xi \in (0,1)$. Assuming that the function γ is differentiable at $\xi \in (0,1)$ and that $\gamma'(\xi) < -1$, we have values x_0, x_1 such that the inequalities $0 \leq x_0 < \xi < x_1 \leq 1$ hold and $\gamma(x_0) = x_1$, $\gamma(x_1) = x_0$.*

It is known (Theorem 8.2, see also [243]) that for anti-ferromagnetic case, there exists a unique translation-invariant Gibbs measure corresponding to the unique solution of the equation $z = \lambda g_m^k(z)$. We let z_* denote this solution.

Proposition 11.2. *For $k \geq 6$ and $\lambda \in (\lambda_{c_1}, \lambda_{c_2})$ the system of equations (11.61) has three solutions $(z_*, z_*), (z_2^*, z_3^*), (z_3^*, z_2^*)$, where $\lambda_{c_i} = b_i^k, i = 1, 2$ and*

$$\begin{aligned} b_1 &= \frac{(k-1-\sqrt{k^2-6k+1})(1-\theta)(\theta+q-1)z_*^{\frac{k-1}{k}}}{2(mz_*+\theta+q-m-1)^2}, \\ b_2 &= \frac{(k-1+\sqrt{k^2-6k+1})(1-\theta)(\theta+q-1)z_*^{\frac{k-1}{k}}}{2(mz_*+\theta+q-m-1)^2}. \end{aligned} \tag{11.62}$$

Proof. For function (11.60) it is easy to see that

(1) $\psi(z_*) = z_*$.

(2) function $\psi(z)$ is defined on $\mathbb{R}_+$, is bounded and differentiable at z_*.

Then by Kesten's lemma (Lemma 11.7), for $\psi'(z_*) < -1$, the equation (11.61) has three solutions of the form $(z_*, z_*), (z_2^*, z_3^*), (z_3^*, z_2^*)$.

The inequality $\psi'(z_*) < -1$ equivalent to the inequality

$$b^2 + (k-1)\frac{(\theta-1)(\theta+q-1)z_*^{\frac{k-1}{k}}}{(mz_*+\theta+q-m-1)^2}b$$

$$+k\frac{(1-\theta)^2(\theta+q-1)^2 z_*^{2\frac{k-1}{k}}}{(mz_*+\theta+q-m-1)^4} < 0,$$

where $b = \sqrt[k]{\lambda}$. Consequently,

$$(b - b_1)(b - b_2) < 0,$$

where b_1, b_2 are defined in (11.62). Proposition is proved. □

We have thus the following theorem.

Theorem 11.10. *Let $|A| = k, k \geq 6$ and $\lambda \in (\lambda_{c_1}, \lambda_{c_2})$. Then for the anti-ferromagnetic Potts model with the special external field (given by (11.55)) there are at last two H_A — weakly periodic (non-periodic) Gibbs measures, where $\lambda_{c_i} = b_i^k, i = 1, 2$.*

Case $\alpha = 0$. In this case the system of equations (11.51) on the invariant set $M_m, m = 1, 2, ..., q-1$ can be reduced to the following system of equations:

$$\begin{cases} z_1 = \left(\frac{(\theta+m-1)z_1+q-m}{mz_1+\theta+q-m-1}\right)^{k-|A|} \left(\frac{(\theta+m-1)z_2+q-m}{mz_2+\theta+q-m-1}\right)^{|A|} \\ z_2 = \left(\frac{(\theta+m-1)z_3+q-m}{mz_3+\theta+q-m-1}\right)^{|A|-1} \left(\frac{(\theta+m-1)z_4+q-m}{mz_4+\theta+q-m-1}\right)^{k+1-|A|} \\ z_3 = \left(\frac{(\theta+m-1)z_2+q-m}{mz_2+\theta+q-m-1}\right)^{|A|-1} \left(\frac{(\theta+m-1)z_1+q-m}{mz_1+\theta+q-m-1}\right)^{k+1-|A|} \\ z_4 = \left(\frac{(\theta+m-1)z_4+q-m}{mz_4+\theta+q-m-1}\right)^{k-|A|} \left(\frac{(\theta+m-1)z_3+q-m}{mz_3+\theta+q-m-1}\right)^{|A|}. \end{cases} \tag{11.63}$$

The following proposition is similar to Proposition 11.1.

Proposition 11.3. *Let $m \in \{1, 2, ..., q-1\}$ be fixed and $\boldsymbol{z} = (z_1, z_2, z_3, z_4)$ is a solution of the system of equations (11.63). If $z_i = z_j$ for some $i \neq j$, then $z_1 = z_2 = z_3 = z_4$.*

Theorem 11.11. *Let $|A| = k$ and $k \geq 6$. If one of the following conditions is satisfied*

(1) $\frac{4k}{k+1+\sqrt{k^2-6k+1}} \leq q < \frac{4k}{k+1-\sqrt{k^2-6k+1}}$ *and* $0 < \theta < \theta_2$*;*
(2) $q \leq \frac{4k}{k+1+\sqrt{k^2-6k+1}}$ *and* $\theta_1 < \theta < \theta_2$*,*

then there are at last $2^q - 2$ weakly periodic (non-periodic) Gibbs measures, where

$$\begin{aligned} \theta_1 &= \frac{4\,k - kq - q - q\sqrt{k^2 - 6\,k + 1}}{4k}, \\ \theta_2 &= \frac{4\,k - kq - q + q\sqrt{k^2 - 6\,k + 1}}{4k}. \end{aligned} \tag{11.64}$$

Proof.
Consider the case $0 < \theta < 1$ and $|A| = k$, then the system of equations (11.63) has the following form

$$\begin{cases} z_1 = (g_m(z_2))^k \\ z_2 = (g_m(z_3))^{k-1}(g_m(z_4)) \\ z_3 = (g_m(z_2))^{k-1}(g_m(z_1)) \\ z_4 = (g_m(z_3))^k . \end{cases} \tag{11.65}$$

Reduce (11.65) to the following system

$$\begin{cases} z_2 = (g_m(z_3))^{k-1} \cdot g_m((g_m(z_3))^k) \\ z_3 = (g_m(z_2))^{k-1} \cdot g_m((g_m(z_2))^k). \end{cases} \tag{11.66}$$

Introducing the notation

$$\varphi(z) = (g_m(z))^{k-1} \cdot g_m((g_m(z))^k), \tag{11.67}$$

the system of equations (11.66) can by written as the following

$$\begin{cases} z_2 = \varphi(z_3) \\ z_3 = \varphi(z_2). \end{cases} \tag{11.68}$$

Note that the following hold

(1) $\varphi(1) = 1$.

(2) the function $\varphi(z)$ is defined on $\mathbb{R}_+$, is bounded and differentiable at $z = 1$.

Therefore by Lemma 11.7, for $\varphi'(1) < -1$, the system of equations (11.68) has three solutions of the form

$$(1, 1), (z_2^*, z_3^*), (z_3^*, z_2^*).$$

The inequality $\varphi'(1) < -1$ equivalent to the inequality

$$k\frac{(\theta-1)^2}{(\theta+q-1)^2} + (k-1)\frac{(\theta-1)}{(\theta+q-1)} + 1 < 0, \tag{11.69}$$

Consequently,

$$2k(\theta - \theta_1)(\theta - \theta_2) < 0,$$

where θ_1, θ_2 are defined in (11.64).

It is clear, that if $k < 5$, then θ_1, θ_2 are complex numbers, if $k = 5$, then $\theta_1 = \theta_2$ and the inequality (11.69) has no solution.

We now assume that $k \geq 6$. We suppose that θ_1, θ_2 are simultaneously negative. Then (11.69) has no solution. If $\theta_1 \leq 0, 0 < \theta_2 < 1$, i.e.,

$$\frac{4k}{k+1+\sqrt{k^2-6k+1}} \leq q < \frac{4k}{k+1-\sqrt{k^2-6k+1}},$$

then the inequality (11.69) has a solution $\theta \in (0, \theta_2)$, which proves item 1 of theorem.

Let $0 \leq \theta_1 < 1, 0 < \theta_2 < 1$, then the inequality

$$q \leq \frac{4k}{k+1+\sqrt{k^2-6k+1}}$$

is satisfied.

The inequality (11.69) has solution $\theta_1 < \theta < \theta_2$. It is easy to see that θ_1, θ_2 less or equal to 1. Thus, for every m, under conditions of theorem we obtain two weakly periodic (non-periodic) Gibbs measures.

From (11.53) we note, that m is number of coordinates other than 1 of the vectors from $\mathbb{R}^{q-1}$. It is obvious that the number of such vectors is equal to $\sum_{m=1}^{q-1} \binom{q-1}{m} = 2^{q-1} - 1$. Consequently, when the conditions of theorem are satisfied we obtain $2(2^{q-1} - 1) = 2^q - 2$ weakly periodic (non-periodic) Gibbs measures. Theorem is proved. □

Remark 11.3. The new Gibbs measures described in Theorem 11.10 and Theorem 11.11 allow to describing an uncountable set of non-periodic Gibbs measures (see Chapter 2 of [243] for such a construction) different from the previously known ones.

Remark 11.4. For the case $m = q - 1$ (see (11.54)), Theorem 11.11 coincides with Theorem 3 in [223].

Chapter 12

Free energies of the Potts model

In this chapter, for the Potts model on the Cayley tree, some explicit formulae of the free energies and entropies (according to vector-valued boundary conditions) are obtained. They include translation-invariant, periodic, Dobrushin-like boundary conditions, as well as those corresponding to weakly periodic Gibbs measures.

12.1 Definitions and equations

Note that on Cayley trees, not only Gibbs measures but also the free energy (and the entropy) depend on the boundary condition (BC). A study of this dependence is given in [78]. Namely, for Ising model, it is shown that for all previously known BCs the free energies exist. Later, in [79] a construction of new Gibbs measures (called alternating Gibbs measures) is presented and their corresponding free energies are given. Moreover, it was proved that free energy of some alternating Gibbs measures may not exist.

In this chapter, following [247], we study free energies of the Potts model on the Cayley tree.

Recall some definitions. Let $\Gamma^k = (V, L, i)$, be a Cayley tree, where V is the set of vertices of Γ^k, L is the set of edges of Γ^k and i is the incidence function associating each edge $l \in L$ with its endpoints $x, y \in V$. If $i(l) = \{x, y\}$, then x and y are called *nearest neighboring vertices*, and we write $l = \langle x, y \rangle$.

As it was mentioned in Chapter 2 there exists a one-to-one correspondence between the set V of vertices of the Cayley tree of order $k \geq 1$ and the group G_k of the free products of $k+1$ cyclic groups $\{e, a_i\}$, $i = 1, ..., k+1$ of the second order (i.e., $a_i^2 = e$, $a_i^{-1} = a_i$) with generators $a_1, a_2, ..., a_{k+1}$.

Here we represent the Hamiltonian of the Potts model by scalar product of vectors: identify the set Φ of spins by the set $\{\sigma_1, ..., \sigma_q\}$, where $\sigma_i \in \mathbb{R}^{q-1}$ such that

$$\sigma_i\sigma_j = \begin{cases} -\frac{1}{q-1}, & \text{if } i \neq j \\ 1, & \text{if } i = j. \end{cases} \tag{12.1}$$

Then we have

$$\delta_{\sigma(x)\sigma(y)} = \frac{q-1}{q}\left(\sigma(x)\sigma(y) + \frac{1}{q-1}\right). \tag{12.2}$$

Using this formula the Hamiltonian of the Potts model can be reduced to the Hamiltonian of the Ising model with q spin values:

$$H(\sigma) = -J \sum_{\langle x,y\rangle \in L} \sigma(x)\sigma(y). \tag{12.3}$$

We fix a basis $\{e_1, \ldots, e_{q-1}\}$ on $\mathbb{R}^{q-1}$, such that $e_i = \sigma_i, i = 1, 2, \ldots, q-1$.

Lemma 12.1. *The following equality holds*

$$\sum_{i=1}^{q} \sigma_i = 0. \tag{12.4}$$

Proof. Let $\sigma_q = \sum_{i=1}^{q-1} \gamma_i e_i = \sum_{i=1}^{q-1} \gamma_i\sigma_i$. Then by (12.1) we get

$$\sigma_q\sigma_j = \sum_{i=1}^{q-1} \gamma_i\sigma_i\sigma_j = \begin{cases} -\frac{1}{q-1}\sum_{\substack{i=1\\ i\neq j}}^{q-1} \gamma_i + \gamma_j, & \text{if } j \leq q-1 \\ -\frac{1}{q-1}\sum_{i=1}^{q-1} \gamma_i, & \text{if } j = q. \end{cases} \tag{12.5}$$

On the other hand (again by (12.1)) we have

$$\sigma_q\sigma_j = \begin{cases} -\frac{1}{q-1}, & \text{if } j \leq q-1 \\ 1, & \text{if } j = q. \end{cases} \tag{12.6}$$

Comparing (12.5) and (12.6) we get the system

$$\begin{aligned} &-\tfrac{1}{q-1}\textstyle\sum_{\substack{i=1\\ i\neq j}}^{q-1} \gamma_i + \gamma_j = -\tfrac{1}{q-1}, \quad j \leq q-1 \\ &-\tfrac{1}{q-1}\textstyle\sum_{i=1}^{q-1} \gamma_i = 1 \end{aligned} \tag{12.7}$$

which has unique solution $\gamma_i = -1$, for all $i \leq q-1$. Therefore, $\sigma_q = -\sum_{i=1}^{q-1} \sigma_i$. This completes the proof. □

Define a finite-dimensional distribution of a probability measure μ in the volume V_n as

$$\mu_n(\sigma_n) = Z_n^{-1} \exp\left\{ -\beta H_n(\sigma_n) + \sum_{x \in W_n} h_x \sigma(x) \right\}, \tag{12.8}$$

where $\beta = 1/T$, $T > 0$ is the temperature, $h_x \in \mathbb{R}^{q-1}$,

$$H_n(\sigma_n) = -J \sum_{\langle x,y \rangle \in L_n} \sigma(x)\sigma(y),$$

and Z_n^{-1} is the normalizing factor, i.e.,

$$Z_n = Z_n(\beta, h) = \sum_{\sigma_n \in \Omega_n} \exp\left(-\beta H_n(\sigma_n) + \sum_{x \in W_n} h_x \sigma(x) \right).$$

The collection of vectors $h = \{h_x \in \mathbb{R}^{q-1}, x \in V\}$ stands for (generalized) BC.

The following limit (if exists) is called *free energy* corresponding to BC h:

$$E(\beta, h) = - \lim_{n \to \infty} \frac{1}{\beta |V_n|} \ln Z_n(\beta, h).$$

We say that probability distributions (12.8) are compatible if for all $n \geq 1$ and $\sigma_{n-1} \in \Phi^{V_{n-1}}$ we have

$$\sum_{\sigma^{(n)} \in \Phi^{W_n}} \mu_n(\sigma_{n-1} \vee \sigma^{(n)}) = \mu_{n-1}(\sigma_{n-1}), \tag{12.9}$$

where $\sigma_{n-1} \vee \sigma^{(n)}$ is the concatenation of the configurations.

The next statement describes the condition on h_x ensuring that $\mu_n(\sigma_n)$ are compatible.

Theorem 12.1. *Measures (12.8) satisfy (12.9) if only if for all $x \in V \setminus \{x^0\}$ the following equation holds:*

$$h_x = \sum_{y \in S(x)} F(h_y, \theta), \tag{12.10}$$

where $F : h = (h_1, \ldots, h_{q-1}) \in \mathbb{R}^{q-1} \to F(h, \theta) = (F_1, \ldots, F_{q-1}) \in \mathbb{R}^{q-1}$ is defined as

$$F_i = \ln \left(\frac{(\theta - 1)e^{h_i} + \sum_{j=1}^{q-1} e^{h_j} + 1}{\theta + \sum_{j=1}^{q-1} e^{h_j}} \right), \qquad \theta = \exp(J\beta).$$

Proof. We already have proof of more general form of this theorem, i.e., Theorem 2.2. Here we shall prove (12.10) for Hamiltonian (12.3). Because, a formula obtained in this proof will be helpful to find a general form of free energy.

Substituting (12.8) in (12.9), in view of (12.2) we get

$$\frac{Z_{n-1}}{Z_n}\sum_{\sigma^{(n)}}\exp\left(J\beta\sum_{x\in W_{n-1}}\left(\sum_{y\in S(x)}\sigma(x)\sigma(y)\right)+\sum_{x\in W_{n-1}}\left(\sum_{y\in S(x)}h_y\sigma(y)\right)\right)$$
$$=\exp\left(\sum_{x\in W_{n-1}}h_x\sigma(x)\right),\quad \sigma(x)\in\Phi.$$

Consequently,

$$\frac{Z_{n-1}}{Z_n}\prod_{x\in W_{n-1}}\prod_{y\in S(x)}\sum_{\sigma(y)}\exp\left(J\beta\sigma(x)\sigma(y)+h_y\sigma(y)\right)$$
$$=\prod_{x\in W_{n-1}}\exp(h_x\sigma(x)),\quad \sigma(x)\in\Phi. \tag{12.11}$$

Fix $x\in W_{n-1}$ and rewrite (12.11) for $\sigma(x)=\sigma_i$, $i=1,\dots,q-1$ and $\sigma(x)=\sigma_q$. Then dividing each of them to the last one we get

$$\prod_{y\in S(x)}\frac{\sum_{\sigma(y)}\exp((J\beta\sigma_i+h_y)\sigma(y))}{\sum_{\sigma(y)}\exp((J\beta\sigma_q+h_y)\sigma(y))}=\frac{\exp(h_x\sigma_i)}{\exp(h_x\sigma_q)},\, i=1,2,...,q-1. \tag{12.12}$$

For any vector $h=(h_1,\dots,h_{q-1})$, using (12.1) we get

$$h\sigma_i=\begin{cases}\frac{q}{q-1}h_i-\frac{1}{q-1}\sum_{j=1}^{q-1}h_j, & \text{if } i=1,\dots,q-1,\\ -\frac{1}{q-1}\sum_{j=1}^{q-1}h_j, & \text{if } i=q.\end{cases}$$

Therefore $h_x\sigma_i-h_x\sigma_q=\frac{q}{q-1}h_{x,i}$, where $h_x=(h_{x,1},\dots,h_{x,q-1})$. Now we change the variables as $\frac{q}{q-1}h_{x,i}\to h_{x,i}$ then we get the equation (12.10).

For a proof of the inverse statement see proof of Theorem 2.2 or [243]. □

In this chapter, we compute free energies which correspond to translation-invariant, periodic, weakly periodic (see Definition 11.3 and some non-periodic BCs (Gibbs measures)).

12.2 Formula of free energy

The following theorem gives a general form of the free energy.

Theorem 12.2. *For BCs satisfying (12.9), the free energy is given by the formula*

$$E(\beta, h) = -\lim_{n\to\infty} \frac{1}{|V_n|} \sum_{x\in V_n} a(x), \tag{12.13}$$

where

$$a(x) = \frac{1}{q\beta} \sum_{i=1}^{q} \ln \left(\sum_{u=1}^{q} \exp\{(J\beta\sigma_i + h_x)\sigma_u\} \right). \tag{12.14}$$

Proof. In view of (12.12) we have

$$\prod_{y\in S(x)} \sum_{u=1}^{q} \exp(J\beta\sigma_i\sigma_u + \sigma_u h_y) = b(x)\exp(\sigma_i h_x), \quad i = 1, \ldots, q.$$

Multiplying all such equalities and using (12.4) we obtain

$$\begin{aligned} b^q(x) &= \prod_{i=1}^{q} \prod_{y\in S(x)} \sum_{u=1}^{q} \exp(J\beta\sigma_i\sigma_u + \sigma_u h_y) \\ &= \prod_{y\in S(x)} \prod_{i=1}^{q} \sum_{u=1}^{q} \exp(J\beta\sigma_i + h_y)\sigma_u, \end{aligned}$$

hence

$$b(x) = \prod_{y\in S(x)} \prod_{i=1}^{q} \left(\sum_{u=1}^{q} \exp(J\beta\sigma_i + h_y)\sigma_u \right)^{1/q}.$$

Let $A_n = \prod_{x\in W_n} b(x)$. It is clear that $Z_n = A_{n-1}Z_{n-1}$. Consequently

$$\begin{aligned} Z_n &= \prod_{x\in W_{n-1}} b(x) Z_{n-1} \\ &= \prod_{x\in W_{n-1}} b(x) \prod_{x\in W_{n-2}} b(x) Z_{n-2} = \ldots = \prod_{x\in V_{n-1}} b(x), \end{aligned}$$

and

$$\ln Z_n = \sum_{x\in V_{n-1}} \ln b(x).$$

Hence we obtain that

$$a(x) = \frac{1}{q\beta} \sum_{i=1}^{q} \ln \left(\sum_{u=1}^{q} \exp\{(J\beta\sigma_i + h_x)\sigma_u\} \right).$$

□

12.3 Concrete free energies

12.3.1 *Translation-invariant case*

Consider translation-invariant set of vectors h_x, i.e.,

$$h_x = h = (h_1, h_2, ..., h_{q-1}) \in \mathbb{R}^{q-1}, \quad \forall x \in G_k.$$

Then from (12.10) we obtain

$$h_i = k \ln\left(\frac{(\theta-1)e^{h_i} + \sum_{j=1}^{q-1} e^{h_j} + 1}{\theta + \sum_{j=1}^{q-1} e^{h_j}}\right), \quad i = 1, \dots, q-1. \tag{12.15}$$

Denoting $z_i = \exp(h_i), i = 1, \dots, q-1$, we get from (12.15)

$$z_i = \left(\frac{(\theta-1)z_i + \sum_{j=1}^{q-1} z_j + 1}{\theta + \sum_{j=1}^{q-1} z_j}\right)^k, \quad i = 1, 2, ..., q-1. \tag{12.16}$$

For $k = 2$ we denote

$$\begin{aligned} x_1(m) &= \frac{\theta - 1 - \sqrt{(\theta-1)^2 - 4m(q-m)}}{2m}, \\ x_2(m) &= \frac{\theta - 1 + \sqrt{(\theta-1)^2 - 4m(q-m)}}{2m}, \end{aligned} \tag{12.17}$$

where

$$\theta \geq \theta_m = 1 + 2\sqrt{m(q-m)}, \quad m = 1, \dots, q-1.$$

It is easy to see that

$$\theta_m = \theta_{q-m} \text{ and } \theta_1 < \theta_2 < \cdots < \theta_{[\frac{q}{2}]-1} < \theta_{[\frac{q}{2}]} \leq q+1.$$

Let $k = 2$, $J > 0$, then the following statements are known (see Chapter 9 or/and [138]).

1. If $\theta < \theta_1$, then the system of equations (12.15) has a unique solution $h_0 = (0, 0, \dots, 0)$;
2. If $\theta_m < \theta < \theta_{m+1}$ for some $m = 1, \dots, [\frac{q}{2}] - 1$, then the system of equations (12.15) has solutions

$$h_0 = (0, 0, \dots, 0), \quad h_{1i}(s), \quad h_{2i}(s), \quad i = 1, \dots, \binom{q-1}{s},$$

$$h'_{1i}(q-s), \quad h'_{2i}(q-s), \quad i = 1, \dots, \binom{q-1}{q-s}, \quad s = 1, 2, \dots, m,$$

where $h_{ji}(s)$, (resp. $h'_{ji}(q-s)$) $j = 1, 2$ is a vector with s (resp. $q-s$) coordinates equal to $2\ln x_j(s)$ (resp. $2\ln x_j(q-s)$) and remaining $q-s-1$ (resp. $s-1$) coordinates equal to 0. The number of such solutions is equal to

$$1 + 2\sum_{s=1}^{m} \binom{q}{s};$$

3. If $\theta_{[\frac{q}{2}]} < \theta \neq q+1$, then there are $2^q - 1$ solutions to (12.15);
4. If $\theta = q+1$ then the number of solutions is as the following

$$\begin{cases} 2^{q-1}, & \text{if } q \text{ is odd} \\ 2^{q-1} - \binom{q-1}{q/2}, & \text{if } q \text{ is even;} \end{cases}$$

5. If $\theta = \theta_m$, $m = 1, \ldots, [\frac{q}{2}]$, $(\theta_{[\frac{q}{2}]} \neq q+1)$ then $h_{1i}(m) = h_{2i}(m)$. The number of solutions is equal to

$$1 + \binom{q}{m} + 2\sum_{s=1}^{m-1}\binom{q}{s}.$$

Thus any solution of (12.15) has the form

$$h = (\underbrace{h_*, h_*, \ldots, h_*}_{m}, 0, 0, \ldots, 0), \quad m \geq 0, \tag{12.18}$$

up to a permutation of coordinates.

In this subsection we shall calculate free energies and entropy

$$S(\beta, h) = -\frac{dE(\beta, h)}{dT},$$

for the set of translation-invariant vectors $h_x = h$, with h given by (12.18).

Case $m = 0$. In this case $h = h_0 = (0, 0, ..., 0) \in \mathbb{R}^{q-1}$. From (12.13) we have

$$\begin{aligned} E_{TI}(\beta, h_0) = -a(x) &= -\frac{1}{q\beta}\sum_{i=1}^{q}\ln\left(\sum_{u=1}^{q}\exp(J\beta\sigma_i\sigma_u)\right) \\ &= -\frac{1}{q\beta}\sum_{i=1}^{q}\ln\left(\exp(J\beta) + (q-1)\exp\left(\frac{J\beta}{1-q}\right)\right) \\ &= -\frac{1}{q\beta}q\ln\left(\exp(J\beta) + (q-1)\exp\left(\frac{J\beta}{1-q}\right)\right) \\ &= -J - \frac{1}{\beta}\ln\left(1 + (q-1)\exp\left(\frac{Jq\beta}{1-q}\right)\right). \end{aligned}$$

Corresponding entropy has the form

$$\begin{aligned} S_{TI}(\beta, h_0) &= -\frac{dE_{TI}(\beta, h_0)}{dT} \\ &= \ln\left(1 + (q-1)\exp\left(\frac{Jq\beta}{1-q}\right)\right) + \frac{Jq\beta\exp\left(\frac{Jq\beta}{1-q}\right)}{1 + (q-1)\exp\left(\frac{Jq\beta}{1-q}\right)}. \end{aligned}$$

Case $m \neq 0$. Using (12.18) we shall calculate free energy:

$$E_{TI}(\beta, m, h_x) = -\lim_{n\to\infty} \frac{1}{|V_n|} \sum_{x\in V_n} a(x)$$
$$= -a(x) = -\frac{1}{q\beta} \sum_{i=1}^{q} \ln\left(\sum_{u=1}^{q} \exp\{(J\beta\sigma_i + h_x)\sigma_u\}\right)$$
$$= -\frac{q-m}{q\beta} \ln\left(m \cdot e^{\left(-\frac{J\beta}{q-1}+\frac{q-m}{q-1}h_*\right)} + e^{\left(J\beta-\frac{m}{q-1}h_*\right)} + (q-m-1)\cdot e^{\left(-\frac{J\beta}{q-1}-\frac{m}{q-1}h_*\right)}\right)$$
$$-\frac{m}{q\beta} \ln\left((m-1)\cdot e^{\left(-\frac{J\beta}{q-1}+\frac{q-m}{q-1}h_*\right)} + e^{\left(J\beta+\frac{q-m}{q-1}h_*\right)} + (q-m)\cdot e^{\left(-\frac{J\beta}{q-1}-\frac{m}{q-1}h_*\right)}\right). \quad (12.19)$$

Taking into account $h_* = 2\ln x_j(m)$, $j = 1, 2$ (where $x_j(m)$ are defined by (12.17)) we calculate corresponding entropy:

$$S_{TI}(\beta, m, h_x) = -\frac{dE_{TI}(\beta, m, h_x)}{dT} = -E_{TI}(\beta, m, h_x)\beta$$
$$+ \frac{J\beta(q-m)}{q\left(m\cdot e^{\left(-\frac{J\beta}{q-1}+\frac{q-m}{q-1}h_*\right)} + e^{\left(J\beta-\frac{m}{q-1}h_*\right)} + (q-m-1)\cdot e^{\left(-\frac{J\beta}{q-1}-\frac{m}{q-1}h_*\right)}\right)}$$
$$\times \left[m(1+(q-m)A)e^{\left(-\frac{J\beta}{q-1}+\frac{q-m}{q-1}h_*\right)} + (1-q-mA)e^{\left(J\beta-\frac{m}{q-1}h_*\right)} + (q-m-1)(1-mA)e^{\left(-\frac{J\beta}{q-1}-\frac{m}{q-1}h_*\right)}\right]$$
$$+ \frac{J\beta m}{q\left((m-1)\cdot e^{\left(-\frac{J\beta}{q-1}+\frac{q-m}{q-1}h_*\right)} + e^{\left(J\beta+\frac{q-m}{q-1}h_*\right)} + (q-m)\cdot e^{\left(-\frac{J\beta}{q-1}-\frac{m}{q-1}h_*\right)}\right)}$$
$$\times \left[(m-1)(1+(q-m)A)e^{\left(-\frac{J\beta}{q-1}+\frac{q-m}{q-1}h_*\right)} + ((m-q)A-1)e^{\left(J\beta-\frac{m}{q-1}h_*\right)} + (q-m)(1-mA)e^{\left(-\frac{J\beta}{q-1}-\frac{m}{q-1}h_*\right)}\right], \quad (12.20)$$

where $A = \dfrac{2e^{J\beta}}{(q-1)(e^{J\beta} - 2me^{\frac{h_*}{2}} - 1)}$.

Remark 12.1.

1. We note that the entropy for $2\ln x_1(m)$ and $2\ln x_2(m)$ can be calculated by the formula (12.20) replacing h_* by $2\ln x_1(m)$ and $2\ln x_2(m)$ respectively.

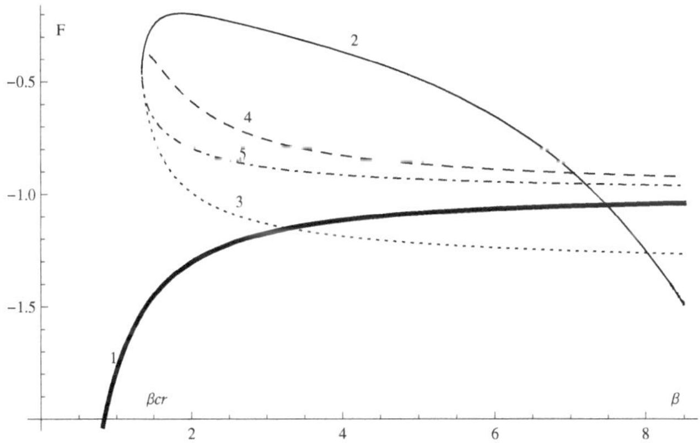

Fig. 12.1 Case $q = 3$. The free energy $F(\beta, h_0)$ the bold solid line (line 1); the free energy $F(\beta, (2\ln x_1(1), 0))$ solid line (line 2); the free energy $F(\beta, (2\ln x_2(1), 0))$ the dotted line (line 3); the free energy $F(\beta, (2\ln x_1(2), 2\ln x_1(2)))$ the dashed line (line 4); the free energy $F(\beta, (2\ln x_2(2), 2\ln x_2(2)))$ the dotted-dashed line (line 5), where $x_i(m), i = 1, 2, m = 1, 2$ defined in (12.17).

2. Notice that under any permutations of the coordinates of the vector h_x the free energy and entropy do not change.

In Fig. 12.1 graphs of the free energies (12.19) are shown.

12.3.2 *Non-translation-invariant BCs*

In this subsection, we consider non-translation-invariant BCs (Gibbs measures) constructed in [94]. Recall this construction: consider the half tree. Namely the root x^0 has k nearest neighbors.

Consider an infinite path $\pi = \{x^0 = x_0 < x_1 < \dots\}$ (the notation $x < y$ meaning that pathes from the root to y go through x).

Take two different solutions h_*^1 and h_*^2 of (12.15) (having the form (12.18)). Associate to this path a collection h^π of vectors given by the condition

$$h_x^\pi = \begin{cases} h_*^1, & \text{if } \ x \prec x_n, \, x \in W_n, \\ h_*^2, & \text{if } \ x_n \preceq x, \, x \in W_n, \end{cases} \tag{12.21}$$

$n = 1, 2, \dots$ where $x \prec x_n$ (resp. $x_n \prec x$) means that x is on the left (resp. right) from the path π (see Fig. 12.2).

Remark 12.2. In (12.21) on vertices of path π we chose the value h_*^2. But in fact (see [94] and [243]) the Gibbs measure corresponding to the path π do not depend on the values of h_x^π, $x \in \pi$.

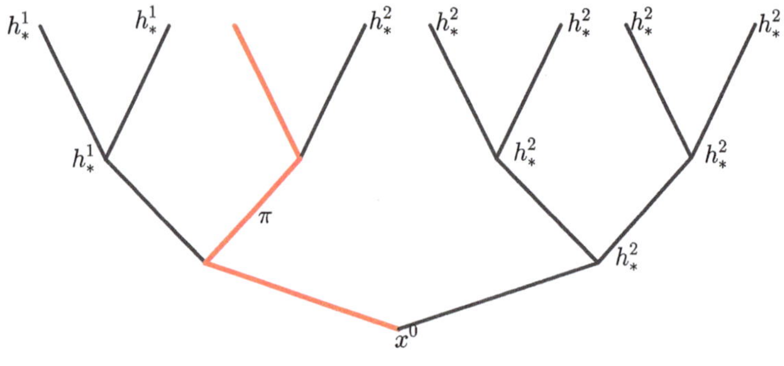

Fig. 12.2 Distribution of values of the function h_x^π on tree.

For a given infinite path π we put

$$\mathcal{W}_n^\pi = \{x \in W_n : x \prec x_n\},$$

where $n = 1, 2, \ldots$.

Lemma 12.2. *For any π there exists the limit*

$$\lim_{n\to\infty} \frac{|\mathcal{W}_n^\pi|}{|W_n|} = a^\pi. \tag{12.22}$$

Proof. Put

$$\Delta_1^\pi = \mathcal{W}_1^\pi; \quad \Delta_n^\pi = \mathcal{W}_n^\pi \cap S(x_{n-1}),$$

where $n \geq 2$.

It is easy to see that

$$|\mathcal{W}_n^\pi| = \sum_{i=1}^{n} |\Delta_i^\pi| \cdot |W_{n-i}|,$$

and $0 \leq |\Delta_i^\pi| \leq k-1$, for any i, π.

Let $a_n^\pi = \frac{|\mathcal{W}_n^\pi|}{|W_n|}$. Then using $|W_n| = k^{n-1}(k+1)$ we get

$$a_{n+1}^\pi - a_n^\pi = \frac{|\mathcal{W}_{n+1}^\pi|}{|W_{n+1}|} - \frac{|\mathcal{W}_n^\pi|}{|W_n|} = \frac{\Delta_{n+1}^\pi}{k^n(k+1)} \geq 0,$$

i.e., the sequence a_n^π is a monotone non-decreasing. It is easy to see that $a_n^\pi < 1$.

Consequently $\lim_{n\to\infty} \frac{|\mathcal{W}_n^\pi|}{|W_n|} = a^\pi$ exists and finite. □

We use the Stolz-Cesáro theorem (see e.g. [142]):

Theorem 12.3. *Let $(a_n)_{n\geq 1}$ and $(b_n)_{n\geq 1}$ be two sequences of real numbers. Assume that $(b_n)_{n\geq 1}$ is a strictly monotone and divergent sequence*[1] *and the*

[1] i.e., strictly increasing and approaching $+\infty$, or strictly decreasing and approaching $-\infty$.

following limit exists:

$$\lim_{n\to\infty}\frac{a_{n+1}-a_n}{b_{n+1}-b_n}=c.$$

Then, the following equality holds

$$\lim_{n\to\infty}\frac{a_n}{b_n}=c.$$

Denote $V_n^l=\cup_{i=1}^n\mathcal{W}_n^\pi$. From Lemma 12.2 by Theorem 12.3 we get

$$\lim_{n\to\infty}\frac{|V_n^l|}{|V_n|}=\lim_{n\to\infty}\frac{|\mathcal{W}_n^\pi|}{|W_n|}=a^\pi. \tag{12.23}$$

Now we calculate free energy for the set of vectors h_x^π. In view of (12.23) we get

$$\begin{aligned}E_G(\beta,m,h_x^\pi)&=-\lim_{n\to\infty}\frac{1}{|V_n|}\sum_{x\in V_n}a(x)\\&=-\frac{1}{q\beta}\lim_{n\to\infty}\frac{|V_n^l|}{|V_n|}\cdot\sum_{i=1}^q\ln\left(\sum_{u=1}^q\exp\{(J\beta\sigma_i+h_*^1)\sigma_u\}\right)\\&\quad-\frac{1}{q\beta}\lim_{n\to\infty}\left(1-\frac{|V_n^l|}{|V_n|}\right)\cdot\sum_{i=1}^q\ln\left(\sum_{u=1}^q\exp\{(J\beta\sigma_i+h_*^2)\sigma_u\}\right)\\&=a^\pi E_{TI}(\beta,m,h_*^1)+(1-a^\pi)E_{TI}(\beta,m,h_*^2),\end{aligned} \tag{12.24}$$

where $E_{TI}(\beta,m,h)$ is defined in (12.19).

Using (12.20) and (12.24) we get the following formula for the corresponding entropy:

$$S_G(\beta,m,h_x^\pi)=a^\pi S_{TI}(\beta,m,h_x^1)+(1-a^\pi)S_{TI}(\beta,m,h_x^2).$$

12.3.3 *Periodic BCs*

In this subsection consider periodic BCs and calculating free energies for them.

We consider the case $q=3$. It is known (see Chapter 12 or [244]) that there are only $G_k^{(2)}$ — periodic Gibbs measures, where $G_k^{(2)}$ is the set of all words of even lengths. The corresponding set of vectors $h=\{h_x\in\mathbb{R}^{q-1}: x\in G_k\}$ has the form

$$h_x=\begin{cases}\mathbf{h}^1, & \text{if } x\in G_k^{(2)},\\ \mathbf{h}^2, & \text{if } x\in G_k\setminus G_k^{(2)},\end{cases}$$

where $\mathbf{h}^1=(h_1^1,h_2^1)$, $\mathbf{h}^2=(h_1^2,h_2^2)$ (see Fig. 11.1).

In view of (12.10) we have

$$\begin{cases} h_1^1 = k\ln\left(\dfrac{\theta\exp(h_1^2)+\exp(h_2^2)+1}{\exp(h_1^2)+\exp(h_2^2)+\theta}\right) \\ h_2^1 = k\ln\left(\dfrac{\exp(h_1^2)+\theta\exp(h_2^2)+1}{\exp(h_1^2)+\exp(h_2^2)+\theta}\right) \\ h_1^2 = k\ln\left(\dfrac{\theta\exp(h_1^1)+\exp(h_2^1)+1}{\exp(h_1^1)+\exp(h_2^1)+\theta}\right) \\ h_2^2 = k\ln\left(\dfrac{\exp(h_1^1)+\theta\exp(h_2^1)+1}{\exp(h_1^1)+\exp(h_2^1)+\theta}\right). \end{cases} \tag{12.25}$$

For $k = 3$, $J < 0$ it is known (see Theorem 11.4 and [123]) that if $0 < \theta < \frac{1}{4}$ then the system (12.25) has at least two solutions of the form: $h = (h^1, h^1, h^2, h^2)$. Now for such periodical set of vectors h_x, we shall describe free energies. From (12.14) we have

$$a(x) = \begin{cases} d(\mathbf{h}^1), \text{ if } x \in G_k^{(2)} \\ d(\mathbf{h}^2), \text{ if } x \in G_k \setminus G_k^{(2)}, \end{cases}$$

where

$$d(\mathbf{h}) = \frac{1}{q\beta}\sum_{i=1}^{q}\left(\sum_{u=1}^{q}\exp\{(J\beta\sigma_i+\mathbf{h})\sigma_u\}\right). \tag{12.26}$$

Denote

$$V_{even,n} = \{x \in V_n : x \in G_k^{(2)}\}, \quad V_{odd,n} = \{x \in V_n : x \in G_k \setminus G_k^{(2)}\}.$$

It is easy to cheek that for $n = 2p$,

$$|V_{even,2p}| = \frac{k^{2p+1}-1}{k-1}, \quad |V_{odd,2p}| = \frac{k^{2p}-1}{k-1},$$

for $n = 2p+1$,

$$|V_{even,2p+1}| = \frac{k^{2p}-1}{k-1}, \quad |V_{odd,2p+1}| = \frac{k^{2p+2}-1}{k-1},$$

also we have

$$|V_n| = \frac{(k+1)k^n-2}{(k-1)}.$$

Using these formulas we calculate free energy:

$$
\begin{aligned}
&E_{per}(\beta, h_x)\\
&= -\lim_{n\to\infty} \frac{|V_{even,n}|d(\mathbf{h}^1) + |V_{odd,n}|d(\mathbf{h}^2)}{|V_n|}\\
&= -\lim_{n\to\infty} \begin{cases} \dfrac{|V_{even,2p}|d(\mathbf{h}^1) + |V_{odd,2p}|d(\mathbf{h}^2)}{|V_{2p}|} & \text{if } n = 2p\\ \dfrac{|V_{even,2p+1}|d(\mathbf{h}^1) + |V_{odd,2p+1}|d(\mathbf{h}^2)}{|V_{2p+1}|} & \text{if } n = 2p+1, \end{cases}\\
&= -\frac{1}{k+1} \begin{cases} kd(\mathbf{h}^1) + d(\mathbf{h}^2) & \text{if } n = 2p,\ p \to \infty\\ d(\mathbf{h}^1) + kd(\mathbf{h}^2) & \text{if } n = 2p+1. \end{cases}
\end{aligned}
$$

From this equality it follows that if $d(\mathbf{h}^1) \neq d(\mathbf{h}^2)$, then for the periodic BCs a free energy does not exist. For $k = q = 3$ and fixed $\theta = \frac{1}{5}$, using a computer analysis one can see that $d(\mathbf{h}^1) \neq d(\mathbf{h}^2)$.

Remark 12.3. It is known (see [78]) that for periodic Gibbs measures of the Ising model on Cayley trees the free energy exists. But for the Potts model we proved that free energy of periodic Gibbs measures may not exist.

Remark 12.4. Although, it was proved in [78] that in absence of external field in the nearest neighbor Ising model on the Cayley tree, the free energy exists for all known BCs, in [79] it was shown that the model exhibits a new temperature-driven phase transitions. These transitions are characterized by a change in the number of Gibbs measures as well as by a drastic change of the behavior of free energies at the new transition points. Moreover, for the Ising model in the presence of an external field the free energies of translation-invariant and some alternating BCs are found. It is shown that free energy of alternating Gibbs measures may not exist, in the sense that it may assume different accumulating points which already appears when a magnetic field is turned on.

12.3.4 *Weakly periodic BCs*

Construct a weakly periodic BC. For $A \subset \{1, 2, ..., k+1\}$ we consider $H_A = \{x \in G_k : \sum_{j\in A} w_j(x)\text{–even}\}$, where $w_j(x)$ is the number of a_j in a word x, $G_k/H_A = \{H_A, G_k \setminus H_A\}$ is a quotient group. For simplicity, we set $H_0 = H_A$, $H_1 = G_k \setminus H_A$. The H_A — weakly periodic sets of vectors

$h = \{h_x \in \mathbb{R}^{q-1} : x \in G_k\}$ has the following form

$$h_x = \begin{cases} h_1, \text{ if } & x_\downarrow \in H_0, \ x \in H_0 \\ h_2, \text{ if } & x_\downarrow \in H_0, \ x \in H_1 \\ h_3, \text{ if } & x_\downarrow \in H_1, \ x \in H_0 \\ h_4, \text{ if } & x_\downarrow \in H_1, \ x \in H_1. \end{cases} \tag{12.27}$$

Here $h_i = (h_{i1}, h_{i2}, ..., h_{iq-1})$, $i = 1, 2, 3, 4$. By (12.10), we have

$$\begin{cases} h_1 = (k - |A|)F(h_1, \theta) + |A|F(h_2, \theta) \\ h_2 = (|A| - 1)F(h_3, \theta) + (k + 1 - |A|)F(h_4, \theta) \\ h_3 = (|A| - 1)F(h_2, \theta) + (k + 1 - |A|)F(h_1, \theta) \\ h_4 = (k - |A|)F(h_4, \theta) + |A|F(h_3, \theta). \end{cases} \tag{12.28}$$

In [223] (see also previous chapter) it was shown that for $|A| = k$, $k \geq 6$ the system of equations (12.28) has at least two (not translation-invariant) solutions, which generate sets of vectors h_x in the form of (12.27), where all coordinates of vectors $h_i, i = 1, 2, 3, 4$ are equal and $h_i \neq h_j$ for $i \neq j$. Now for such weakly periodic sets of vectors h_x, we calculate the corresponding free energy.

We introduce the following

$$\begin{aligned} \mathcal{A}_n &= |\{\langle x, y\rangle \in L_n : x \in H_0, \ y = x_\downarrow \in H_0\}|, \\ \mathcal{B}_n &= |\{\langle x, y\rangle \in L_n : x \in H_0, \ y = x_\downarrow \in H_1\}|, \\ \mathcal{C}_n &= |\{\langle x, y\rangle \in L_n : x \in H_1, \ y = x_\downarrow \in H_0\}|, \\ \mathcal{D}_n &= |\{\langle x, y\rangle \in L_n : x \in H_1, \ y = x_\downarrow \in H_1\}|, \end{aligned}$$

where L_n is a set of edges in V_n.

It is known from [78] that

$$\lim_{n\to\infty} \frac{\mathcal{A}_n}{|V_n|} = \lim_{n\to\infty} \frac{\mathcal{D}_n}{|V_n|} = \frac{1}{2(k+1)},$$
$$\lim_{n\to\infty} \frac{\mathcal{B}_n}{|V_n|} = \lim_{n\to\infty} \frac{\mathcal{C}_n}{|V_n|} = \frac{k}{2(k+1)}.$$

Using these formulas we calculate free energy:

$$\begin{aligned} E_{WP}(\beta, q, h_x) &= -\lim_{n\to\infty} \frac{1}{|V_n|}(\mathcal{A}_n d(h_1) + \mathcal{B}_n d(h_2) + \mathcal{C}_n d(h_3) + \mathcal{D}_n d(h_4)) \\ &= \frac{1}{2(k+1)} [E_{TI}(\beta, q-1, h_1) + kE_{TI}(\beta, q-1, h_2) \\ &\qquad + kE_{TI}(\beta, q-1, h_3) + E_{TI}(\beta, q-1, h_4)], \end{aligned}$$

where $E_{TI}(\beta, m, h_x)$ is defined in (12.19). Now the entropy is

$$\begin{aligned} S_{WP}(\beta, q, h_x) = \frac{1}{2(k+1)} &(S_{TI}(\beta, q-1, h_1) + kS_{TI}(\beta, q-1, h_2) \\ &+ kS_{TI}(\beta, q-1, h_3) + S_{TI}(\beta, q-1, h_4)), \end{aligned}$$

where $S_{TI}(\beta, m, h_x)$ is defined in (12.20).

12.4 Comments and references for Part III

This part of the book mainly based on works [28], [33], [75], [78], [80], [90], [94], [95], [98], [110], [111], [122], [123]-[126], [138], [139], [171], [176], [187], [223]-[227], [236], [243]-[245], [247], [249], [254], [255]. Many of these papers written after 2013 and are related to the Potts model on trees.

In my opinion, one of nice results of the theory of Gibbs measures on trees is Theorem 8.3, which gives full description of TISGMs for q-state Potts model. Namely, it states that there are $[q/2]$ distinct critical temperatures changing the number of TISGMs and there can be up to $2^q - 1$ such measures. Before of this result it was only known one critical temperature and q TISGMs [94], [95].

Applications of results of [176] allowed (in Chapter 10) give conditions of extremality of all TISGMs. Remaining results of this part are related to recently obtained results on boundary configurations, non-zero external fields, periodic and weakly periodic Gibbs measures and free energies of the Potts model.

A modern direction of the theory of Gibbs measures is non-Archimedean (p-adic) version of the measures. In p-adic case the Hamiltonian and measures have p-adic values [133]. This is now very actively developing area (see for example [46], [147] and references therein). In particular, it was already showed that the p-adic Potts model is completely different from the real case see [5], [193]-[197].

Another direction is related to the quantum Potts model. Such a model considered in [3] (see also [306] and references therein) and motivated that quantum critical points in quasiperiodic magnets can realize new universality classes, with critical properties distinct from those of clean or disordered systems. A quantum phase transitions separating ferromagnetic and paramagnetic phases in the quasiperiodic q-state Potts model in $2+1D$ is studied. It is shown that the critical behavior is largely independent of q, and is controlled by an infinite-quasiperiodicity fixed point.

For future reading see [11], [15], [25], [28], [33]-[35], [52], [55], [59], [64], [71], [75], [81], [88], [89], [96], [129], [140], [141], [158], [173], [184], [206], [209], [211], [228], [232], [269], [281], [286].

Bibliography

[1] K. Adam, D. Zöllner, and D.P. Field, 3D microstructural evolution of primary recrystallization and grain growth in cold rolled single-phase aluminum alloys. *Modelling Simul. Mater. Sci. Eng.* **26** (2018), 035011, 16 pages.

[2] S. Adams, R. Briceo, B. Marcus, Brian, and R. Pavlov, Representation and poly-time approximation for pressure of $\mathbb{Z}^2$ lattice models in the non-uniqueness region. *J. Stat. Phys.* **162**(4) (2016), 1031–1067.

[3] U. Agrawal, S. Gopalakrishnan, and R. Vasseur, Romain Quantum criticality in the 2D quasiperiodic Potts model. *Phys. Rev. Lett.* **125**(26) (2020), 265702, 6 pp.

[4] F.S. de Aguiar, L.B. Bernardes, and S. Goulart Rosa, Jr., Metastability in the Potts model on the Cayley tree. *J. Stat. Phys.* **64**(3-4) (1991), 673–682.

[5] M.A.K. Ahmad, L. Liao, and M. Saburov, Periodic p-adic Gibbs measures of q-state Potts model on Cayley trees I: The chaos implies the vastness of the set of p-adic Gibbs measures. *J. Stat. Phys.* **171**(6) (2018), 1000–1034.

[6] B. Alberts, A. Johnson, J. Lewis, M. Raff, K. Roberts, and P. Walter, *Molecular Biology of the Cell* (4th edition). New York: Garland Science; 2002.

[7] R. Allena, M. Scianna, and L. Preziosi, A cellular Potts model of single cell migration in presence of durotaxis. *Math. Biosci.* **275** (2016), 57–70.

[8] G. Altan-Bonnet, A. Libchaber, and O. Krichevsky, Bubble dynamics in double-stranded DNA. *Phys. Rev. Lett.* **90**(13) (2003), 138101.

[9] V. Andasari, R.T. Roper, M.H. Swat, and M.A.J. Chaplain, Integrating intracellular dynamics using CompuCell3D and bionetsolver: Applications to multiscale modelling of cancer cell growth and invasion. *PLoS ONE* **7**(3) (2012), e33726.

[10] A.R.A. Anderson, M.A.J. Chaplain, and K.A. Rejniak, *Single-Cell-Based Models in Biology and Medicine.* Mathematics and Biosciences in Interaction. Birkhäuser Verlag, Basel, 2007.

[11] É.O. Aquino, F.W. Lima, A.D. Araújo, and F.R.N. Costa, Potts model in one-dimension on directed small-world networks. *J. Stat. Phys.* **171**(6) (2018), 1112–1121.

[12] B.-R. Aurélien, *Mobile Spins on Lattice as Model for Liquid Crystals and Topological Excitations and Skyrmions*. Ph.D. Thesis, University of Cergy-Pontoise, Paris, France, 15 October 2018.

[13] J.C.A. Barata, and D.H.U. Marchetti, Griffiths' singularities in diluted Ising models on the Cayley tree. *J. Stat. Phys.* **88**(1-2) (1997), 231–268.

[14] R.J. Baxter, *Exactly Solved Models in Statistical Mechanics*, Reprint of the 1982 original. Academic Press, Inc. [Harcourt Brace Jovanovich, Publishers], London, (1989).

[15] L. Beaudin, A review of the Potts model. *Rose-Hulman Undergraduate Math. Jour.* **8**(1) (2007), Article 13, 25 pages.

[16] L. Beaudin, J. Ellis-Monaghan, G. Pangborn, and R. Shrock, A little statistical mechanics for the graph theorist. *Discrete Math.* **310**(13-14) (2010), 2037–2053.

[17] V. Beffara, and H. Duminil-Copin, The self-dual point of the two-dimensional random-cluster model is critical for $q \geq 1$. *Probab. Theory Relat. Fields* **153**(34) (2012), 511–542.

[18] J. Besag, On the statistical analysis of dirty pictures. *J. Roy. Statist. Soc. Ser. B* **48**(3) (1986), 259–302.

[19] C.J. Benham, Elastic model of supercoiling. *Proc. Natl. Acad. Sci. U.S.A.* **74** (1977), 2397–2401.

[20] C.J. Benham, An elastic model of the large-scale structure of duplex DNA. *Biopolymers* **18** (1979), 609–623.

[21] S.N. Bernshtein, Solution of a mathematical problem connected with the theory of heredity. *Uch. Zap. Nauchno-Issled. kaf. Ukr. Otd. Mat.* **1** (1924), 83–115.

[22] A. Blanca, Z. Chen, D. Štefankovič, and E. Vigoda, Hardness of identity testing for restricted Boltzmann machines and Potts models. *J. Mach. Learn. Res.* 22 (2021), Paper No. 152, 56 pp.

[23] M. Blatt, S. Wiseman, and E. Domany, Superparamagnetic clustering of data. *Phys Rev Lett.* **76**(18) (1996), 3251–3254.

[24] P.M. Bleher, J. Ruiz, and V.A. Zagrebnov, On the purity of the limiting Gibbs state for the Ising model on the Bethe lattice. *J. Stat. Phys.* **79** (1995), 473–482.

[25] P. Bleher, M. Lyubich, and R. Roeder, Lee-Yang-Fisher zeros for the DHL and 2D rational dynamics, II. Global pluripotential interpretation. *J. Geom. Anal.* **30**(1) (2020), 777–833.

[26] S.E.M. Boas, Y. Jiang, R.M.H. Merks, S.A. Prokopiou, and E.G. Rens, Cellular Potts model: Applications to vasculogenesis and angiogenesis. Probabilistic cellular automata, 279–310. *Emerg. Complex. Comput.* **27**, Springer, Cham, 2018.

[27] T. Bodineau, Translation invariant Gibbs states for the Ising model. *Probab. Theory Relat. Fields* **135**(2), (2006), 153–168.

[28] L.V. Bogachev, and U.A. Rozikov, On the uniqueness of Gibbs measure in the Potts model on a Cayley tree with external field. *J. Stat. Mech. Theory Exp.* **7** (2019), 073205, 76 pp.

[29] M. Bolanowski, and A. Paszkiewicz, The use of statistical signatures to detect anomalies in computer network. *Lect. Notes Electr. Eng.* **324** (2015), 251–260.

[30] C. Borgs, *Statistical physics expansion methods in combinatorics and computer sciencc*, http://rcscarch.microsfort.com/ borgs/CBMS.pdf, 2004.

[31] S. Bornholdt, Expectation bubbles in a spin model of markets: Intermittency from frustration across scales. *Int. J. Mod. Phys.* **12**(5) (2001), 667–674.

[32] S. Bornholdt, A q-spin Potts model of markets: Gain-loss asymmetry in stock indices as an emergent phenomenon. *Phys. A* **588** (2022), Paper No. 126565.

[33] G.I. Botirov, and U.A. Rozikov, The Potts model with competing interactions on the Cayley tree: the contour method. *Theoret. Math. Phys.* **153**(1) (2007), 1423–1433.

[34] G.I. Botirov, and U.A. Rozikov, On q-component models on the Cayley tree: The general case. *J. Stat. Mech. Theory Exp.* **10** (2006), P10006, 8 pp.

[35] G.I. Botirov, and U.U. Qayumov, Ground states for the Potts model with competing interactions and a countable set of spin values on a Cayley tree. *Theoret. Math. Phys.* **209**(2) (2021), 1633–1642.

[36] A. Bovier, *Statistical mechanics of disordered systems. A mathematical perspective.* Cambridge Series in Statistical and Probabilistic Mathematics, 18. Cambridge University Press, Cambridge, 2006.

[37] Z.S. Boxonov, and U.A. Rozikov, A dynamical system of temperature-dependent sex linked inheritance. *Uzbek Math. J.* **2** (2018), 17–31.

[38] Y. Boykov, O. Veksler, and R. Zabih, Fast approximate energy minimization via graph cuts. *IEEE Trans. Pattern Anal. Machine Intell.* **23**(11) (2001), 1222–1239.

[39] T. Brookings, J.M. Carlson, and J. Doyle, Three mechanisms for power laws on the Cayley tree. *Phys. Rev. E.* **72** (2005), 056120, 18 pp.

[40] R.K. Bullough, Y. Chen, and J.T. Timonen, Thermodynamics of Toda lattice models: application to DNA. *Phys. D* **68**(1) (1993), 83–92.

[41] R. Bundschuh, Unified approach to partition functions of RNA secondary structures. *J. Math. Biol.* **69** (2014), 1129–1150.

[42] F. Caltagirone, G. Parisi, and T. Rizzo, Dynamical critical exponents for the mean-field Potts glass. *Phys. Rev. E* **85** (2012), 051504.

[43] E. Carlon, *Thermodynamics of DNA microarrays.* Stochastic models in biological sciences, 229–233. Banach Center Publ., 80, Polish Acad. Sci., Warsaw, 2008.

[44] J. Cavender, Taxonomy with confidence. *Math. BioSci.* **40** (1978), 271–280.

[45] G. Celeux, F. Forbes, and N. Peyrard, *EM-based image segmentation using Potts models with external field.* [Research Report] RR-4456, INRIA. 2002. inria-00072132.

[46] A. Chambert-Loir, J. Nicaise, and J. Sebag, *Motivic integration.* Progress in Mathematics, 325. Birkhüser/Springer, New York, 2018.

[47] V. Chandola, A. Banerjee, and V. Kumar, Anomaly detection: A survey. *ACM Comput. Surv.* **41**(3) (2009), 58 pages.

[48] S. Chatterjee, M. Mangeat, R. Paul, and H. Rieger, Flocking and reorientation transition in the 4-state active Potts model. *EPL (Europhysics Letters)* **130**(6) (2020), 66001, (8 pages).

[49] N. Chen, J.A. Glazier, and M.S. Alber, *A Parallel Implementation of the Cellular Potts Model for Simulation of Cell-based Morphogenesis.* Cellular automata, 58–67, Lecture Notes in Comput. Sci., 4173, Springer, Berlin, 2006.

[50] N. Chen, J.A. Glazier, J. A. Izaguirre, and M.S. Alber, A parallel implementation of the cellular Potts model for simulation of cell-based morphogenesis. *Comput. Phys. Commun.* **176** (2007), 670–681.

[51] I. Chio, C. He, A. Ji, and R.K.W. Roeder, Limiting measure of Lee-Yang zeros for the Cayley tree. *Comm. Math. Phys.* **370**(3) (2019), 925–957.

[52] L. Cioletti, and R. Vila, Graphical representations for Ising and Potts models in general external fields. *J. Stat. Phys.* **162**(1) (2016), 81–122.

[53] F. Claudio, M. Giancarlo, and Z. Claudio (Eds.), *DNA computing.* Lecture Notes in Computer Science, 3384. Springer-Verlag, Berlin, 2005.

[54] S. Cohen-Addad, R. Höhler, and O. Pitois, Flow in foams and flowing foams. *Ann. Rev. Fluid Mech.* **45** (2013), 241–267.

[55] L. Coquille, H. Duminil-Copin, D. Ioffe, and Y. Velenik, On the Gibbs states of the noncritical Potts model on $\mathbb{Z}^2$. *Probab. Theory Related Fields* **158**(1-2) (2014), 477–512.

[56] L. Coquille, Examples of DLR states which are not weak limits of finite volume Gibbs measures with deterministic boundary conditions. *Jour. Stat. Phys.* **159**(4) (2015), 958-971.

[57] M. Daniel, and M. Vanitha, Bubble solitons in an inhomogeneous, helical DNA molecular chain with flexible strands. *Phys. Rev. E* **84** (2011), 031928.

[58] C. Daskalakis, E. Mossel, and S. Roch, Evolutionary trees and the Ising model on the Bethe lattice: a proof of Steel's conjecture. *Probab. Theory Related Fields* **149**(1-2) (2011), 149–189.

[59] A. Dembo, A. Montanari, A. Sly, and N. Sun, The replica symmetric solution for Potts models on d-regular graphs. *Comm. Math. Phys.* **327** (2014), 551–575.

[60] R.L. Dobrushin, The description of a random field by means of conditional probabilities and conditions of its regularity. *Theory Probab. Appl.* **13** (1968), 197–224.

[61] C. Dombry, A probabilistic study of DNA denaturation. *J. Stat. Phys.* **120**(3-4) (2005), 695–719.

[62] C. Domb, and R. B. Potts, Order-disorder statistics. IV. A two-dimensional model with first and second interactions. *Proc. Roy. Soc. London Ser. A* **210** (1951), 125–141.

[63] J.M. Dresch, M.A. Thompson, D.N. Arnosti, and C. Chiu, Two-layer mathematical modeling of gene expression: incorporating DNA-level information and system dynamics. *SIAM J. Appl. Math.* **73**(2) (2013), 804–826.

[64] H. Duminil-Copin, V. Sidoravicius, and V. Tassion, Continuity of the phase transition for planar random-cluster and Potts models with $1 \leq q \leq 4$. *Comm. Math. Phys.* **349**(1) (2017), 47–107.

[65] H. Duminil-Copin, A. Raoufi, and V. Tassion, Sharp phase transition for the random-cluster and Potts models via decision trees. *Ann. of Math.* (2) **189**(1) (2019), 75–99.

[66] R. Durrett, *Probability Models for DNA Sequence Evolution* (2nd edition). Probability and its Applications (New York). Springer, New York, 2008.

[67] S.W. Englander, N.R. Kallenbach, A.J. Heeger, J.A. Krumhansl, and A. Litwin, Nature of the open state in long polynucleotide double helices: Possibility of soliton excitations. *Proc. Natl. Acad. Sci. U.S.A.* **77** (1980), 7222–7226.

[68] A.C.D. van Enter, G. Iacobelli, and S. Taati, Potts model with invisible colors: Random-cluster representation and Pirogov-Sinai analysis. *Rev. Math. Phys.* **24**(2) (2012), 1250004, 42 pp.

[69] R.C. Erson, and J. Stewart, A general model for the genetic analysis of pedigree data. *Hum. Hered.* **21** (1971), 523–542.

[70] D.K. Faddeev, and I.S. Sominskii, *Problems in higher algebra.* Freeman, San Francisco, 1965.

[71] O.N. Feldheim, and Y. Spinka, Long-range order in the 3-state antiferromagnetic Potts model in high dimensions. *J. Eur. Math. Soc. (JEMS)* **21**(5) (2019), 1509–1570.

[72] M.E. Fisher, The renormalization group in the theory of critical phenomena. *Rev. Modern Phys.* **46** (1974), 597–616.

[73] M. Formentin, and C. Külske, A symmetric entropy bound on the non-reconstruction regime of Markov chains on Galton-Watson trees. *Electron. Commun. Probab.* **14** (2009), 587–596.

[74] G. Franzese, and A. Coniglio, Phase transitions in the Potts spin-glass model. *Phys. Rev. E* **58**(3) (1998), 2753–2759.

[75] S. Friedli, and Y. Velenik, *Statistical Mechanics of Lattice Systems. A Concrete Mathematical Introduction.* Cambridge University Press, Cambridge, 2018.

[76] F. Friedrich, A. Kempe, V. Liebscher, and G. Winkler, Complexity penalized M-estimation: Fast computation. *J. Comput. Graph. Statist.* **17**(1) (2008), 201–224.

[77] Y. Fu, and P.W. Anderson, Application of statistical mechanics to NP-complete problems in combinatorial optimisation. *J. Phys. A* **19**(9) (1986), 1605–1620.

[78] D. Gandolfo, M.M. Rakhmatullaev, U.A. Rozikov, and J. Ruiz, On free energies of the Ising model on the Cayley tree. *Jour. Stat. Phys.* **150**(6) (2013), 1201–1217.

[79] D. Gandolfo, F.H. Haydarov, U.A. Rozikov, U.A., and J. Ruiz, New phase transitions of the Ising model on Cayley trees. *J. Stat. Phys.* **153**(3) (2013) 400–411.

[80] D. Gandolfo, M.M. Rahmatullaev, and U.A. Rozikov, Boundary conditions for translation-invariant Gibbs measures of the Potts model on Cayley trees. *J. Stat. Phys.* **167**(5) (2017), 1164–1179.

[81] D. Gandolfo, C. Maes, J. Ruiz, and S. Shlosman, Glassy states: The free Ising model on a tree. *J. Stat. Phys.* **180**(1-6) (2020), 227–237.
[82] R.N. Ganikhodzhaev, Quadratic stochastic operators, Lyapunov functions and tournaments. *Russian Acad. Sci. Sbornik Math.* **76** (1993), 489–506.
[83] R.N. Ganikhodzhaev, A chart of fixed points and Lyapunov functions for a class of discrete dynamical systems. *Math. Notes* **56** (1994), 1125–1131.
[84] R.N. Ganikhodzhaev, F.M. Mukhamedov, and U.A. Rozikov, Quadratic stochastic operators and processes: Results and open problems. *Inf. Dim. Anal. Quant. Prob. Rel. Fields* **14**(2) (2011), 279–335.
[85] N.N. Ganikhodjaev, An application of the theory of Gibbs distributions to mathematical genetics. *Doklady Math.* **61**, (2000), 321–323.
[86] N.N. Ganikhodjaev, and U.A. Rozikov, On disordered phase in the ferromagnetic Potts model on the Bethe lattice. *Osaka J. Math.* **37**(2) (2000), 373–383.
[87] N.N. Ganikhodjaev, and U.A. Rozikov, On quadratic stochastic operators generated by Gibbs distributions. *Regul. Chaotic Dyn.* **11**(4) (2006), 467–473.
[88] N.N. Ganikhodjaev, The Potts model on $\mathbb{Z}^d$ with countable set of spin values. *J. Math. Phys.* **45**(3) (2004), 1121–1127.
[89] N.N. Ganikhodjaev, and U.A. Rozikov, The Potts model with countable set of spin values on a Cayley tree. *Lett. Math. Phys.* **75**(2) (2006), 99–109.
[90] N.N. Ganikhodzhaev, and U.A. Rozikov, Description of periodic extreme Gibbs measures of some lattice models on a Cayley tree. *Theoret. Math. Phys.* **111**(1) (1997), 480–486.
[91] N.N. Ganikhodzhaev, and U.A. Rozikov, The classes of normal subgroups of finite index of group representation of the Cayley tree. *Uzbek Math. Jour.* **4** (1997), 31–39 (Russian).
[92] N.N. Ganikhodzhaev, and U.A. Rozikov, Pirogov-Sinai theory with new contours for symmetric models. *Int. J. Geom. Methods Mod. Phys.* **5**(4) (2008), 537–546.
[93] N.N. Ganikhodjaev, Exact solution of an Ising model on the Cayley tree with competing ternary and binary interactions. *Theor. Math. Phys.* **130**(3) (2002), 419–424.
[94] N.N. Ganikhodjaev, On pure phases of the three-state ferromagnetic Potts model on the second-order Bethe lattice. *Theor. Math. Phys.* **85**(2) (1990), 1125–1134.
[95] N.N. Ganikhodzhaev, On pure phases of the ferromagnetic Potts model Bethe lattices. *Dokl. AN Uzbekistan.* No. 6-7 (1992), 4–7.
[96] A. Galanis, D. Štefankovič, E. Vigoda, and L. Yang, Ferromagnetic Potts model: Refined #BIS-hardness and related results. *SIAM J. Comput.* **45**(6) (2016), 2004–2065.
[97] O. Garet, Central limit theorems for the Potts model. *Math. Phys. Electron. J.* **11** (2005), Paper 4, 27 pp.
[98] H.O. Georgii, *Gibbs Measures and Phase Transitions* (2nd edition). de Gruyter Studies in Mathematics, 9. Walter de Gruyter, Berlin, 2011.

[99] R. Gheissari, *Dynamics of 2D Potts and Random-Cluster models.* Thesis (Ph.D.). New York University, 2019, 320 p.

[100] G.P. Gladyshev, and Yu. A. Ershov, Principles of the thermodynamics of biological systems. *J. Theoret. Biol.* **94**(2) (1982), 301–343.

[101] G.P. Gladyshev, Classical thermodynamics, tandemism and biological evolution. *J. Theoret. Biol.* **94**(2) (1982), 225–239.

[102] J. A. Glazier, M.P. Anderson, and G. Grest, Coarsening in the two-dimensional soap froth and the large-Q Potts model: A detailed comparison. *Philos. Mag. B.* **62** (1990), 615–645.

[103] L.A. Goldberg, M. Jalsenius, R. Martin, and M. Paterson, Improved mixing bounds for the anti-ferromagnetic Potts model on $\mathbb{Z}^2$. *LMS J. Comput. Math.* **9** (2006), 1–20.

[104] F. Graner, and J. Glazier, Simulation of biological cell sorting using a two-dimensional extended Potts model. *Phys. Rev. Lett.* **69**(13) (1992), 2013–2017.

[105] L. Grans-Samuelsson, L. Liu, Y. He, J.L. Jacobsen, and H. Saleur, The action of the Virasoro algebra in the two-dimensional Potts and loop models at generic Q. *J. High Energy Phys.* **10** (2020), 109, 52 pp.

[106] G. Grimmett, *The random-cluster model.* Grundlehren der Mathematischen Wissenschaften [Fundamental Principles of Mathematical Sciences], 333. Springer-Verlag, Berlin, 2006.

[107] P.J. Green, and S. Richardson, Hidden Markov models and disease mapping. *J. Amer. Statist. Assoc.* **97** (2002), 1055–1070.

[108] W.M. Haddad, *A Dynamical Systems Theory of Thermodynamics.* Princeton Series in Applied Mathematics. Princeton University Press, Princeton, NJ, 2019.

[109] O. Häggström, The random-cluster model on a homogeneous tree. *Probab. Theory and Relat. Fields* **104** (1996), 231–253.

[110] O. Häggström, and C. Külske, Gibbs properties of the fuzzy Potts model on trees and in mean field. *Markov Proc. Rel. Fields* **10**(3) (2004), 477–506.

[111] Y. Higuchi, Remarks on the limiting Gibbs states on a $(d+1)$-tree. *Publ. RIMS, Kyoto Univ.* **3** (1977), 335–348.

[112] S. Hohng, R.Zhou, M.K. Nahas, J. Yu, K. Schulten, D.M.J. Lilley, and T. Ha, Fluorescence-force spectroscopy maps two-dimensional reaction landscape of the holliday junction. *Science* **318** (2007), 279–283.

[113] R. Holliday, A mechanism for gene conversion in fungi. *Genet Res.* **5** (1964), 282–304.

[114] M.A. Idiart, and Y. Levin, Rupture of a liposomal vesicle. *Phys. Rev. E.* **69**(6) (2004), 061922.

[115] F.H. Jafarpour, The application of the Yang-Lee theory to study a phase transition in a non-equilibrium system. *J. Stat. Phys.* **113**(1–2) (2003), 269–281.

[116] R.D. Jenks, Quadratic differential systems for interactive population models. *J. Differ. Equations* **5** (1969), 497–514.

[117] J.L. Jensen, and A-M.K. Pedersen, Probabilistic models of DNA sequence evolution with context dependent rates of substitution. *Adv. in Appl. Probab.* **32**(2) (2000), 499–517.

[118] I. Kanter, Potts-glass models of neural networks. *Phys. Rev. A* (3) **37**(7) (1988), 2739–2742.

[119] A. Karolak, D.A. Markov, L.J. McCawley, and K.A. Rejniak, Towards personalized computational oncology: From spatial models of tumour spheroids, to organoids, to tissues. *J. R. Soc. Interface* **15** (2018), 20170703, 16 pages.

[120] A. Katok, and B. Hasselblatt, *Introduction to the Modern Theory of Dynamical Systems.* Encyclopedia of Mathematics and its Applications, 54. Cambridge University Press, Cambridge, 1995.

[121] H. Kesten, Quadratic transformations: A model for population growth. *I. Adv. Appl. Probab.* **2** (1970), 1–82.

[122] H. Kesten, and B.P. Stigum, Additional limit theorem for indecomposable multi-dimensional Galton-Watson processes. *Ann. Math. Statist.* **37** (1966), 1463–1481.

[123] R.M. Khakimov, On the existence of periodic Gibbs measures for the Potts model on a Cayley tree. (Russian) *Uzbek. Mat. Zh.* **3** (2014), 134–142.

[124] R.M. Khakimov, New Periodic Gibbs Measures for q-state Potts Model on a Cayley Tree. *Jour. Siberian Federal Univ. Math. Phys.* **7**(3) (2014), 297–304.

[125] R.M. Khakimov, and M.T. Makhammadaliev, Translation invariance of periodic Gibbs measures for the Potts model on the Cayley tree. *Theoret. Math. Phys.* **199**(2) (2019), 726–735.

[126] R.M. Khakimov, and F.Kh. Khaidarov, Translation-invariant and periodic Gibbs measures for the Potts model on a Cayley tree. *Theoret. Math. Phys.* **189**(2) (2016), 1651–1659.

[127] N.M. Khatamov, Holliday junctions for the HC (cycle) Blume-Capel model of DNA. *Uzbek Math. Jour.* **1** (2021), 87–97.

[128] N. M. Khatamov, Holliday junctions in the Blume-Capel model of DNA. *Theoret. Math. Phys.* **206**(3) (2021), 383–390.

[129] N.M. Khatamov, and G. T. Madgoziev, Gibbs measures for a generalized Potts model with the interaction radius two on a Cayley tree. *Theoret. Math. Phys.* **183**(3) (2015), 836–845.

[130] A.Yu. Khrennikov, *Non-Archimedean Analysis: Quantum Paradoxes, Dynamical Systems and Biological Models.* Kluwer Acad. Publ., Dordreht, 1997.

[131] R. Kindermann, and J.L. Snell, Markov random fields and their applications. *Contemporary Mathematics* **1** (1980).

[132] S. Kjelstrup, and D. Bedeaux, *Non-equilibrium thermodynamics of heterogeneous systems.* Series on Advances in Statistical Mechanics, 16. World Scientific Publishing Co. Pte. Ltd., Hackensack, NJ, 2008.

[133] N. Koblitz, *p-adic Numbers, p-adic Analysis, and zeta-Functions.* Springer, Berlin, 1977.

[134] R. Kotecký, and S.B. Shlosman, First-order phase transitions in large entropy lattice models. *Comm. Math. Phys.* **83** (1982), 493–515.
[135] J. Krishnan, R. Torabi, A. Schuppert, and E. Di Napoli, A modified Ising model of Barabsi-Albert network with gene-type spins. *J. Math. Biol.* **81**(3) (2020), 769–798.
[136] E. Kropff, and A. Treves, The storage capacity of Potts models for semantic memory retrieval. *J. Stat. Mech. Theory Exp.* **8** (2005), 08010, 19 pp.
[137] V.K. Kuetche, Ab initio bubble-driven denaturation of double-stranded DNA: Self-mechanical theory. *J. Theoret. Biol.* **401** (2016), 15–29.
[138] C. Külske, U.A. Rozikov, and R.M. Khakimov, Description of the translation-invariant splitting Gibbs measures for the Potts model on a Cayley tree. *J. Stat. Phys.* **156**(1) (2014), 189–200.
[139] C. Külske, and U.A. Rozikov, Fuzzy transformations and extremality of Gibbs measures for the Potts model on a Cayley tree. *Random Structures Algorithms* **50**(4) (2017), 636–678.
[140] C. Külske, *Stochastic Processes on Trees*, Preprint, Ruhr-University, Bochum, 2017.
[141] C. Külske, and D. Meiβner, Stable and metastable phases for the Curie-Weiss-Potts model in vector-valued fields via singularity theory. *J. Stat. Phys.* **181**(3) (2020), 968–989.
[142] S.H. Kung, Sums of integer powers via the Stolz-Cesáro theorem. *Math. Assoc. America.* **40** (2009), 42–44.
[143] L. Laanait, A. Messager, and J. Ruiz, Phases coexistence and surface tensions for the Potts model. *Commun. Math. Phys.* **105**(4) (1986), 527–545.
[144] O.E. Lanford, and D. Ruelle, Observables at infinity and states with short range correlations in statistical mechanics. *Comm. Math. Phys.* **13** (1969), 194–215.
[145] C.R. Laramy, M.N. O'Brien, and C.A. Mirkin, Crystal engineering with DNA. *Nat. Rev. Mater* **4** (2019), 201–224.
[146] D.A. Lavis, *Equilibrium Statistical Mechanics of Lattice Models. Theoretical and Mathematical Physics.* Springer, Dordrecht, 2015.
[147] A. Le Ny, L. Liao, and U.A. Rozikov, *p*-adic boundary laws and Markov chains on trees. *Lett. Math. Phys.* **110**(10) (2020), 2725–2741.
[148] J.L. Lebowitz, and C. Maes, *Entropy: A dialogue.* Entropy, 269-276. Princeton Ser. Appl. Math., Princeton Univ. Press, Princeton, NJ, 2003.
[149] T.D. Lee, and C.N. Yang, Statistical theory of equations of state and phase transitions. II. Lattice gas and Ising model. *Phys. Rev.* **87**(2) (1952), 410–419.
[150] S. Lehnert, *Biomolecular Action of Ionizing Radiation.* Series in Medical Physics and Biomedical Engineering. Taylor-Francis, Boca Raton, FL, USA, 2008.
[151] L. Lenzini, F.D. Patti, Francesca, S. Lepri, R. Livi, and S. Luccioli, Thermodynamics of DNA denaturation in a model of bacterial intergenic sequences. *Chaos Solitons Fractals* **130** (2020), 109446, 8 pp.
[152] R. M. Levy, A. Haldane, and W. F. Flynn, Potts Hamiltonian models of protein co-variation, free energy landscapes, and evolutionary fitness. *Curr. Opin. Struct. Biol.* **43** (2017), 55–62.

[153] Q. Li, X. Wang, F. Liang, F. Yi, Y. Xie, A. Gazdar, and G. Xiao, A Bayesian hidden Potts mixture model for analyzing lung cancer pathology images. *Biostatistics* **20**(4) (2019), 565–581.

[154] E.H. Lieb, and A.D. Sokal, A general Lee-Yang theorem for one-component and multicomponent ferromagnets. *Comm. Math. Phys.* **80**(2) (1981), 153–179.

[155] E.H. Lieb, and J. Yngvason, The physics and mathematics of the second law of thermodynamics. *Phys. Rep.* **310**(1) (1999), 96 pp.

[156] E.H. Lieb, and J. Yngvason, A second look at the second law of thermodynamics. Prospects in mathematical physics, 127–157, *Contemp. Math.*, 437, *Amer. Math. Soc.*, Providence, RI, 2007.

[157] A.D. Linkevich, A sequential pseudo-inverse learning rule for networks of formal neurons. *J. Phys. A* **25**(15) (1992), 4139–4146.

[158] W. Liu, and N. Ning, Information reconstruction on an infinite tree for a 4×4-state asymmetric model with community effects. *J. Stat. Phys.* **177**(3) (2019), 438–467.

[159] D.M. Lopes, and J.C.M. Mombach, Two-dimensional wetting transition modeling with the Potts model. *Braz. J. Phys.* **47** (2017), 672–677.

[160] M. Löwe, and F. Vermet, The capacity of q-state Potts neural networks with parallel retrieval dynamics. *Statist. Probab. Lett.* **77**(14) (2007), 1505–1514.

[161] Yu. I. Lyubich, Basic concepts and theorems of the evolutionary genetics of free populations. *Russian Math. Surveys* **26** (1971), 51–123.

[162] Yu. I. Lyubich, A topological approach to a problem in mathematical genetics. *Russian Math. Surveys* **34** (1979), 60–66.

[163] C. Maes, and V.K. Vande, The fuzzy Potts model. *J. Phys. A* **28**(15) (1995), 4261–4270.

[164] W. Magnus, A. Karrass, and D. Solitar, *Combinatorial group theory: Presentations of groups in terms of generators and relations* (2nd revised edition). Dover, New York, 1976.

[165] M. Mak, F. Spill, K. Roger, and M. Zaman, Single-cell migration in complex microenvironments: Mechanics and signaling dynamics. *J. Biomech. Eng.* **138**(2) (2016), 021004, 8 pages.

[166] M. Mandel, and J. Marmur, Use of Ultraviolet Absorbance-Temperature Profile for Determining the Guanine plus Cytosine Content of *DNA. Meth. Enzymol.* **12** (2) (1968), 198–206.

[167] M. Manghi, and N. Destainville, Physics of base-pairing dynamics in DNA. *Phys. Rep.* **631** (2016), 1–41.

[168] J. Malo, J.C. Mitchell, C. Vénien-Bryan, J.R. Harris, H. Wille, D.J. Sherratt, and A.J. Turberfield, Engineering a 2D protein-DNA crystal. *Angew. Chem. Int. Ed.* **44** (2005) 3057–3061.

[169] V.A. Malyshev, and R.A. Minlos, *Gibbs random fields. Cluster expansions.* Mathematics and its Applications (Soviet Series), 44. Kluwer Academic Publishers Group, Dordrecht, 1991.

[170] E. Marinari, S. Mossa, and G. Parisi, Glassy Potts model: A disordered Potts model without a ferromagnetic phase. *Phys. Rev. B* **59** (1999), 8401.

[171] J.B. Martin, *Reconstruction thresholds on regular trees.* Discrete random walks (Paris, 2003), 191–204, Discrete Math. Theor. Comput. Sci. Proc., AC, Assoc. Discrete Math. Theor. Comput. Sci., Nancy, 2003.

[172] P.P. Martin, *Potts models and related problems in statistical mechanics.* Series on Advances in Statistical Mechanics, **5**. World Scientific, Singapore, 1991.

[173] P.P. Martin, and S.F. Zakaria, Zeros of the 3-state Potts model partition function for the square lattice revisited. *J. Stat. Mech. Theory Exp.* **8** (2019), 084003, 25 pp.

[174] T. Martínez-Seara, and A. Delshams, *Splitting measure for invariant manifolds in plane fields.* (Spanish) Proceedings of the XIth Congress on Differential Equations and Applications/First Congress on Applied Mathematics, 1989, 329–333, Univ. Málaga, Málaga, 1990.

[175] D.H. Martirosian, Translation invariant Gibbs states in the q-state Potts model. *Commun. Math. Phys.* **105**(2), (1986), 281–290.

[176] F. Martinelli, A. Sinclair, and D. Weitz, Fast mixing for independent sets, coloring and other models on trees. *Random Struct. Algoritms* **31** (2007), 134–172.

[177] F. Martinelli, A. Sinclair, and D. Weitz, Glauber dynamics on trees: Boundary conditions and mixing time. *Comm. Math. Phys.* **250**(2) (2004), 301–334.

[178] H. Meyer-Ortmanns, Immigration, integration and ghetto formation. *Internat. J. Modern Phys. C* **14**(3) (2003), 311–320.

[179] M. Mézard, and A. Montanari, Reconstruction on trees and spin glass transition. *J. Stat. Phys.* **124** (2006), 1317–1350.

[180] M. Mezard, G. Parisi, and M. A. Virasoro, *Spin Glass Theory and Beyond.* World Scientific, Singapore, 1987.

[181] C. Micheletti, and D. Marenduzzo, *Modelling DNA as a Flexible Thick Polymer: DNA Elasticity and Packaging Thermodynamics.* Physical and numerical models in knot theory, 127–147, Ser. Knots Everything, 36, World Sci. Publ., Singapore, 2005.

[182] R.A. Minlos, *Introduction to Mathematical Statistical Physics.* University lecture series, AMS, v.19, 2000.

[183] B. M. Mognetti, P. Cicuta, and L. Di Michele, Programmable interactions with biomimetic DNA linkers at fluid membranes and interfaces. *Rep. Progr. Phys.* **82**(11) (2019), 116601, 34 pp.

[184] R.H.M. Morais, J.P. Santos, S. Oliveira, R.G.B. Mendes, D.S. Rosa, R.M. Francisco, and F.C. Sá Barreto, Magnetocaloric effect in the Potts model based on the effective-field theory. *Phys. Lett. A* **424** (2022), Paper No. 127844.

[185] E. Mossel, Reconstruction on trees: Beating the second eigenvalue. *Ann. Appl. Probab.* **11**(1) (2001), 285–300.

[186] E. Mossel, and Y. Peres, Information flow on trees. *Ann. Appl. Probab.* **13**(3) (2003), 817–844.

[187] E. Mossel, *Survey: Information flow on trees.* Graphs, morphisms and statistical physics, 155–170, DIMACS Ser. Discrete Math. Theoret. Comput. Sci., 63, Amer. Math. Soc., Providence, RI, 2004.

[188] E. Mossel, On the impossibility of reconstructing ancestral data and phylogenies. *J. Comput. Biol.* **10**(5) (2003), 669–678.

[189] E. Mossel, and M. Steel, A phase transition for a random cluster model on phylogenetic trees. *Math. Biosci.* **187**(2) (2004), 189–203.

[190] U. Mückstein, H. Tafer, J. Hackermüller, S.H. Bernhart, P.F. Stadler, and I.L. Hofacker, Thermodynamics of RNA-RNA binding. *Bioinformatics* **22**(10) (2006), 1177–1182.

[191] F.M. Mukhamedov, and N.N. Ganikhodjaev, *Quantum Quadratic Operators and Processes.* Lecture Notes in Mathematics, 2133. Springer, Cham, 2015.

[192] F.M. Mukhamedov, U.A. Rozikov, and J.F. Mendes, On contour arguments for the three state Potts model with competing interactions on a semi-infinite Cayley tree. *J. Math. Phys.* **48**(1) (2007), 013301, 14 pp.

[193] F.M. Mukhamedov, and U.A. Rozikov, On Gibbs measures of p-adic Potts model on the Cayley tree. *Indag. Math., New Ser.* **15** (2004), 85–100.

[194] F.M. Mukhamedov, and U.A. Rozikov, On inhomogeneous p-adic Potts model on a Cayley tree. *Infin. Dimens. Anal. Quantum Probab. Relat. Top.* **8** (2005), 277–290.

[195] F.M. Mukhamedov, U.A. Rozikov, and J.F.F. Mendes, *On phase transitions for p-adic Potts model with competing interactions on a Cayley tree.* p-adic mathematical physics, 140150, AIP Conf. Proc., 826, Amer. Inst. Phys., Melville, NY, 2006.

[196] F.M. Mukhamedov, and O.N. Khakimov, On periodic Gibbs measures of p-adic Potts model on a Cayley tree. *p-Adic Numbers Ultrametric Anal. Appl.* **8**(3) (2016), 225–235.

[197] F.M. Mukhamedov, and O.N. Khakimov, Phase transition and chaos: P-adic Potts model on a Cayley tree. *Chaos Sol. Fractals* **87** (2016), 190–196.

[198] E. Müller-Hartmann, Theory of the Ising model on a Cayley tree. *Z. Phys. B* **27** (1977), 161–168.

[199] J. Murray, *Mathematical Biology, Section Temperature-dependent Sex Determination (TSD): Crocodilian Survivorship*, pp. 119–144. Springer-Verlag, New York. 2002.

[200] A. Murua, and F. A. Quintana, Semiparametric Bayesian regression via Potts model. *J. Comput. Graph. Statist.* **26**(2) (2017), 265–274.

[201] M. Naim, V. Boboeva, C.J. Kang, and A. Treves, Reducing a cortical network to a Potts model yields storage capacity estimates. *J. Stat. Mech. Theory Exp.* **4** (2018), 043304, 35 pp.

[202] F. R. Nardi, and A. Zocca, Tunneling behavior of Ising and Potts models in the low-temperature regime. *Stochastic Process. Appl.* **129**(11) (2019), 4556–4575.

[203] L. Nicolao, and M. Ostilli, Critical states in political trends. How much reliable is a poll on Twitter? *Phys. A* **533** (2019), 121920, 18 pp.

[204] V. T. Ngo, P.-T. Nguyen, and H.T. Diep, Statistical physics approach to liquid crystals: Dynamics of mobile Potts model leading to Smectic phase, phase transition by Wang-Landau method. *Entropy* **22**(11) (2020), Paper No. 1232, 19 pp.

[205] E.P. Normatov, and U.A. Rozikov, A description of harmonic functions using the properties of the group representation of the Cayley tree. *Math. Notes.* **79**(3-4) (2006), 399–407.

[206] M. Ostilli, and F. Mukhamedov, 1D three-state mean-field Potts model with first- and second-order phase transitions. *Phys. A* **555** (2020), 124415, 10 pp.

[207] S. Ott, and Y. Velenik, Potts models with a defect line. *Comm. Math. Phys.* **362**(1) (2018), 55–106.

[208] N.B. Ouchi, J.A. Glazier, J-P. Rieu, A. Upadhyaya, and Y. Sawada, Improving the realism of the cellular Potts model in simulations of biological cells. *Phys. A* **329**(3-4) (2003), 451–458.

[209] D. Panchenko, Free energy in the Potts spin glass. *Ann. Probab.* **46**(2) (2018), 829–864.

[210] A. Paszkiewicz, Modeling and analysis of anomalies in the network infrastructure based on the Potts model. *Entropy.* **23**(8) (2021), Paper No. 949, 13 pp.

[211] U. Paun, G method in action: normalization constant, important probabilities, and fast exact sampling for Potts model on trees. *Rev. Roumaine Math. Pures Appl.* **65**(2) (2020), 101–128.

[212] A.K. Pedersen, C. Wiuf, and F.B. Christiansen, A codon-based model designed to describe lentiviral evolution. *Mol. Biol. Evol.* **15** (1998), 1069–1081.

[213] J.K. Percus, *Mathematics of genome analysis.* Cambridge Studies in Mathematical Biology, 17. Cambridge University Press, Cambridge, 2002.

[214] F. Peruggi, F. di Liberto, and G. Monroy, The Potts model on Bethe lattices. *I. General results. J. Phys. A* **16**(4) (1983), 811–827.

[215] F. Peruggi, F. di Liberto, and G. Monroy, Phase diagrams of the q-state Potts model on Bethe lattices. *Phys. A* **141**(1) (1987), 151–186.

[216] L. Pilz, *Gibbs measures for cell systems.* Stochastic modelling in biology (Heidelberg, 1988), 150–165, World Sci. Publ., Teaneck, NJ, 1990.

[217] M. Plischke, and B. Bergesen, *Equilibrium Statistical Mechanics* (3rd edition). World Scientific, Singapore, 2006.

[218] I.J. Polmear, *Light Alloys from Traditional Alloys to Nanocrystals* (4th edition) Elsevier Butterworth-Heinemann, Burlington, MA, 2006.

[219] R.B. Potts, Some generalized order-disorder transformations. *Proc. Cambridge Philos. Soc.* **48** (1952), 106–109.

[220] C. Preston, *Gibbs States on Countable Sets.* Cambridge University Press, London, (1974).

[221] V.V. Prasolov, *Polynomials.* Springer-Verlag Berlin Heidelberg, 2004.

[222] B. Radhakrishnan, G.B. Sarma, and T. Zacharia, Modeling the kinetics and microstructural evolution during static recrystallization-Monte Carlo simulation of recrystallization. *Acta Mater.* **46** (1998), 4415–4433.

[223] M.M. Rahmatullaev, The existence of weakly periodic Gibbs measures for the Potts model on a Cayley tree. *Theor. Math. Phys.* **180**(3) (2014), 1019–1029.

[224] M.M. Rahmatullaev, A weakly periodic Gibbs measure for the ferromagnetic Potts model on a Cayley tree. *Sib. Math. J.* **56**(5) (2015), 929–935.

[225] M.M. Rahmatullaev, On weakly periodic Gibbs measures of the Potts model with a special external field on a Cayley tree. *Zh. Mat. Fiz. Anal. Geom.* **12**(4) (2016), 302–314.

[226] M.M. Rahmatullaev, On weakly periodic Gibbs measures for the Potts model with an external field on a Cayley tree. *Ukrainian Math. J.* **68**(4) (2016), 598–611.

[227] M.A. Rasulova, On periodic Gibbs measures for the Potts-SOS model on a Cayley tree. *Theoret. Math. Phys.* **199**(1) (2019), 586–592.

[228] J. Rau, *Statistical Physics and Thermodynamics. An Introduction to Key Concepts.* Oxford University Press, Oxford, 2017.

[229] J. Reichardt, and S. Bornholdt, Detecting fuzzy community structures in complex networks with a Potts model. *Phys. Rev. Lett.* **93** (2004), Art. 218701, 4 pp.

[230] E. A. Rietman, S. Taylor, H.T. Siegelmann, M.A. Deriu, M. Cavaglia, and J.A. Tuszynski, Using the Gibbs function as a measure of human brain development trends from fetal stage to advanced age. *Int. J. Mol. Sci.* **21** (2020), 1116, 17 pp.

[231] V. Riva, and J. Cardy, Holomorphic parafermions in the Potts model and stochastic Loewner evolution. *J. Stat. Mech. Theory Exp.* **12** (2006), P12001, 19 pp.

[232] N.F. Robertson, M. Pawelkiewicz, J.L. Jacobsen, and H. Saleur, Integrable boundary conditions in the antiferromagnetic Potts model. *J. High Energy Phys.* **5** (2020), 144, 34 pp.

[233] H.E. Rodriguez, P. Barrat-Charlaix, and M. Weigt, Toward inferring Potts models for phylogenetically correlated sequence data. *Entropy* **21**(11) (2019), Paper No. 1090, 20 pp.

[234] J.A. Rose, R.J. Deaton, and A. Suyama, Statistical thermodynamic analysis and design of DNA-based computers. *Nat. Comput.* **3**(4) (2004), 443–459.

[235] B.Rosental, Z. Kozhekbaeva, N. Fernhoff, J.M. Tsai, and N. Traylor-Knowles, Coral cell separation and isolation by fluorescence-activated cell sorting. *BMC Cell Biology* **18**, 30 (2017).

[236] U.A. Rozikov, On pure phase of the anti-ferromagnetic Potts model on the Cayley tree. *Uzbek Math. J.* **1** (1999), 73–77 (Russian).

[237] U. A. Rozikov, On q-component models on Cayley tree: Contour method. *Lett. Math. Phys.* **71**(1) (2005), 27–38.

[238] U.A. Rozikov, and Y.M. Suhov, Gibbs measures for SOS models on a Cayley tree. *Infin. Dimens. Anal. Quantum Probab. Relat. Top.* **9** (2006), 471–488.

[239] U.A. Rozikov, A constructive description of ground states and Gibbs measures for Ising model with two-step interactions on Cayley tree. *J. Stat. Phys.* **122**(2) (2006), 217–235.

[240] U.A. Rozikov, A contour method on Cayley trees. *J. Stat. Phys.* **130**(4) (2008), 801–813.

[241] U.A. Rozikov, and F.T. Ishankulov, Description of periodic p-harmonic functions on Cayley trees. *Nonlinear Diff. Equations Appl.* **17**(2) (2010), 153–160.

[242] U.A. Rozikov, and J.P. Tian, Evolution algebras generated by Gibbs measures. *Lobachevskii Jour. Math.* **32**(4) (2011), 270–277.

[243] U.A. Rozikov, *Gibbs Measures on Cayley Trees.* World Sci. Publ., Singapore, 2013.

[244] U.A. Rozikov, and R.M. Khakimov, Periodic Gibbs measures for the Potts model on the Cayley tree. *Theor. Math. Phys.* **175**(2) (2013), 699–709.

[245] U.A. Rozikov, Gibbs measures on Cayley trees: Results and open problems. *Rev. Math. Phys.* **25**(1) (2013), 1330001, 112 pp.

[246] U.A. Rozikov, Tree-hierarchy of DNA and distribution of holliday junctions. *Jour. Math. Biol.* **75**(6-7) (2017), 1715–1733.

[247] U.A. Rozikov, and M.M. Rahmatullaev, On free energies of the Potts model on the Cayley tree. *Theor. Math. Phys.* **190**(1) (2017), 98–108.

[248] U.A. Rozikov, *Holliday junctions for the Potts model of DNA.* Algebra, complex analysis, and pluripotential theory, 151–165. Springer Proc. Math. Stat., **264**, Springer, Cham, 2018.

[249] U.A. Rozikov, R.M. Khakimov, and F.Kh. Khaidarov, Extremality of translation invariant Gibbs measures for Potts model on Cayley tree. *Theor. Math. Phys.* **196**(1) (2018), 1043–1058.

[250] U.A. Rozikov, *An Introduction to Mathematical Billiards.* World Sci. Publ., Hackensack, NJ, 2019.

[251] U.A. Rozikov, *Population Dynamics: Algebraic and Probabilistic Approach.* World Sci. Publ. Singapore. 2020, 460 pp.

[252] U.A. Rozikov, Thermodynamics of interacting system of DNAs. *Theor. Math. Phys.* **206**(2) (2021), 174–183.

[253] U.A. Rozikov, Thermodynamics of DNA-RNA renaturation. *Int. J. Geom. Methods Mod. Phys.* **18**(6) (2021), 2150096, 14 pp.

[254] U.A.Rozikov, Gibbs measures of Potts model on Cayley trees: A survey and applications. *Reviews in Math. Phys.* **33**(10) (2021), 2130007, 58 pages.

[255] U.A. Rozikov, M.M. Rahmatullaev, and R.M. Khakimov, Periodic Gibbs measures for the Potts model with translation-invariant and periodic external fields on the Cayley tree. *Theor. Math. Phys.* **210**(1) (2022), 135–153.

[256] S. Sanyal, and J.A. Glazier, Viscous instabilities in flowing foams: A Cellular Potts Model approach. *J. Stat. Mech. Theory Exp.* **10** (2006), 10008, 9 pp.

[257] J. Salas, and A.D. Sokal, Absence of phase transitions for the antiferromagnetic Potts model via the Dobrushin uniqueness theorem. *J. Stat. Phys.* **86** (1997), 551–579.

[258] D. Savage, X. Zhang, X. Yu, P. Chou, and Q. Wang, Anomaly detection in online social networks. *Social Networks* **39** (2014), 62–70.

[259] T.C. Schelling, Dynamic models of segregation. *J. Math. Sociology* **1** (1971), 143–186.

[260] O. Schramm, Conformally invariant scaling limits: An overview and a collection of problems. In: *Proceedings of the International Congress of Mathematicians* (Madrid, August 22–30, 2006), Vol. I, 513–543. Eur. Math. Soc., Zürich, 2007.
[261] C. Schulze, Potts-like model for ghetto formation in multi-cultural societies. *Int. J. Modern Phys. C.* **16**(3) (2005), 351–355.
[262] M. Scianna, and L. Preziosi, *Cellular Potts Models. Multiscale Extensions and Biological Applications.* Chapman and Hall/CRC Mathematical and Computational Biology Series. CRC Press, Boca Raton, FL, 2013.
[263] A. Shahraz, A. Borhan, and K.A. Fichthorn, Wetting on physically patterned solid surfaces: The relevance of molecular dynamics simulations to macroscopic systems. *Langmuir* **29**(37) (2013), 11632–11639.
[264] A. Shahraz, A. Borhan, and K.A. Fichthorn, Kinetics of droplet wetting mode transitions on grooved surfaces: Forward flux sampling. *Langmuir* **30**(51) (2014), 15442–15450.
[265] D. Sherrington, and S. Kirkpatrick, Solvable model of a spin glass. *Phys. Rev. Lett.* **35** (1975), 1792–1796.
[266] A.N. Shiryaev, *Probability* (2nd edition). Graduate Texts in Mathematics, vol. 95. Springer, New York, 1996.
[267] Ya. G. Sinai, *Theory of Phase Transitions: Rigorous Results.* Pergamon, Oxford, 1982.
[268] T.N. de Silva, and V. Sivised, *A Statistical Mechanics Perspective for Protein Folding from q-state Potts Model.* arXiv:1709.04813.
[269] A. Sly, Reconstruction for the Potts model. *Ann. Probab.* **39**(4) (2011), 1365–1406.
[270] A.P. Solon, and J. Tailleur, Flocking with discrete symmetry: The two-dimensional active Ising model. *Phys. Rev. E* (3) **92**(4) (2015), 042119, (18 pages).
[271] F. Spitzer, Markov random fields on an infinite tree. *Ann. Probab.* **3** (1975), 387–398.
[272] J. Srolvitz, M.P. Anderson, G.S. Grest, and P.S. Sahni, Grain growth in two-dimensions. *Scr. Metall.* **17** (1983), 241–246.
[273] H.E. Stanley, *Introduction to Phase Transitions and Critical Phenomena,* Oxford University Press, Oxford, 1971.
[274] M. Steel, and M. Charleston, Five surprising properties of parsimoniously colored trees. *Bull. Math. Biol.* **57** (1995), 367–375.
[275] Yu.M. Suhov, and U.A. Rozikov, A hard-core model on a Cayley tree: An example of a loss network. *Queueing Syst.* **46**(1/2) (2004), 197–212.
[276] L. Sun, Y.F. Chang, and X. Cai, A discrete simulation of tumor growth concerning nutrient concentration. *Int. J. Modern Phys. B* **18** (2004), 2651–2657.
[277] D. Swigon, The Mathematics of DNA Structure, Mechanics, and Dynamics. In: *IMA Volumes in Mathematics and Its Applications,* vol. 150. Springer, New York, NY, 2009, pp. 293–320.
[278] A. Szabó, and R.M.H. Merks, Cellular Potts modeling of tumor growth, tumor invasion, and tumor evolution. *Front. Oncol.* **3** (2013), Art. 87, 12 p.

[279] T. Takaishi, Simulations of financial markets in a Potts-like model. *Int. J. Modern Phys. C* **16**(08) (2005), 1311–1317.

[280] T. Takaishi, Analysis of spin financial market by GARCH model. *J. Phys. Conf. Ser.* **454** (2013) 012041.

[281] D.R. Tan, C.D. Li, W.P. Zhu, and F.J. Jiang, A comprehensive neural networks study of the phase transitions of Potts model. *New J. Phys.* **22** (2020), 063016, 18 pages.

[282] F. Tanaka, A. Kameda, M. Yamamoto, and A. Ohuchi, *Nearest-neighbor thermodynamics of DNA sequences with single bulge loop.* DNA computing, 170–179, Lecture Notes in Comput. Sci., 2943, Springer, Berlin, 2004.

[283] R. Tahir-Kheli, *General and Statistical Thermodynamics.* Graduate Texts in Physics. Springer, Heidelberg, 2012.

[284] P. Tautu, *Interacting Biological Cell Systems.* Stochastic modelling in biology (Heidelberg, 1988), 50-90, World Sci. Publ., Teaneck, NJ, 1990.

[285] V. Tikare, and J.D. Cawley, Application of the Potts model to simulation of Ostwald ripening. *J. Amer. Ceramic Soc.* **81**(3) (1998), 485–491.

[286] P.N. Timonin, Statistics of geometric clusters in Potts model: Statistical mechanics approach. *Proc. A.* **476**(2240) (2020), 215–228.

[287] C. Thompson, *Mathematical Statistical Mechanics.* Princeton Univ Press. 1972.

[288] S. Turner, and J.A. Sherratt, Intercellular adhesion and cancer invasion: A discrete simulation using the extended Potts model. *J. Theoret. Biol.* **216**(1) (2002), 85–100.

[289] S.M. Ulam, Problems in Modern Math. Wiley, New York, 1964.

[290] R. Varro, Gonosomal algebra. *J. Algebra* **447**(1) (2016), 1–30.

[291] A. Voss-Böhme, *Cellular Potts models for interacting cell populations: Mathematical foundation, challenges, and future prospects,* Probabilistic cellular automata, 311–325, Emerg. Complex. Comput., 27, Springer, Cham, 2018.

[292] M.J. de Vries, Technological processes. In: *Teaching about Technology. Contemporary Issues in Technology Education.* Springer, 2016.

[293] F. Wagner, D. Grensing, and J. Heide, New order parameters in the Potts model on a Cayley tree. *J. Phys. A* **34**(50) (2001), 11261–11272.

[294] J. Wang, J. Wang, and H.E. Stanley, Multiscale multifractal DCCA and complexity behaviors of return intervals for Potts price model. *Phys. A* **492** (2018), 889–902.

[295] F. Wang, K. Yamasaki, S. Havlin, and H.E. Stanley, Scaling and memory of intraday volatility return intervals in stock markets. *Phys. Rev. E* **73** (2006), 026117.

[296] W. Wang, C. Wu, and J. Deng, Piecewise constant signal and image denoising using a selective averaging method with multiple neighbors. *Inverse Probl. Imaging* **13**(5) (2019), 903–930.

[297] J. Watson, F.A. Hays, and P.S. Ho, Definitions and analysis of DNA Holliday junction geometry. *Nucleic Acids Res* **32** (2004), 3017–3027.

[298] D. Weaire, F. Bolton, P. Molho, and J. A. Glaziers, Investigation of an elementary model for magnetic froth. *J. Phys.: Condens. Matter* **3**(13) (1991), 2101–2115.

[299] A.Weinmann, M. Storath, and L. Demaret, The L^1-Potts functional for robust jump-sparse reconstruction. *SIAM J. Numer. Anal.* **53**(1) (2015), 644–673.

[300] F.Y. Wu, The Potts model. *Rev. Modern Phys.* **54** (1982), 235–268.

[301] F.Y. Wu, *Exactly solved models: A journey in statistical mechanics. Selected papers with commentaries (1963–2008).* With an introduction by C. N. Yang and a review of Wu's research by J.-M. Maillard. World Scientific Publishing Co. Pte. Ltd., Hackensack, NJ, 2009.

[302] P. Xie, Model for RuvAB-mediated branch migration of Holliday junctions. *J. Theor. Biology.* **249** (2007) 566–573.

[303] Y. Xing, and J. Wang, Nonlinear complexity behaviors of agent-based $3D$ Potts financial dynamics with random environments. *Phys. A* **492** (2018), 824–836.

[304] D. Xiong, and H. Zhao, q-state Potts-glass neural network based on pseudoinverse rule. *J. Phys. A* **43**(44) (2010), 445001, 9 pp.

[305] K. Xu, and J. Wang, Weighted fractional permutation entropy and fractional sample entropy for nonlinear Potts financial dynamics. *Phys. Lett. A* **381**(8) (2017), 767–779.

[306] H. Yi, First- and second -order quantum phase transitions of a q-state Potts model in fractal lattices. *Phys. Rev. E* **96**(6) (2017), 062105, 6 pp.

[307] M. Zahradnik, A short course on the Pirogov-Sinai theory. *Rendiconti Math. Serie VII* **18** (1998), 411–486.

[308] M.I. Zakharevich, The behavior of trajectories and the ergodic hypothesis for quadratic mappings of a simplex. *Russian Math. Surveys.* **33** (1978), 207–208.

[309] M. Zoli, Twist versus nonlinear stacking in short DNA molecules. *J. Theoret. Biol.* **354** (2014), 95–104.

[310] B.J. Zubillaga, A.L.M. Vilela, C. Wang, K.P. Nelson, and H.E. Stanley, A three-state opinion formation model for financial markets. *Phys. A* **588** (2022), Paper No. 126527.

Index

Lightning Source UK Ltd.
Milton Keynes UK
UKHW050624270822
407875UK00005B/37

9 789811 251238